Wildlife Biology

Wildlife Biology

Edited by **Martin Winter**

SYRAWOOD
PUBLISHING HOUSE
New York

Published by Syrawood Publishing House,
750 Third Avenue, 9th Floor,
New York, NY 10017, USA
www.syrawoodpublishinghouse.com

Wildlife Biology
Edited by Martin Winter

International Standard Book Number: 978-1-68286-321-3 (Hardback)

Printed in China.

Contents

Preface

Wildlife biology is gaining significance all over the world. The study of all species of animals and their habitats has fascinated scientists for many years. This book covers the existing theories and concepts in the field of wildlife biology and also sets ground for new research. It discusses some innovative techniques and modules to record and comprehend the physiology, demographics and related aspects of the varied species on the planet. It further delves into the management and conservation practices for harboring sustainable habitats. This book aims to provide a better understanding of the advancing field of wildlife biology through diverse case studies and ground-breaking researches. This book will benefit students and professionals alike.

After months of intensive research and writing, this book is the end result of all who devoted their time and efforts in the initiation and progress of this book. It will surely be a source of reference in enhancing the required knowledge of the new developments in the area. During the course of developing this book, certain measures such as accuracy, authenticity and research focused analytical studies were given preference in order to produce a comprehensive book in the area of study.

This book would not have been possible without the efforts of the authors and the publisher. I extend my sincere thanks to them. Secondly, I express my gratitude to my family and well-wishers. And most importantly, I thank my students for constantly expressing their willingness and curiosity in enhancing their knowledge in the field, which encourages me to take up further research projects for the advancement of the area.

Editor

1

ACTIVITIES TOWARDS PRIMATE CONSERVATION IN PANAMA

P. Méndez-Carvajal[1,*], I. Ruiz-Bernard[1], G. De León[1], Y. González[1], E. Miranda[1], L. Loría[1], G. Berguido[1], A. Cortes[1] & E. Soto[1]

[1] Fundación Pro-Conservación de los Primates Panameños (FCPP), 0816-07905, Panamá, República de Panamá.

* Corresponding author e-mail: fcprimatespanama@gmail.com

Keywords
Primate Conservation; Panama.

Abstract
Neotropical Primates are threatened due to deforestation and hunting pressure. Nowadays there are many subspecies of primates living in reduced areas. These subspecies are poorly studied due to factors such as remote locations, political situations or the expensive logistics required. Although population surveys are important to obtain baseline information for the conservation of critically endangered primates, there are some opportunities where population censuses could be used in conjunction with rapid conservation strategies. In this report we encourage scientists to use collateral techniques that could positively impact the local community while other studies are underway. Some of those activities are the involvement of biology students from the target country, the use of informative material to spread the scientific information acquired, and raising the awareness of community leaders so that a long term conservation program can be established in the future.

Introduction

Panama is home of 13 subspecies of non-human primates, two mantled howler monkeys *Alouatta palliata palliata* and *A. p. aequatorialis* [16,18], the Coiba and Azuero howler monkey *A. coibensis coibensis* and *A. c. trabeata* [12, 13, 14, 15], four spider monkeys, the *Ateles geoffroyi panamensis*, *A. g. azuerensis*, *A. g. grisescens* and *A. fusciceps rufiventris*, these considered the most endangered primates in Panama [11, 13]. Followed by two capuchin monkeys *Cebus imitator* and *C. capucinus*, both have been identified as Regionally Endangered by FCPP, due to the fact that these species are directly hunted by poachers in order to minimize the damage they cause in crops [13, 15]. Panama also has the Central American Squirrel monkey *Saimiri oerstedii oerstedii*, Regionally Endemic for Chiriqui province in Panama [16], the tamarin *Saguinus geoffroyi* and the only nocturnal monkey found in Mesoamerica, the *Aotus zonalis*, turning Panama in the most diverse country of Mesoamerica in terms of primates [12, 13, 18, 20]. However, primate conservation actions have been recently initiated with the FCPP in 2001 [9, 21]. Most of the subspecies are considered Vulnerable, Endangered, and Critically Endangered, due to anthropogenic pressures including: exposure to crop spraying, pet trade, land exploitation for agricultural activities, cattle ranching, hunting, mining, wrong tourism activities, expansion of teak plantations, and most recently real estate investments [12, 14]. Due to these reasons, FCPP has developed an educational program to conserve the Panamanian primates while doing long term monitoring surveys around the country. Our objectives have been to create good attitudes and practices from the human community to mitigate deceasing of the primate's population living in the fragments forest and living fences out to the protected areas.

We consider being in touch with the communities improve our communication with the local people and impact in a positive way to the wildlife population living near to these habitats in Panama. We summarized here some activities that have been important to develop interest in people to study, protect and conserve the non-human primates in Panama. These activities have been generating data through before-after questionaries' for each town visited (e.g. Chiriqui province, Azuero Peninsula, Chucanti Reserve, between others) [10, 11, 12, 13, 15, 16]. We would like to encourage native biologist or primatologist from Latin-American countries to promote similar plans in order to protect their endangered species and obtain accurate data from their countries. These initiatives in Panama have been implemented in the following ways:

Training of students from the Biology School of the University of Panama

This training is free and includes teaching different techniques for surveying primates, specifically for *Alouatta*, *Ateles* and *Cebus* genus. The training includes how to collect field data on primate group structure, mapping, compass use, and calculating distances and height for habitat description. After five hours of theory, the project offers the students the opportunity to practice these methods in the field acting as volunteers in five to ten day surveys. Field training includes the use of GPS, recorder equipment to obtain vocalization data, fecal sampling and other related activities. The goal of this initiative is to develop the interest of senior biology students in primate research. At the same time, the students will be supporting our project with their fieldwork.

Educational activities at the primary and secondary schools

Educational talks have been developed to offer basic information to children in the regional schools. Over all deforested areas where remaining primate's populations live close to the villages and are at a higher risk [19, 21]. In these presentations, we talk about primate biodiversity, the primates, and the characteristics that humans share with them as part of the same taxonomic Order [3,16]. The talks also touch on basic information about primate food, their importance in the ecosystem as seed dispersers, pollinators, and how they contribute to the survival of other wild animals (e.g., deer, pacas, agouties, peccaries, and others). We normally conclude with ten minutes of questions or drawing activities, where we measure the knowledge of the students after each talk and answer any remaining questions that they might have [21]. We give each school posters with information of the subspecies primates and how they can help to minimize their risk of extinction. The initiative also gives away bookmarks, t-shirts and stickers, with the image of the most endangered non-human primates in the region and a message on the back of how the locals can help the primates which are a part of their heritage and responsibility [5, 6, 7].

Educational activities around natural reserves and villages

For those protected areas, we have contacted members of the Environmental Authority of Panama (ANAM) and their forest rangers every year before and during our field work [1, 11, 12]. Our visits with the forest rangers include an open discussion where we exchange information about our results, what we do in Panama for primates and, their experiences sighting monkeys and conservation issues in the area. We give them new information and t-shirts with conservation messages or any other informative material. Our goal is focused on obtaining information on the ranger's needs and how they think they can improve protection of the natural areas [17, 19]. That information is used to evaluate the effectiveness of the parks and reserves according to the Action Plan for Mesoamerican Primates document [20], and make recommendations to the ANAM's directors for future management of the parks and reserves of the zone.

We are interested in knowing locals point of view about having primates near to their houses, as well as the past history of the occurrence of the primates and local extinctions of already isolated populations. Our goal is to involve local people and have them participate in our activities, obtaining information about the history of their fauna, native plant knowledge, and the presence and problems related to the non-human primates [7, 19, 21].

Road signs, youtube videos, and newspaper publications

The information obtained is passed to the general public beyond the scientific community by putting signs along the roads, posting videos with conservation messages on the internet and publishing articles in national newspapers. Panamanian primates can be found in gallery forests along some secondary roads and near to bridges, we have posted some Spanish-English signs on the appropriate sides, to warn people not to feed the monkeys and to drive in high speed. Feeding monkeys was practiced in the past, in some areas where people feel sympathy for them [12, 21], however, it caused mortality as monkeys road-killed and being poisoned by locals, arguing that the monkeys (any species) could damage their crops, or plantations, and to protect their cattle from the larvae of the botfly *Cuterebra baeri* that they see on the howler monkeys [12]. The consistency of our visits every year and volunteers visiting the communities are now helping to protect the monkeys in those vulnerable areas, changing the perception of them in a positive way [21].

Secondary school students, who have easy access to internet, are now less familiar with their native monkeys living in their own town. FCPP has posted videos on *youtube.com*, with environmental education subjects [12, 21]. The idea is to keep the cyber students and the general public informed about the fauna and flora in our region as an optional way to inform Panamanians about our activities. Publications of general information related to this project are also part of our goals, so we are publishing articles related to conservation in national newspapers [21]. This information is accessible to the local community where we work and encourage the conservation of the fauna and flora in the country.

Outcomes

We have further evidence of positive effects of the program we are conducting on the local people, based on our recent questionnaires, the people are showing an increase of 30% in basic knowledge related to the monkeys, and owners of agricultural lands are avoiding the unnecessary cutting of trees, as happened before [2, 5, 6, 10, 22]. The primate's knowledge of the land owners had a further improvement of 71% in 2011 compared with the first pre-survey (42% for 2001) [10, 12, 21]. Students trained by our program have assisted visiting primatologist researchers in Panama, conducting their own research, and some others have been part of our environmental education team, giving talks in secondary schools around Panama, this technique has led to an increase of the number of Panamanian biologists interested in studying native primates. FCPP address more than 300 local students per year in the entire country, and forest rangers are directly in communication with FCPP to protect National Parks and Reserves [12, 21]. Since 2001 to the present we have published more than ten newspaper articles related to the critical situation of the monkeys in the area, jaguar conservation issues, and two more demanding the ANAM to stop logging activities in the Maje Mountain Chain in Darien province, where deforestation is being increasing yearly [13].

After being in the whole educational roll by our side, government and National Institutions in Panama, have been approached in support to our efforts. Recently, the Secretary of Science, Technology and Innovation of the Republic of Panama (SENACYT) in collaboration with FCPP helped us printing out educational guides for school teachers [21]. The document "Guía Didáctica por la Conservación de los Primates de Azuero" prepared by FCPP incorporated dynamics and topics related to the primates and to the yearly study plan for the Panamanian communities. Measurements of the effectiveness of these conservation activities include the continual monitoring and evaluation of primate populations and vegetation in old and new forest patches [10, 12]. At the moment, all the primate' groups that have been sighted since 2001 have survived and their habitats have been almost untouched [12, 13, 15, 21], this could be considered an important achievement of our project. Future projects include the measurement and comparison of forest expansion and regeneration; that will be done using advanced techniques of GIS. Around 12 YouTube videos have been released in the web to promote conscience to the Panamanian primates, so far with an average of visits of 1,100.

Recommendations for national scientists

The direct access to the community and interaction with the local people call our attention for other species that could be positively impacted by our work (e.g. bats, wildcats), helping in the end, with the complete system of flora and fauna [1, 2, 3, 4, 19]. The participation of the local biologists, local people and the support from national and international institutions is crucial to keep up our long term efforts of education and conservation, trying to stop hunting for monkeys [6, 7, 9, 12, 16, 21].

Based on the Action Plan for the Mesoamerican Primates [20], the Primate Specialist

Group has classified the two most important steps toward the primate's conservation. There are the Study Priorities (E) and the Conservation Actions (AC). Taking this document as a guide, the FCPP has achieved the identification and monitoring populations in protected areas, estimation of densities and population growth, the actual distribution, long-term evaluations for primate communities, and viability of populations [21]. We have accomplished the promotion of new primatologists to study and continue long-term projects, the advice to the governmental authorities in the primate's knowledge. Basically, what we have seen while working in the communities is that people reject the idea of conservation not because they do not care about the species but because they lack of information and they think they cannot use their own land to obtain their resources for subsistence, also found in [1, 3, 4, 6, 7, 8, 17, 21, 22] as implications of restricting the use of land without incentives for the locals [3, 21]. For areas already deforested but with certain amount of fauna, we recommend a permanent environmental education rather than creating more natural reserves because new natural reserves will not be functional without an educational plan [7, 21]. Our yearly monitoring of primates and our educational efforts in the area encourage the cooperation of the locals and help us to stay vigilant to any changes [7, 9, 21, 22]. Our idea moreover, is to encourage the use of living fences [10], which is already a cultural practice, and promote the connectivity of patches of forest with gallery forest and reserves already established [19, 20, 21].

Conclusions

The conservation activities for primates in Panama have been developed with the strong support of local volunteers and of national and international organizations not settled in Panama. Efforts for conservation do not have to start with a huge amount of money; they could start with real compromise and voluntary efforts from native scientists, to work for the conservation of their species and ecosystems. For visiting-conservationist scientists, the goal should be to train and/or support native scientists to develop conservation actions, which will be the best way to protect their resources in the future of that country (having in mind that their resources are still our resources as we all are affected by world deforestation). Panama has suffered a lack of local scientists interested in the study of primatology, even being the first place where Neotropical Primates were studied and primatology developed in Latinamerica [5, 9, 21]. Even with 100 years of permanent presence of scientists from other countries, established in Panama [5], its shows that conservation really need to be attached with locals, and visitor scientists should incorporate in their schedule a better plan to do a better impact with their host country while doing their research [3, 7, 21]. This will be positive for the visitors and also for the natural resources that normally is on need to generate scientific information to conserve the tropics and the humanity.

Acknowledgements

We thank the Panama National Environmental Authority (ANAM), particularly would like to thank Elena Méndez, Marta C. de Méndez, Kenia Sánchez, Somaly Silva, Valeria Franco, Jorge Garzón, Robert Duarte, Alejandro Garrido, for their support as volunteers. Special thanks to the Panamanian journalists Angel

Santos Barrios, Cynthia Sánchez, Thays Domínguez, Rocio Martins, Herasto Reyes, Ismael Hernández, José Poveda, Milciades Pinzón, Omar Rodríguez, Sugeiris Mitre, Víctor Dagoberto Torres, Vielka Corro Rios, and Yelena Rodríguez. We are grateful to Aida Bustamante Ho, Corri Waitt, Juan Carlos Serio-Silva, Alonso Santos-Murgas, Robert Horwich, Robert Horsburgh and Lynda Dillon for their valuable support. This project has been possible thanks to Primates Conservation Inc., Idea Wild, Distribuidora Panatex, S.A., Ford Motor Company, Mohamed bin-Zayed Species Conservation Fund, Rufford Small Grants Foundation and Wild Futures. We thank Barbara Rethore and Julien Chapuis from Conserve-Action, and Timothy Bearder from Oxford BBC radio station for filming and promote our efforts of conservation in Panama.

References

Five "key references", selected by the authors, are marked below (Three recommended (●) and two highly recommended (●●) papers).

1. ANAM (Autoridad Nacional del Ambiente). 1999. Estrategia Nacional Ambiental. Análisis de la Situación Ambiental de Panamá. Ministerio de Economía y Finanzas, Banco Interamericano de Desarrollo.
2. Castro, G. 2004. Para una Historia Ambiental Latinoamericana, Instituto Cubano del Libro Editorial de Ciencias Sociales, Impreso en Colombia.
3. Cowlishaw, G. and Dunbar, R. 2000. Primate Conservation Biology. University of Chicago Press, USA.
4. Gligo, N. 1999. La dimensión ambiental en el desarrollo de América Latina, Comisión Económica para América Latina, Chile.
5. Heckadon-Moreno, S. 1998. Naturalistas del Istmo de Panamá. Instituto Smithsonian de Investigaciones Tropicales, Editorial Santillana, Costa Rica.
6. Heckadon-Moreno, S. 2001. Panamá: Puente Biológico. Instituto Smithsonian de Investigaciones Tropicales-Panamá. Imprelibros-Colombia.
7. ● Horwich, R.H. 1996. Effective solutions for howler conservation. International Journal of Primatology. 19(3): 579-598. http://dx.doi.org/10.1023/A:1020368624122
8. MacKinnon, K., G. Hatta, H. Halim, and A. Mangalik. 1996. The ecology of Kalimantan. Periplus Editions, Singapore.
9. Méndez-Carvajal, P.G. 2007. Historical Perspectives of Primatology in Panama: From The Panama Canal to the Present Day. Canopy 6 (1): 4-6.
10. Méndez-Carvajal, P.G. 2008. Living fences: a farmer strategy that keeps the Azuero primates surviving in fragmented habitats. Canopy 6 (2): 9-11.
11. Méndez-Carvajal, P.G. and I. Ruiz-Bernard. 2009. Estudio Poblacional del mono araña de Azuero (*Ateles geoffroyi azuerensis*), Península de Azuero, Panamá. Tecnociencia 11(1): 24-44.
12. ● Méndez-Carvajal, P.G. 2011. Population size, Distribution and Conservation Status of Howler Monkeys (*Alouatta coibensis trabeata*) and Spider Monkeys (*Ateles geoffroyi azuerensis*) on the Azuero Peninsula, Panama. Primate Cons. 26:1-12.
13. ● Méndez-Carvajal, P.G. 2012. Preliminary Primate Survey at the Chucanti Nature Reserve, Darien Province, Republic of Panama. Mesoamericana 16(3), 22-29.
14. Méndez-Carvajal, P.G. 2012. Estudio de diversidad de mamíferos en cuatro hábitats de transición asociados a una plantación de Teca (*Tectona grandis*) dentro de la Cuenca del Canal de Panamá, Las Pavas, Chorrera, Panamá. Tecnociencia 14(2):55-83.

15. Méndez-Carvajal, P.G. 2012. Population Study of Coiba Howler Monkeys (*Alouatta coibensis coibensis*) and Coiba Capuchin Monkeys (*Cebus capucinus imitator*), Coiba Island National Park, Republic of Panama. J Primatol 1:104.

16. ●● Miranda-Jiménez, E.G. and Méndez-Carvajal, P.G. 2012. Conservation of Chiriquí's Squirrel monkey (*Saimiri oerstedii oerstedii*), Chiriqui Province, western Panama. Canopy Journal. 13(1):24-26.

17. Pimbert, M.P, and J.N. Pretty. 1995. 'Parks, people and professionals. Putting "participation" into protected area management', UNRISD-IIED-WWF Discussion Paper, UNRISD No. 57.

18. Rylands, A., Groves, C., Mittermeier, R., Cortes-Ortiz, L., and Hines, J. 2006. "Taxonomy and Distributions of Mesoamerican Primates". In Estrada, A., Garber, P., Pavelka, M. & Luecke, L. New Perspectives in the Study of Mesoamerican Primates. (pp. 29–80), New York: Springer. http://dx.doi.org/10.1007/0-387-25872-8_3

19. ●● Rabinowitz, A. 2003. Manual de Capacitación para la Investigación de Campo y la Conservación de la Vida Silvestre. Wildlife Conservation Society, Bronx, New York.

20. Rodríguez-Luna, E., L. Cortes-Ortiz, R. Mittermeier and A. Rylands. 1996. Plan de Acción para los Primates Mesoamericanos. Grupo Especialista de Primates-Sección Neotropical. Xalapa, Veracruz, México. Borrador de Trabajo. 102 pp. (+19 Apéndice).

21. Ruiz-Bernard, I., Y. González, G. De León, and P. G. Méndez-Carvajal. 2010. Integración de programas de educación ambiental en proyectos de conservación de primates. Congreso Mesoamericano de Biología y Conservación, p.118. San José, Costa Rica, 8–12 November 2010. Abstract.

22. Suárez, J.O. 1981. Hombres y Ecología en Panamá. Ed. Universitaria, Instituto Smithsonian de Investigaciones Tropicales, Panamá.

2

Validation of a Citizen Science-Based Model of Coyote Occupancy with Camera Traps in Suburban and Urban New York, USA

C.M. Nagy[1,*], M.E. Weckel[1], A. Toomey[2], C.E. Burns[3] & J. Peltz[4]

[1] Mianus River Gorge Preserve, 167 Mianus River Rd., Bedford, NY 10506 USA; Sackler Institute of Comparative Genomics, American Museum of Natural History, Central Park West at 79th Street., New York 10024 USA.

[2] Lancaster Environment Centre, Lancaster University, United Kingdom LA1 4YQ.

[3] 402 Ranch View Rd., Santa Cruz, CA, 95064 USA.

[4] College Internship in Suburban Ecology, Mianus River Gorge Preserve, Bedford, NY, 10506 USA.

* Corresponding author email: cmwnagy@gmail.com.

Keywords

Canis latrans;
Citizen Science;
Human-Wildlife Interaction;
Model Validation;
Suburban;
Urban.

Abstract

We evaluated the accuracy of a previously published model of coyote (*Canis latrans*) sightings in suburban Westchester County, New York. This model was originally developed using citizen reports of coyote sightings to predict the probability of a human-coyote interaction based on proximity to habitat features. Because the data were obtained from surveys, researchers could not separate patterns of site occupancy by coyotes from possible patterns of detection by respondents. Nevertheless, the model could be an indicator of site occupancy within the suburban matrix. We sought to evaluate the predictive power of the human-coyote interaction model with data gathered via a more rigorous method. To build a set of validation sites, we surveyed 11 parks in Westchester County and one park in Bronx County, NY with camera traps between April and October of 2010. The probability of photographing a coyote in a single trap-night was 0.06 ± 0.12 and all sites had >0.9 probability of detecting a coyote at least once given the total trap-nights at each site. During validation, we also added four additional sites that had been surveyed by other researchers with camera traps as additional "present" sites. Predictions of coyote presence or absence based on the human-coyote interaction model for these 16 validation sites were compared to the observed survey results. The model, which contained distances to forest, grassland, and pooled medium and high development performed well in predicting the observed data (kappa = 0.75 ± 0.17, Area-Under-Curve of Receiver-Operator-Characteristic plots = 0.90). The model appears to sufficiently predict coyote occupancy in a suburban-urban landscape and will form the basis of for development of a more comprehensive model of coyote distribution in the New York City metropolitan area. Furthermore, its accuracy illustrates how citizen science can provide reliable estimates of wildlife-habitat patterns in urban areas.

Introduction

In the past few decades, coyotes (*Canis latrans*) have greatly expanded their range east from the deserts and plains of the Western United States into nearly the entire midwest and east coast, and have colonized many urban centers outside of their historical range such as Chicago and Boston. In New York State, the species has colonized the suburban and urban areas of Westchester County, and most recently individuals have been observed in New York City, although there has yet to be any breeding populations found in the island-bound sections of New York City or Nassau and Suffolk Counties on Long Island [1]. Public and expert opinion regarding urban coyotes ranges from curiosity and interest to concern for the safety of people, pets, and property [2-7]. To facilitate human-coyote coexistence [8], wildlife professionals are especially interested in predicting where coyotes will establish themselves and where coyotes and humans are likely to come into contact. To this end, Weckel et al. [9] developed and tested a citizen science-based landscape model of human-coyote interaction (HCI) from 1553 survey responses from Westchester County residents regarding whether they had seen a coyote on their property between 2003 and 2006. The 2001 National Landcover Dataset (NLCD) [10] information was used as habitat covariates to model the probability of such an observation using logistic regression with 777 of the surveys. Weckel et al.'s model contained distance to forest, distance to grassland and agricultural land, and distance to pooled medium development and high development (>50% paved surface). The predicted HCI probabilities from this final model were positively associated with the actual observations from the remaining 776 surveys ($r = 0.88$, $P < 0.001$ [9]).

The term "citizen science" is used to describe methodologies that incorporate non-expert volunteers as participants in research on tasks ranging from data collection to full project development [11-13]. As with any method, citizen science has its own inherent limitations depending on research objectives and requirements, although in many cases, the quality of data collected via citizen science methods has held up when validated by data collected by professionals [14-16]. However, in the case of the HCI model, the use of volunteer-submitted surveys about anecdotal sightings made true inferences about coyote site occupancy impossible and a criticism of Weckel et al.'s [9] HCI model was that it potentially modeled detection patterns (i.e., landscape features where it was easiest to detect a coyote) as opposed to occupancy (i.e., those landscape features where coyotes actually are). The model thus may describe patterns of detection, patterns of occupancy, or a combination of both and the researchers restricted their conclusions and recommendations regarding their model to predicting HCI only, not true coyote habitat use or selection.

However, the HCI model did describe trends that were similar to what is known about suburban/urban coyote habitat use [17-23] and may in fact predict biological patterns in urban/suburban areas. The objective of the present study was to compare predictions of the HCI model to an independent dataset of coyote sightings collected with a method that could account for detection rate and sampling effort. Specifically, we used infrared camera traps to measure coyote distribution in Westchester County parks and compared the observed distribution to predictions based on the HCI model. If the HCI model sufficiently predicted new data of true coyote sightings using more controlled methods – as opposed to just detection by human residents – then we could infer that the HCI model predictions do reflect coyote occupancy. The HCI model could then be used to model coyote occupancy in the suburban matrix (areas between protected areas) and serve as the foundation for further modeling of coyote movement and dispersal through the New York metropolitan area. This study also provided another test of citizen science as a framework for making reliable predictions of wildlife distribution.

Materials and methods

Westchester County is a 1,295 km^2 suburban county immediately north of New York City. There is a distinct urban-rural gradient from the large cities in the southernmost sections of the County to smaller towns and villages in the north. In 2010, population densities on U.S. Census tracts estimated 87 people/km^2 in the northernmost sections of the County to 20,812 people/km^2 in the city of Mt. Vernon on the border of New York City [24]. We surveyed parks along the urban-rural gradient, yet concentrated our efforts south of Interstate 287 (Fig. 1). This 8-lane highway is an east-west dividing line between the generally exurban-rural areas in the north and the more densely populated and urbanized southern portion. Since only a limited number of sites could be surveyed, we chose sites that had some amount of undeveloped wooded area (and thus had a reasonable chance of being occupied by coyotes) but still attempted to include a diversity of parks that varied in regard to total size, forested area, and percent forest (Table 1). Within each park we restricted camera locations to undeveloped wooded areas of each park, i.e., we avoided playgrounds, picnic areas, golf courses, etc., at the request of park managers and to minimize vandalism and theft. These undeveloped areas were largely reclaimed agricultural fields that are presently deciduous forests approximately 50 to 100 years old. The field work and data analysis was conducted with the assistance of volunteers as part of an Earthwatch Institute [25] project. Earthwatch is an international organization that supports ecological research by including private donors as volunteer researchers. Donors help fund specific research projects and then assist researchers in field work, data collection, and analysis.

In the spring and summer of 2010, we deployed Moultrie GameSpy 4.0sec infrared flash camera traps in 11 Westchester County, NY parks and 1 park in Bronx, NY. Camera density should be smaller than the target animal's home range to minimize the risk of missing animals "in between" camera placements. We placed cameras approximately every 0.25km^2, which is smaller than the smallest recorded urban coyote home ranges of 0.5 – 3.0 km^2; [19,20,23]). Each park was surveyed for 4 - 6 weeks. The number of cameras in each park was thus determined by park area, though we placed a minimum of two cameras per park. Cameras were placed at a height of approximately 0.5m and were baited with fatty acid scent discs [26]. Cameras were never placed on human trails owing to a high chance for theft and vandalism. While we attempted to place cameras near coyote sign or on game trails, our priority was to minimize camera loss.

We calculated a detection rate (p) for our cameras by dividing the number of successful detections by the total number of trap-nights (TR) for each camera

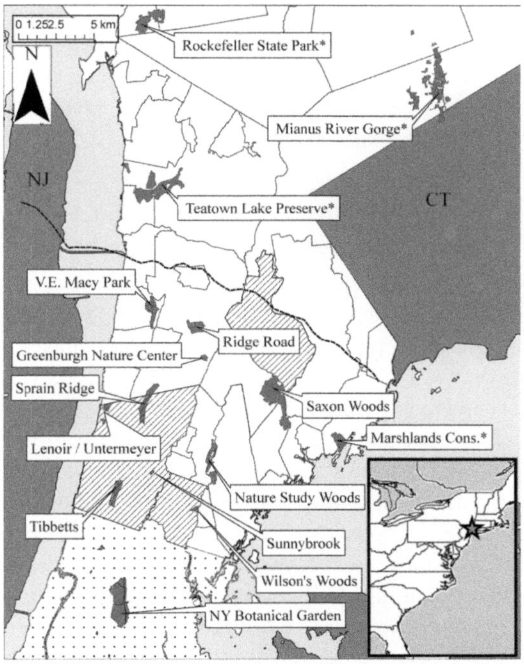

Fig. 1: Map of Westchester and Bronx Counties, NY, USA, showing study sites for 2010 camera trapping and municipal boundaries. The dashed line indicates Interstate 287. Sites marked with an asterisk were surveyed in previous studies and used for model validation only. Hatched areas indicate urban centers in Westchester County: White Plains (northernmost parcel), Yonkers (southwest), and Mt. Vernon (south-central). Dotted pattern indicates portions of Bronx and Manhattan Counties of New York City. Star marker in inset map shows general location of study area in the northeast USA.

Table 1: Total areas (TA), undeveloped areas (UA; primarily deciduous forest), and percent undeveloped (% UA) for sites surveyed for coyotes, Westchester County, NY, 2010.

Site	TA (ha)	UA (ha)	% UA	$HCIz^2$	$SD_{HCIz}{}^3$
Greenburgh Nature Center	13.16	10.47	80%	0.29	0.02
Lenoir Park	16.49	10.49	64%	0.46	0.17
Nature Study Woods	90.26	38.31	42%	0.40	0.07
New York Botanical Garden	101.17	45.86	45%	0.26	0.08
Ridge Road Park	69.77	51.33	74%	0.48	0.03
Saxon Woods Park	365.54	193.45	53%	0.45	0.06
Sunnybrook Park	4.65	0.86	18%	0.14	0.01
Sprain Ridge Park	100.11	90.64	91%	0.33	0.05
Tibbetts Brook Park	63.46	28.13	44%	0.11	0.01
Untermeyer Park	6.16	2.52	41%	0.36	0.01
V.E. Macy Park	84.10	55.49	66%	0.37	0.06
Wilson's Woods Park	12.35	0.64	5%	0.13	0.01
Marshlands Conservatory[1]	66.59	28.85	43%	0.51	0.05
Mianus River Gorge Preserve[1]	312.22	286.87	92%	0.82	0.07
Rockefeller State Park[1]	283.25	248.23	88%	0.53	0.03
Teatown Reservation[1]	142.80	125.99	88%	0.81	0.03

[1] Separate Westchester, New York camera surveys added for validation only
[2] HCIz: Predicted probability of coyote presence based on the mean values of HCI model parameters of all raster cells in the park; optimized threshold for categorizing predicted presence was >0.36.
[3] SD_{HCIz}: Standard deviation of model-predicted probabilities of all raster cells in the park.

deployment that photographed at least one coyote. This per trap-night rate was used to determine if those parks in which no coyotes were detected were likely true absences (i.e., coyotes were absent from the site) or false absences (i.e., coyotes were present at the site but were not photographed) by calculating a final detection rate given the total number of trap-nights in each park,

$$p_{\text{final}} = 1 - (1\text{-}p)^{\text{TR}}$$

If this final detection rate was >0.85, we were confident that we had sampled sufficiently to regard parks with no detections as truly "coyote-absent" for the purposes of HCI model validation.

For model validation purposes, we defined sampling units as whole parks, as opposed to designating individual cameras as independent sites. The reasons for this decision were based on the fact that multiple cameras in a particular park may not have been completely independent, and thus it would be inappropriate to model site occupancy at the scale of individual cameras. We were also more concerned with the applicability of the HCI model in predicting large-scale patterns of occupancy (e.g., whole parks versus locations within parks). Conversely, although most of our parks were quite far apart, it was still possible that individual coyotes could access and use multiple parks across the study area. However, management decisions will occur at the park level and thus obtaining predictions at this scale was useful. Additionally, since parks were distinct "greenspaces" within a dense urban matrix, we thought that the use of any particular park could be reasonably assumed to be an independent selection on the part of a coyote.

We were able to increase the size of our validation set by including results from concurrent camera trapping studies performed in four other Westchester County Parks. These sites – all located in the more northern portions of Westchester County – had confirmed detections of coyotes and provided validation sites from more rural areas.

The final HCI model recommended by Weckel et al. [9] consisted of distance to all forest types (dFOR), distance to grassland and agricultural land (dGRASS), and distance to pooled medium development and high development (>50% paved surface; dMHD), denoted as [dFOR + dGRASS + dMHD]. For a given location (i.e., raster cell), the HCI probability is a function of the distances to the closest 30 x 30m patch of each respective habitat type, based on the 2006 NLCD [27]. Since we defined our sampling unit as individual parks, we had to extrapolate HCI values – which occur on a 30 x 30m scale – to a single probability for each park. To obtain whole-park HCI's from the numerous raster cells that occur within every park, we calculated park averages for all covariates by averaging the raster values for each land use type within each park. A zonal HCI (HCIz) could then be calculated for each park using Weckel et al.'s [9] parameter values (Intercept = -0.141; β_{dFOR} = -0.00126; β_{dGRASS} = -0.00047; β_{dMHD} = 0.000881) and the average covariate values. HCIz was then used to investigate the accuracy of the model on predicting our observed camera data at each site. If the HCI model accurately predicted the camera survey data, we could be confident that the model predicts coyote site occupancy rather than detection patterns exhibited by Westchester residents who answered the original survey. All spatial calculations were performed using ARCGIS 9.3 Spatial Analyst [28].

When validating a species-occupancy model, one compares predictions to observed presences and absences at a set of validation sites. Model predictions, however, are continuous probabilities and must be converted to a set of binary presence-absence predictions. Thus, a threshold probability is necessary; sites with probabilities above this threshold are classified as predicted presences and those below the threshold are categorized as absences. Historically, a threshold of 0.5 has been used, but there are no particular reasons why 0.5 is optimal and in fact this threshold may be inappropriately arbitrary [29,30]. In determining the threshold probability for this study, we followed the recommendations of Freeman and Moisen [30], in which the observed prevalence (the proportion of presences/total sites) is maintained in the final predictions.

We then compared our model predictions to our observations. We calculated the proportion-correctly-classified (PCC; correct predictions/total sites), sensitivity (correctly predicted presences/all predicted presences), specificity (correctly predicted absences/all predicted absences), the true skill statistic (TSS) [31], and Cohen's kappa (model accuracy corrected for chance [30,32,33]) of the HCI model on the validation set using R statistical software and the PresenceAbsence package [34]. We also calculated the area-under-curve (AUC) for receiver-operator characteristic (ROC) plots of the predictions versus the observations [35].

Results

The 12 parks were surveyed for 1449 trap-nights total (Table 2) and coyotes were detected in 5 out of 12 parks. Despite our attempts to standardize camera density, five cameras were lost to theft and vandalism. This reduced our camera density in some parks below our goal of ⩾ 4 cameras/km^2, although no park had less than 1 camera/km^2 (Table 2). Cameras did not malfunction otherwise.

Table 2: Coyote presence and model predictions (HCIz) for camera trap data and coyote HCI models (Weckel *et al.* 2010), Westchester County and Bronx, New York, 2010.

Site	Trap-Nights	Camera Density (#/km^2)	# Detections[2]	Classified Correctly?
Greenburgh Nature Center	93	23.17	1	No
Lenoir Park	66	6.06	0	No
Nature Study Woods	112	4.94	5	Yes
New York Botanical Garden	485	7.41	0	Yes
Ridge Road Park	93	4.36	1	Yes
Saxon Woods Park	224	2.57	6	Yes
Sunnybrook Park	33	21.49	0	Yes
Sprain Ridge Park	33	1.00	0	Yes
Tibbetts Brook Park	119	3.84	0	Yes
Untermeyer Park	33	16.15	0	Yes
V.E. Macy Park	102	3.56	8	Yes
Wilson's Woods Park	56	16.20	0	Yes
Marshlands Conservatory[1]	X	X	X	Yes
Mianus River Gorge Preserve[1]	X	X	X	Yes
Rockefeller State Park[1]	X	X	X	Yes
Teatown Reservation[1]	X	X	X	Yes
Proportion Correct Presences (Sensitivity)				0.89
Proportion Correct Absences (Specificity)				0.86
Proportion Correctly Classified (PCC)				0.88

[1] Separate Westchester, New York camera surveys added for validation only.
[2] # Detections: Total number of coyote detections, defined as photographs taken on separate days.

Coyotes were found in parks that had lower zonal distances to forest and grassland and greater distances from medium-high development (Table 3). Detection rate for cameras in parks where a coyote was photographed at least once was 0.066 ± 0.012 per trap night, and all parks had at least a 0.90 chance of detecting a coyote (if one was present) given the total trap nights we surveyed. Using an optimized threshold of 0.36, the HCI model correctly classified 14 of the 16 validation sites (87.5%), and had kappa and TSS scores of 0.75 ± 0.17, substantially higher than chance (i.e., "substantial support" as described by Landis and Koch [33]). The AUC value of the ROC plot of the model was 0.90, substantially higher than chance (0.50). While a few of the parks fell very close to the threshold value of 0.36 (Table 2), parks that were occupied had a mean HCIz of 0.52 ± 0.26 (mean ± SD) and mean HCIz of unoccupied parks was 0.18 ± 0.14.

Table 3: Zonal distances between sites where coyotes were detected or not detected and nearest habitat types, Westchester County and Bronx, NY, 2010.

Habitat Type	Coyote-Detected Sites (SD)	Coyote-Undetected Sites (SD)
Forest (m)	162 (78)	239 (112)
Grassland (m)	1188 (360)	1206 (263)
Mid-High Development (m)	442 (160)	975 (238)

Discussion

The primary objective of this study was to determine if Weckel et al.'s [9] HCI model could predict actual coyote distribution in suburban/urban parks, rather than simply being an index of how likely residents tend to observe coyotes. By all measures, the HCI model appears to predict occupied and unoccupied parks well, with kappa and TSS scores above 0.70. Cohen's kappa and/or TSS are more appropriate measures of a model's accuracy than the raw proportion of correctly predicted observations; these indices reflect the correct proportion above what would be expected by chance. Kappa and TSS range from -1 to +1, with values $\leqslant$ 0 indicating agreement between predictions and observations equivalent or worse than chance, and 1.0 representing perfect agreement. There is some debate regarding whether kappa or TSS is the superior metric (see [30,31]); in this study, TSS and kappa were identical. Our third measure of model accuracy, the AUC of a ROC plot, shows the overall accuracy of a model at varying thresholds [29,30,32, but see 36]. An ROC plot displays true positives (sensitivity) vs. false positives (1 – specificity) across a large number of threshold values. A model that performs well will asymptote at 100% sensitivity at low levels of 1 – specificity. Thus, the AUC proportion shows how a model performs independent of threshold compared to randomly assigning observations (AUC = 0.5), as better models will generally have larger AUC's (e.g., >0.8 or >0.9). The HCI model's AUC was also very high (0.90).

The HCI model predicts well-documented patterns of coyote habitat use in suburban areas: coyotes are usually found closer to forest and farther from high-density development [17-23, 37). In particular, occupied parks had much higher mean distances to development than unoccupied parks, even though most parks were actually surrounded by medium- or high-density development (Table 3). This difference then is likely an indication of a positive relationship between coyote occupancy and park size and/or park linearity. HCIz values were calculated as an average of all raster cells in the park; thus, parks with more interior cells would have higher zonal distances to the surrounding development than smaller or more linear parks. For example, Tibbets Brook Park in the city of Yonkers has a few substantial wooded parcels and was close to other parks that were occupied by coyotes (Saxon Woods, White Plains; and Nature Study Woods, New Rochelle) but is comparatively linear, and raster cells in its wooded areas would still have relatively low distances to development.

The parks that had HCIz values very close to the threshold of 0.36 – Sprain Ridge, Untermeyer, and V.E. Macy – were some of the smaller and more urban parks that still had a high percentage of wooded area. V.E. Macy, the occupied park of the three, was the least linear and thus had the highest distance to medium-high development (distances to forest and grasslands were similar among the three) despite its similar location in the urban-suburban matrix. Even though these sites were ultimately classified correctly, the similar predicted HCIz values for some of our smaller, more urban parks may indicate that we are nearing the predictive extent of a suburban coyote-based model as we sample sites with resident coyotes that have acclimated to very urban settings. This could be the case for the coyote(s) detected at the very small and urban Greenburgh Nature Center, the only occupied park that was incorrectly predicted as occupied. In other large cities where coyotes have established themselves, researchers have observed changes in home range size [37-39], activity cycles [19,20], and variable use of non-natural areas [22,37,38].

We acknowledge that the use of a single average score (HCIz) to describe the complex spatial characteristics and habitat quality of parks is likely a simplification and more descriptive predictor variables would increase the accuracy of any predictive model. However, it was necessary to use a single, park-wide score to validate Weckel et al.'s [9] HCI model (which was based on three "distance-to" covariates). Additionally, the utility of a model by land managers is affected as much by its ease-of-use as its accuracy. Given the agreement between the predictions and observations in this study, a zonal average is useful for making baseline predictions while only requiring freely-available NLCD layers and basic knowledge of geographical analysis and software. Additionally, with the possible exception of Lenoir Park, predicted probabilities for all raster cells did not vary much from the HCIz score calculated for the whole park (Table 1). We are currently incorporating park- or area-specific measures such as perimeter/area ratios, potential corridors, and individual forest stand size into future candidate models that will be used in New York City (NYC) and hopefully provide better predictions by better describing landscape variability.

In addition to model testing and documenting the distribution of coyotes in southern Westchester, we hoped to pilot test field methods and the HCI model for use in an ongoing long-term survey of NYC parks. Given the spotty occurrence of coyotes in the NYC area, methods that allow for passive sampling (such as cameras) and utilize the millions of residents (such as citizen science programs) are needed at this time to cast a sufficiently wide net to detect the few coyotes at present that make their way through NYC. While residents' observations have alerted us and NYC officials to a number of coyotes in NYC, our cameras also detected coyotes in some parks that have not been reported by the public (Nagy and Weckel, unpublished data). Cameras have been frequently used to efficiently sample urban coyotes and other rare or wide-ranging mammal species in numerous other situations [40-44], including urban areas [45,46] and can be used to efficiently monitor a large area. In addition to our camera trapping effort, we are also designing scat collection and genetic mark-recapture studies and plan to use neighborhood surveys to record additional coyote sightings and to explore stakeholder opinions regarding coyotes and other urban wildlife in NYC.

An assumption of this analysis was that the study parks were independent sites, i.e., coyote presence at any of our parks was an indication of the suitability of that park, rather than its proximity to another suitable park. We believe this assumption held for most of our sites, but the same individual coyotes may have used multiple parks, particularly those that were relatively close (e.g., Lenoir and Untermeyer, see Figure 1) and thus some sites may not have been entirely independent. Indeed, a fundamental question concerning how coyotes colonize urban New York is to what extent geographically separate habitat patches can be considered connected in terms of the potential for coyote movement. We are currently employing year-long camera monitoring and genetic analysis of collected scat to address questions pertaining to activity patterns and population structure on the level of parks and individual coyotes, as well as to what extent existing coyote models are applicable in the extreme urban landscape of NYC.

The movement of coyotes from Westchester County into NYC and Long Island offers a unique opportunity to view the range expansion of a large predator across a suburban-to-urban-to-suburban spectrum, as the coyotes in Westchester County must cross through NYC in order to eventually enter the more suburban landscape of central and eastern Long Island. The arrival of a top predator to these habitats will have important ecological and wildlife management implications for the region. Thus, coyote researchers in the area should include education, outreach, and citizen involvement components into their research. As with any method citizen science has its own challenges or limitations (such as training volunteer researchers, the reliability or bias of data collected by non-experts, and potential difficulty in adapting technically complex methods to a citizen science framework [see 13-16]), that must be taken into account by researchers. However, the advantages of citizen science may include greater amounts of data over a large geographic area [13,47], access to private properties that make up the majority of the landscape in developed areas and are often the habitats specifically of interest to (sub)urban ecologists [9,48], and fostering of a sense of involvement and partnership amongst researchers and stakeholders [13,15, 47,49-52]. In building their model, Weckel et al. [9] used volunteer-submitted sightings because the study objectives required a large dataset gathered on private properties across an entire county. The objectives of this validation study required more systematic and verifiable sampling; however, we still made use of volunteer assistance during our field work and the early stages of analysis. In both cases, the volunteer-based portion of the research was an essential asset and advantage to the work, not a peripheral element. Including stakeholders in the research process can help educate the public about the necessities and/or benefits of particular conservation plans and management necessities as well as educate researchers as to the values and perspectives of others [6, 11-13,49,50,52].

Currently, coyotes are well-distributed in the southernmost portions of Westchester and the Bronx (the only mainland portion of NYC). However, they are very rare or absent in the island-bound boroughs of Manhattan, Queens, and Brooklyn (C. Nagy and M. Weckel, unpublished data). As a result, tracking and developing predictions regarding this range expansion will require long-term, year-round monitoring across a large area. Further surveys of the wildlife-related opinions and acceptance capacity of NYC residents would also inform management efforts and policy decisions. We encourage wildlife scientists working in urbanized areas to enlist residents and interested volunteers as observers and field assistants, and cooperate with volunteer-supported/grass-roots organizations that can assist with access, outreach, data collection, and funding.

Acknowledgements

We thank the Earthwatch staff who supported this project and especially the dedicated Earthwatch volunteers for their hard work and contributions. We would also like to thank D. Burg for his advice and assistance in the field and two anonymous reviewers for their thoughtful comments. This work was funded in part by Earthwatch Institute, through a grant made to C.E. Burns.

References

Five "key references", selected by the authors, are marked below (Three recommended (●) and two highly recommended (●●) papers).

1. Toomey, A., M. Weckel, C. Nagy, L. J. Gormezano, and S. Silver. 2012. The Last Frontier: The Eastern Coyote in New York City. Wildlife Professional 6: 54-57.
2. Stevens, T.H., T. A. More, and R. J. Glass. 1994. Public attitudes about coyotes in New England. Soc Natur Resour 7:57-66.
DOI:10.1080/08941929409380844
3. Reiter, D.K., M.W. Brunson, and R.H. Schmidt. 1999. Public attitudes towards wildlife damage management and policy. Wildlife Soc B 27:746-758.
4. Gompper, M. 2002. Top carnivores in the suburbs? Ecological and conservation issues raised by colonization of North-eastern North America by coyotes. BioScience 52:185-190.
DOI:10.1641/0006-3568(2002)052[0185:TCITSE]2.0.CO;2
5. Vaske, J.J. and M.D. Needham. 2007. Segmenting public beliefs about conflict with coyotes in an urban recreation setting. Journal of Park and Recreation Administration 25:79-98.
6. Wieczorek Hudenko, H., D. J. Decker, and W. F. Siemer. 2008. Stakeholder insights into the human-coyote interface in Westchester County, New York. Human Dimensions Research Unit 08-1. Cornell University, Ithaca.
7. Lukasik, V.M. and S.M. Alexander. 2011. Human-coyote interactions in Calgary, Alberta. Human Dimensions of Wildlife 16:114-127.
DOI:10.1080/10871209.2011.544014
8. Peterson, M.N., J.L. Birckhead, K. Leong, M. J. Peterson, T.R. Peterson. 2010. Rearticulating the myth of human-wildlife conflict. Conservation Letters 3:74-82.
DOI:10.1111/j.1755-263X.2010.00099.x
9. ●● Weckel, M. E., D. Mack, C. Nagy, R. Christie, A. Wincorn. 2010. Using citizen science to map human-coyote interaction in suburban Westchester County, NY. J Wildl Manage74: 1163-1171.
DOI:10.2193/2008-512

10. Homer, C. C. Huang, L. Yang, B. Wylie and M. Coan. 2004. Development of a 2001 National Landcover Database for the United States. Photogramm. Eng Remote Sensing 70: 829-840.
11. Couvet, D., F. Jiguet, R. Julliard, H. Levrel, and A. Teyssedre. 2008. Enhancing citizen contributions to biodiversity science and public policy. Interdiscipl Science Reviews 33: 95–103.
DOI:10.1179/030801808X260031
12. Silvertown, J. 2009. A new dawn for citizen science. Trends Ecol Evol 24: 467-471.
DOI:10.1016/j.tree.2009.03.017
13. Dickinson, J.L., B. Zuckerberg, and D.N. Bonter. 2010. Citizen science as an ecological research tool: challenges and benefits. Ann Rev Ecol Evol Syst. 41:149-172.
DOI:10.1146/annurev-ecolsys-102209-144636
14. Lee, T., M.S. Quinn, D. Duke. 2006. Citizen science, highways, and wildlife: using a web-based GIS to engage citizens in collecting wildlife information. Ecol Soc 11:11.
15. Delaney, D.G., C.D. Sperling, C.A. Adams, B. Leung. 2008. Marine invasive species: validation of citizen science and implications for national monitoring networks. Biol Invasions 10:117-128.
DOI:10.1007/s10530-007-9114-0
16. Cohn, J.P. 2008. Citizen science: Can volunteers do real research? BioScience 58:192-197.
DOI:10.1641/B580303
17. Atkinson, K. T. and D. M. Shackleton. 1991. Coyote ecology in a rural-urban environment. Can Field Nat 105: 49–54.
18. Quinn, T. 1995. Using public sighting information to investigate coyote use of urban habitat. J Wildl Manage 59: 238–245.
DOI:10.2307/3808936
19. Grinder, M.I. and P.R. Krausman. 2001. Home range, habitat use, and nocturnal activity of coyotes in an urban environment. J Wildl Manage 65: 887-898.
DOI:10.2307/3803038
20. Tigas, L. A., D. H. Van Vuren, and R. M. Sauvajot. 2002. Behavioral responses of bobcats and coyotes to habitat fragmentation and corridors in an urban environment. Biol Conserv 108: 299–306.
DOI:10.1016/S0006-3207(02)00120-9
21. Riley, S.P.D., R. M. Sauvajot, T.K. Fuller, E.C. York, D.A. Kamradt, C. Bromley, and R.K. Wayne. 2003. Effects of urbanization and habitat fragmentation on bobcats and coyotes in southern California. Conserv Biol 17: 566-576.
DOI:10.1046/j.1523-1739.2003.01458.x
22. Way, J. G., I.M. Ortega, E.G. Strauss. 2004. Movement and activity patterns of eastern coyotes in a coastal, suburban environment. Northeast Nat 11: 237–254.
DOI:10.1656/1092-6194(2004)011[0237:MAAPOE]2.0.CO;2
23. Fox, C. and C. Papouchis. 2005. Coyotes in Our Midst: Coexisting with an Adaptable and Resilient Carnivore. Born Free USA, Sacramento.
24. United States Census Bureau. 2010. Available via http://2010.census.gov/2010census/. Cited 26 April 2012.
25. Earthwatch Institute, US. Boston. Available via http://www.earthwatch.org. Cited 21 July 2012.
26. U.S. Department of Agriculture, Animal and Plant Health Inspection Service. Potacello Supply Depot, Pocatello.

27. Fry, J., G. Xian, S. Jin, J. Dewitz, C. Homer, L. Yang, C. Barnes, N. Herold, and J. Wickham. 2011. Completion of the 2006 National Land Cover Database for the Conterminous United States. Photogramm. Eng Remote Sensing 77: 858-864.
28. ESRI, Inc. 2009. ArcGIS Desktop 9.3.1
29. ● Fielding, A.H., and J.F. Bell. 1997. A review of methods for the assessment of prediction errors in conservation presence/absence models. Environ Conserv 24: 38-49. DOI:10.1017/S0376892997000088
30. ●● Freeman, E.A. and G. G. Moisen. 2008. A comparison of the performance of threshold criteria for binary classification in terms of predicted prevalence and kappa. Ecol Model 217: 48-58. DOI:10.1016/j.ecolmodel.2008.05.015
31. Allouche, O., A. Tsoar, and R. Kadmon. 2006. Assessing the accuracy of species distribution models: Prevalence, kappa and the true skill statistic (TSS). J Appl Ecol 43: 1223-1232. DOI:10.1111/j.1365-2664.2006.01214.x
32. ● Manel, S., H.C. Williams, and S.J. Ormerod. 2001. Evaluating presence-absence models in ecology: the need to account for prevalence. J Appl Ecol 38: 921-931. DOI:10.1046/j.1365-2664.2001.00647.x
33. Landis, J.R. and G.G. Koch. 1977. The measurement of observer agreement for categorical data. Biometrics 33: 159–174. DOI:10.2307/2529310
34. Freeman, E.A. and G. Moisen. 2008. PresenceAbsence: An R package for presence absence analysis. J Stat Softw 23: 1-31.
35. Fielding, A.H., and J.F. Bell. 1997. A review of methods for the assessment of prediction errors in conservation presence/absence models. Environ Conserv 24: 38-49. DOI:10.1017/S0376892997000088
36. Lobo, J. M., A. Jiménez-Valverde, and R. Real. 2008. AUC: a misleading measure of the performance of predictive distribution models. Global Ecol Biogeogr 17: 145–151. DOI:10.1111/j.1466-8238.2007.00358.x
37. ● Gehrt, S.D., C. Anchor, and L.A. White. 2009. Home range and landscape use of coyotes in a metropolitan landscape: Conflict or coexistence? J Mammal 90: 1045-1057. DOI:10.1644/08-MAMM-A-277.1
38. Grubbs, S.E. and P.R. Krausman. 2009. Use of urban landscape by coyotes. Southwest Nat 54:1-12. DOI:10.1894/MLK-05.1
39. Riley, S.P.D., R. M. Sauvajot, T.K. Fuller, E.C. York, D.A. Kamradt, C. Bromley, and R.K. Wayne. 2003. Effects of urbanization and habitat fragmentation on bobcats and coyotes in southern California. Conserv Biol17: 566-576. DOI:10.1046/j.1523-1739.2003.01458.x
40. Atwood, T.C., H.P. Weeks, and T.H. Gehring. 2004. Spatial ecology of coyotes along a suburban-to-rural gradient. J Wildl Manage 68: 1000-1009. DOI:10.2193/0022-541X(2004)068[1000:SEOCAA]2.0.CO;2
41. Weckel, M., W. Giuliano, and S. Silver. 2006. Jaguar (*Panthera onca*) feeding ecology: distribution of predator and prey through time and space. J Zool 270:25-30. DOI:10.1111/j.1469-7998.2006.00106.x
42. Kelly, M.J. and E.L. Holub. Camera trapping of carnivores: trap success among camera types and across species, and habitat selection by species, on Salt Pond Mountain, Giles County, Virginia. Northeastern Nat 15:249-262. DOI:10.1656/1092-6194(2008)15[249:CTOCTS]2.0.CO;2

43. Gompper, M.E., R.W. Kays, J.C. Ray, S.D. LaPoint, D.A. Bogan, and J.R. Cryan. 2006. A comparison of noninvasive techniques to survey carnivore communities in northeastern North America. Wildlife Soc B 34:1142-1151.
DOI:10.2193/0091-7648(2006)34[1142:ACONTT]2.0.CO;2
44. Sarmento, P.B., J. P. Cruz, C. I. Eira, C. Fonseca. 2010. Habitat selection and abundance of common genets *Genetta genetta* using camera capture-mark-recapture data. Eur J Wildl Res 56:59-66.
DOI:10.1007/s10344-009-0294-z
45. Harmsen, B.J., R.J. Foster, C.P. Doncaster. 2011. Heterogeneous capture rates in low density populations and consequences for capture-recapture analysis of camera-trap data. Popul Ecol. 53:253-259.
DOI:10.1007/s10144-010-0211-z
46. Hegglin, D., F. Bontadina, S. Gloor, J. Romer, U. Müller, U. Breitenmoser, and P. Deplazes. 2004. Baiting red foxes in an urban area: a camera trap study. J Wildl Manage 68:1010-1017.
DOI:10.2193/0022-541X(2004)068[1010:BRFIAU]2.0.CO;2
47. Ordeñana, M.A., K.R. Crooks, E.E. Boydston, R.N. Fisher, L.M. Lyren, S. Siudyla, C.D. Haas, S. Harris, S.A. Hathaway, G.M. Turschak, A.K. Miles, and D. H. Van Vuren. 2010. Effects of urbanization on carnivore species distribution and richness. J Mammal 91:1322-1331.
DOI:10.1644/09-MAMM-A-312.1
48. Devictor, V., R.J. Whittaker, and C. Beltrame. 2010. Beyond scarcity: Citizen science programmes as useful tools for conservation biogeography. Divers Distrib 16: 354-362.
DOI:10.1111/j.1472-4642.2009.00615.x
49. Webster, C.M., and S. DeStefano. 2004. Using public surveys to determine the distribution of greater roadrunners in urban and suburban Tuscon, Arizona. Proceedings of the 4th International Urban Wildlife Symposium.
50. Evans, C., E. Abrams, R. Reitsma, K. Roux, L. Salmonsen, and P. Marra. 2005. The neighborhood nestwatch program: Participant outcomes of a citizen-science ecological research project. Conserv Biol 19: 589-594.
DOI:10.1111/j.1523-1739.2005.00s01.x
51. Kransy, M. E., and R. Bonney. 2005. Environmental education through citizen science and participatory action research. Pages 292–320 *in* E. Johnson and M. Mappin, ed. Environmental education and advocacy: changing perspectives of ecology and education. Cambridge University Press, Cambridge.
52. Cooper, C.B., J. Dickinson, T. Phillips, and R. Bonney. 2007. Citizen science as a tool for conservation in residential ecosystems. Ecol Soc 12: 11.
53. Bonney, R., C. Cooper, J. Dickinson, S. Kelling, T. Phillips, K.V. Rosenberg, J. Shirk. 2009. Citizen science: a developing tool for expanding science knowledge and scientific literacy. Bioscience 59: 977-984.
DOI:10.1525/bio.2009.59.11.9

3

EVALUATION OF SOME HABITAT IMPROVEMENT ACTIONS IN A MEDITERRANEAN AREA THROUGH OBSERVATION OF REARED PHEASANT AND RED-LEGGED PARTRIDGE LAND USES

F. Santilli[1,*], G. Paci[2] & M. Bagliacca[2]

[1] Game & Wildlife Management Consulting, Campiglia Marittima (Li) Italy; email: perdix@teletu.it.

[2] Department of Veterinary Science, University of Pisa, Italy; email: gisella.paci@unipi.it and marco.bagliacca@unipi.it

* Corresponding author email: perdix@teletu.it.

Keywords

Red-legged partridge; *Alectoris rufa*; Pheasant; *Phasianus colchicus*; Habitat Improvement Actions; Feeders; Hedgerows.

Abstract

Red-legged partridge (*Alectoris rufa*) and pheasant (*Phasianus colchicus*) have suffered population decline in the last decades. Research suggested that decline is mainly related to the intensification of agricultural techniques and land abandonment. We evaluated habitat use and distribution of reared red-legged partridge pairs and cock pheasants in an estate representative of most of the agricultural Mediterranean areas from 2007 to 2012. Particular attention has been addressed to the effect of artificial habitat improvement actions (HIAs) and other management options such as supplemental feeding and hedgerows maintenance. We compared habitat use with its availability and analyzed spatial distribution by comparing points with pheasant cocks and partridge pairs with random ones.

HIAs were used more than their availability by both species and resulted to have an important role in their distribution like hedgerows. Feeders resulted to have a strong effect on pheasant distribution, but not on pairs of red-legged partridges.

Our results show that habitat management could be a fundamental tool for game-bird conservation, though the effects of feeders on red-legged partridge distribution need to be further investigated.

Introduction

Last century has witnessed a marked decline of red-legged partridge (*Alectoris rufa*) all through its range [1-4], including Italy [5]. In Tuscany (Central Italy), this species has become extinct at the beginning of the 20th century with the exception of a population still found in Elba Island. Several attempts to reestablish wild populations have been recently conducted, though not all of them turned out to be successful, probably due to the low quality of captive-reared birds [6-8]. In the last years, bag records of pheasants have shown a dramatic decline despite no significant variation in releasing rate and hunting pressure. In agreement with the trend observed in Italy, wild pheasant populations have declined since 1960s both in Europe and USA [9-13], although this is often masked by the repeated release of captive-reared birds.

This general decline may be variously explained. Still, the main reason behind it is the impact of agricultural intensification and subsequent land abandonment [14-16]. The increase of field size and monocultural farming, the reduction of grassy field margins, hedgerows and tree-rows, the large employment of herbicides and pesticides, and a general reduction of permanent cover have caused a dramatic loss of habitat heterogeneity [15-18].

Game birds management in Italy and other European countries is often based on restocking with captive-reared birds. This practice, however, is controversial and mostly ineffective [19-25], especially when it is not coupled with habitat management and proper predators control. Many studies suggest that the management of agricultural areas might be most effective for game-birds conservation when combined with a reduced predation pressure [26-29]. Habitat management is the basis for any recovery program of wildlife populations: not only may releasing alone be ineffective, but it may also pose significant sanitary and genetic risks.

For this reason, a deeper understanding of the habitat preference of these species may contribute to identify areas suitable for reintroduction of red-legged partridges, the possible spatial competition between the two galliformes, and habitat management strategies to preserve both species. Currently, habitat management choices mostly applied to improve food availability, shelter, and nesting covers are: crops for game (cereal or other unharvested crops, hay-fields mowed only outside the breeding season), overwinter stubble maintenance, beetle banks, and conservation headlands [30-33]. Supplemental feeding as well is a technique widely used especially to enhance game birds populations [34-37].

We evaluated habitat use of red-legged partridge pairs and cock pheasants in an estate of southern Tuscany which can be considered representative of many farmland areas of Central Italy where cereal cultivation has been substituted by fallow fields and trellis system vineyards. Most importantly, we aimed at evaluating the effectiveness of the habitat improvement actions (HIAs) and other management options such as supplemental feeding and hedgerows maintenance. In fact, all these improvements are often planned only on a theoretical basis, but the scientific evaluation of their effect on the behaviour of the released animals needs to be verified.

Study area

Study area is located in Grosseto province (South-western Tuscany, Italy; 10°59'E, 49°55'N). Climate can be defined as warm Mediterranean [38] and the annual mean temperature is 15.7°C (max=30°C in July; min=3°C in January). Average rainfall is 655 mm per year. The area covers 2.9 km^2 and is part of a wider hunting estate of 4.1 km^2. Of the study area, 16.6% is covered by Olm oaks (*Quercus ilex*), while 83.4% by arable land. This last is composed of crops (winter wheat and beans, and sunflowers; 22.0%), fallow fields (24.5%), vineyards (39.2%), and fields of olive tree groves (4.0%). Vineyards are cultivated with trellis driving system. Soil is regularly ploughed and grasses are controlled with herbicides. The remaining fields (10.3%) are formed by HIAs (the minimum requested by the provincial regulation is 3% of the agricultural area). These areas are composed of winter cereals, spring crops (sorghum and maize) and hay fields (lucerne and clover mixed with grasses). Cereal and spring crops remain unharvested until following winter or spring, while hay fields are mowed once a year in late March.

About 400 juvenile pheasants and 100 juvenile red-legged partridges, both captive-reared (artificial incubation), are released by the estate managers every summer for hunting purpose. Birds are purchased in a local farm and gradually released in summer using two open top pens of about 1 ha each. Pheasants are released in the pens at 80-90 days of age, partridges at 120-140 days. As releases started in 2005, the partridge and pheasant populations are the result of a mix of released and wild-born birds. The main population parameters of the two populations are reported in Tables 1 and 2.

Table 1: Main population parameters of pheasants in the study area from 2008 to 2012.

Year	Pheasants released	Pheasants shot	Spring cocks	Spring hens	Spring density	Broods	Broods production rate	Broods size (arithmetical mean)
2008	400	25	18	25	26.6	3	12.0	6.5
2009	400	55	20	15	25.2	5	33.3	5.8
2010	400	34	20	35	32.1	7	20.0	4.5
2011	400	70	20	31	30.7	8	25.8	3.3
2012	400	69	16	64	38.1	10	15.6	5.0

Table 2: Main population parameters of red-legged partridges in the study area from 2007 to 2012.

Year	Red-legged partridges released	Red-legged partridges shot	Spring pairs	Spring pairs density	Broods	Brood production rate	Broods size (arithmetical mean)
2007	300	10	20	6.9	11	60.0	4.6
2008	300	5	18	6.2	11	61.1	5.3
2009	100	7	32	11.0	14	43.8	4.2
2010	100	11	19	6.6	6	31.6	6.7
2011	100	10	25	8.6	10	40.0	7.2
2012	100	12	17	5.9	10	58.8	6.1

Supplemental food is provided all year round by means of 33 feeders set in the study area. Feeders contain a mix of grains (wheat, sorghum and cracked corn) and are put on wooden pallets at least 1.3 high to prevent wild boar and porcupine depredation. Pheasants can pick up on these feeders quite easily [36], while this is more difficult for partridges. Nonetheless, photo trapping showed red-legged partridges feeding on them too.

Every year, about 50 pheasants and 10 partridges are shot during the hunting season. Fox culling is practiced, and corvids (mainly magpies) are controlled by Larsen traps in accordance with Tuscan legislation.

Methods

Pairs of red-legged partridges were monitored from 2007 to 2012, while pheasants from 2008 to 2012. Counts were carried out by means of mapping methods [39,40]. Searching effort was equally distributed by a net of 4 transects totalling 21.9 km in length which covered the whole study area (Fig. 1). Each transect (average length 5.5 km) was completely conducted once a week from 15 March to 15 May and it took about 2 h to complete it. Surveys were carried out during the first 3 h after dawn and the last 3 h before dusk by two trained people (a gamekeeper and a technician) using a 4x4 vehicle. Playback recalls were not used since they could negatively affect red-legged partridge detection in low-density populations [41].

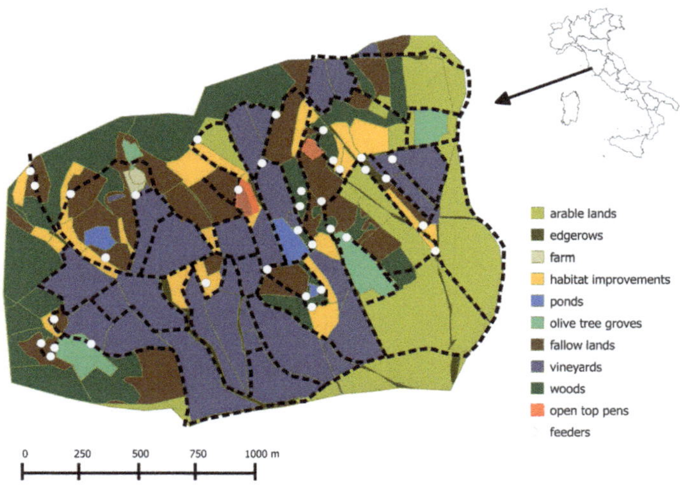

Figure 1: Study area. The dashed line represents the net of transects.

Each observation reporting the type of habitat was registered using a hand hold GPS. Observations of pairs were considered as single observations to avoid their dependence. As for pheasants, we considered only territorial cocks because they can be detected more easily than hens and no-territorial cocks. By so doing, the risk of underestimating the presence of pheasants in closer habitats was reduced. Surface areas covered by different types of habitat were measured from aerial photographs at 1:10.000 using a GIS software (QGIS 1.8.0).

The investigation of the two species' habitat preference was carried out using two different procedures. First, we used $\chi 2$ analysis for independent samples to test the hypothesis that partridges and pheasants occupy different habitats in relation to their availability [42]. When $\chi 2$ reached the minimum significant value ($P<0.05$), the hypothesis was rejected and Bonferroni simultaneous confidence intervals for observed occupying proportions were calculated through the formula:

$$P_i \pm Z_{\alpha 2k} \sqrt{Pi\,(1-P_i)/n} \text{ [43-45]}$$

where P_i is the observed occupying proportion for the ith habitat, Z is the upper standard normal table value corresponding to a probability tail area of $\alpha/2k$ with $\alpha=0.05$ or 0.01 and k=no. of considered habitats, and n is the total number of observations. When the expected occupying proportions felt outside calculated intervals, differences between expected and observed occupying proportions were considered significant. Second, we verified the null hypothesis that the observed distribution of birds was not different from a random distribution. For this purpose, we created a number of random points equal to the number of observations. Then, we measured the distance between each observation and random point from the nearest kind of habitat including the variables that were difficult to analyze with the forward method (*i.e.* hedgerows, woodland edges, water points and feeders). We compared the two distributions by Mann-Whitney test, and we performed Logistic Regression Analysis between points with partridge pairs and pheasant cocks and an equal number of random points, using the distances from the nearest habitat as predictor variables [46]. Logistic Stepwise regression was first performed using the forward method of selection and then the backward one; the probability to enter was set to 0.25 and that to leave to 0.10. Pheasant or partridge presence was coded as 1, while absence as 0. In the end, we compared the distances between feeders and the nearest HIAs with the distances between feeders and the equal number of random points in order to verify the relationship between them (Unpaired t test).

Results

Distances between feeders and HIAs (60.7, SE 9.77) turned out to be significantly shorter than those between feeders and random points (113.6, SE 18.71) ($P<0.05$). Habitat use was significantly different from what expected both for red-legged partridge pairs and pheasant cocks. Partridge pairs showed a high preference for HIAs, while woods and fallow fields were significantly avoided. Nonetheless, the avoidance of woods and fallow fields in our study may be the result of birds being hardly detectable in close habitats. The use of the other habitats did not differ from their availability (Table 3). Pheasant cocks as well showed a higher use of HIAs than the other habitats (Table 4). As compared with random points, partridge pairs were significantly closer to hedgerows, HIAs, and fallow fields (Table 5), while pheasant cocks were significantly closer to HIAs, hedgerows, feeders, fallow fields, and woodland edges (Table 6).

Table 3: Habitat use by red-legged partridge pairs (pooled years: 2007-2012).

Habitats	Availability (km^2)	Expected occupying proportions (km^2)	Observed occupying proportions (km^2)
Arable crops	0.18	0.19	0.22
HIAs	0.09	0.09	0.23**
Olive tree groves	0.03	0.03	0.05
Fallow fields	0.20	0.20	0.14*
Vineyards	0.33	0.32	0.36
Woods	0.17	0.17	<0.01**
Total	2.69	-	n=124

HIAs, habitat improvement actions. *$P<0.05$; **$P<0.01$.

Table 4: Habitat use by pheasant cocks (pooled years: 2007-2012).

Habitats	Availability (km^2)	Expected occupying proportions (km^2)	Observed occupying proportions (km^2)
Arable crops	0.18	0.19	0.17
HIAs	0.09	0.09	0.28**
Olive tree groves	0.03	0.03	0.03
Fallow fields	0.20	0.21	0.14
Vineyards	0.33	0.32	0.28
Woods	0.17	0.16	0.11
Total	2.69		n=94

HIAs, habitat improvement actions. **$P<0.01$.

Logistic regression analysis between partridge pairs observation and random points showed a negative effect of distances from HIAs, hedgerows, vineyards and fallow fields, and a positive one from woods, arable crops, and olive tree groves (Table 7). For pheasants, a negative effect of distances was registered from HIAs, hedgerows, feeders, woods, and fallow fields, and a positive one from arable crops (Table 8).

Table 5: Average distance (±SE) of partridge pairs between random points and habitat variables (Mann-Whitney U test).

Habitat variables	Mean distances		U	P
	Partridge pairs	Random points		
HIAs	148±12.7	181±12.9	6531	<0.05
Hedgerows	40±3.1	64±4.9	5246	<0.0001
Feeders	188±12.1	212±12.5	6802	n.s.
Vineyards	64±9.3	78±11.8	7248	n.s.
Arable crops	244±21.3	200±21.1	6763	n.s.
Fallow fields	97±11.4	127±11.0	6544	<0.05
Olive tree groves	303±14.8	287±14.5	7153	n.s.
Water points	354±19.2	376±18.4	6974	n.s.
Woodland edges	268±21.2	225±224.8	6698	n.s.
Releasing points	1334±521	1535±507	6698	n.s.

HIAs, habitat improvement actions; n.s., not significant.

Table 6: Average distance (±SE) of pheasant cocks between random points and habitat variables (Mann-Whitney U test).

Habitat variables	Mean distances		U	P
	Pheasant cocks	Random points		
HIAs	133±16.0	182±14.7	3293	<0.01
Hedgerows	50±4.2	75±6.7	3290	<0.01
Feeders	159±13.1	221±15.4	3230	<0.01
Vineyards	69±13.2	83±14.2	4212	n.s.
Arable crops	221±25.3	212±24.6	4155	n.s.
Fallow fields	91±10.5	137±14.3	3623	<0.05
Olive tree groves	303±18.1	291±17.3	4198	n.s.
Water points	358±23.6	389±22.9	4022	n.s.
Woodland edges	138±17.0	245±22.7	3210	<0.01
Releasing points	1337±550	1466±515	6698	n.s.

HIAs, habitat improvement actions; n.s., not significant.

Table 7: Results of logistic regression analysis on distances between points with partridge pairs and random ones.

Habitat variables	Estimate	SE	P
HIAs	+0.004	0.0012	<0.001
Hedgerows	+0.017	0.0038	<0.0001
Vineyards	+0.003	0.0014	<0.05
Arable crops	-0.002	0.0008	<0.01
Fallow fields	+0.004	0.0013	<0.01
Olive tree groves	-0.002	0.0010	<0.05
Woodland edges	-0.002	0.0007	<0.05
Constant	+0.712	0.372	<0.05
Log likelihood=147.4; R_{adj}^2=0.142			
Area under curve=0.745; model χ^2=P<0.0001			

HIAs, habitat improvement actions.

Table 8: Results of logistic regression analysis on distances between points with pheasant cocks and random ones.

Habitat variables	Estimate	SE	P
HIAs	+0.003	0.0012	<0.05
Hedgerows	+0.009	0.0035	<0.05
Feeders	+0.003	0.0013	<0.05
Arable crops	-0.003	0.0010	<0.01
Fallow fields	+0.004	0.0016	<0.01
Woodlands edges	+0.003	0.0011	<0.001
Constant	+2.019	0.443	<0.001
Log likelihood=107.13; R_{adj}^2=0.178			
Area under curve=0.766; model χ^2=P<0.0001			

HIAs, habitat improvement actions.

What did not contribute significantly to the model fit in partridges were feeders and water points, while ti was water, olive and vine points in pheasants. In both species the releasing points were discarded from the model during the logistic stepwise regression. Regression models [Rsquare (U)] explained 14.2% of the variance for partridges pairs and 16.0% of pheasants. The area under curve was 0.75 using pairs as the positive level and 0.77 using cock.

Discussion

Patterns of habitat use and selection recorded for red-legged partridges and pheasants in the area under study are consistent with others known for these game birds. Also, despite their importance [47], availability of water and water sources were not considered as a significant variable. This happened probably because drop irrigation was distributed in all vineyards and feeders (equipped with drinkers) provided water during summer. The effect of the releasing point was forgot by both species, probably on account of the time interval between release and counts (birds are released in summer and counts are carried out the next spring) and the small dimension of the area (2.9 km^2) with only two releasing points.

Partridges avoided the dense cover of woods that may also favor predation by mammals [48,49]. Fallow lands were avoided too, even if pairs seemed to stay close to these fields, as confirmed by the logistic regression model. This finding is only seemingly contradictory. Like gray partridges, red-legged are *birds of edges* which are attracted by field margins offering protection, like hedgerows [50,51]. Fallow fields are probably used only in part, but their presence seems to be important to provide cover and protection to aerial predators [52,53]. This also explains the strong effect of hedgerows on the distribution of the spring pairs. Other studies reported hedgerows as the most selected habitat by these species [8,54,55]. Conversely, the role of permanent crops as vineyards and olive tree groves in red-legged partridge presence and distribution remains unclear. Borralho *et al.* [50] found a positive effect of olive tree groves on the probability detection of red-legged partridges, but their use is probably related to the kind of cover at ground level. Vargas *et al.* [56] found that olive groves are a very important nesting habitat, but in summer they may represent an *ecological trap* when the herbaceous cover is dramatically reduced due to agricultural work. However that may be, in our case study olive groves represent a very small part of the study area and do not allow reliable assumptions. Traditional vineyards are known to positively act on red-legged partridge populations [57,58]. The highest abundance of partridges within the Mediterranean part of France was found in agricultural areas highly dominated by vineyards and/or cereal crops [57]. The habitat importance of vineyards is probably more related to its structure as shelter habitat, which ensures a good protection for partridges rather than to its food resource. In our study vineyards were the most important crops for partridges.

The effect of woodland edges, hedgerows and fallow lands on pheasant distribution in the study area was consistent with the ecology of this game bird, while permanent crops, vineyards and olive tree groves did not affect the pheasant positions. In fact, these habitats were excluded by the stepwise selection of the best parameters which affected the pheasant positions with respect to the random points.

Pheasants are primarily birds of woodland edge. In winter they spend most of the time within 30 m from the open ground. Woodland edges also are important features of pheasant breeding territories in spring, although some birds establish their territory along hedgerows, ditches, or in areas of rough ground [9]. In northern Italy, Nelli *et al.* [46] found that re-afforestation has an important role in determining pheasant male distribution.

Supplemental feeding is a popular and easy tool in game birds management. In our study, artificial feeding points seemed to have a strong effect on pheasant distribution, whereas this effect was weaker on partridge behaviour as feeders were discarded by stepwise selection. Possibly, the *suspended* feeders are less suitable for partridges, although these birds were seen feeding on them. Feeders were established mainly for pheasants and they might have been in wrong places for partridges (*i.e.* near woods). In addition, a competition between the two species for the use of feeders cannot be excluded. Despite supplemental feeding is a very common practice, the use of hoppers by partridges needs further investigation, at least when pheasants are present. Hoppers may save the time commonly spent for the feeding activities, thus reducing the risk of predation [29], yet some possible disadvantages cannot be ignored. Feeding with wheat which is low in fiber and protein may reduce the functionality of the internal organs of partridges (heart and caeca) [59].

Both species confirmed to show a high selection of HIAs. Many studies showed that these actions may improve density and/or reproductive success of wild galliformes [13,46,60-64], especially if they are scattered in the landscape [65]. In the study area, different types of crops (winter cereals, spring crops and hay fields) were cultivated side by side, so they could offer both food (seeds and insects) and shelter. Furthermore, HIAs were not treated with herbicides so that they yielded more weeds than seeded crops. Seeds used for HIAs are probably a good and safe food source for birds since they are not treated with pesticides: recently, pesticides-treated seeds have been shown to represent a risk for farmland birds [66]. HIAs remain unharvested and mowing (for grasslands) is postponed until the end of the nesting season (late July) providing food and cover when there is no plentiful supply [67].

Conclusions

In a landscape dominated by intensive agriculture (here vineyards), habitat improvements were confirmed to act as key factors. Indeed, HIAs were used significantly more than favorable crops such as vineyards, having a positive impact not only on game-birds but also on the biodiversity of the agricultural zones [68-70]. However, since feeders were often located close to the HIAs, we cannot exclude that the effect of HIAs on the distribution of the game-birds in our study area was affected by the presence of feeders (and vice versa), even if there was no correlation between distances from feeders and distances from HIAs.

References

Five "key references", selected by the authors, are marked below (Three recommended (●) and two highly recommended (●●) papers).

1. Cramp, S. & Simmons, K.E.L., 1980. The birds of the Western Pale-arctic. Volume II. Oxford University Press, Oxford, UK.
2. Potts, G.R., 1980. The effects of modern agriculture, nest predation and game management on the population ecology of partridges (*Perdix perdix* and *Alectoris rufa*). Ecol. Res. 2:2-79.
3. Aebischer, N.J.& Potts, G.R., 1994. Red-legged partridge. In: G.M. Tucker and M.F. Heath (eds.) Birds in Europe. Their conservation status. Birdlife International, Cambridge, UK, pp. 214-215.
4. Blanco-Aguiar, J.A., Virgòs, E., Villafuerte, R., 2004. Perdiz Roja (*Alectoris rufa*). In: A. Madrono, C. Gonzàlez, & J.C. Atienza (eds.) Libro rojo de las aves de Espana. Direcciòn General para la Biodiversidad-SEO/BirdLife, Madrid, Spain, pp. 182-185.

5. Spanò, S., 2010. La pernice rossa. Il Piviere edizioni, Alessandria, Italy.
6. Meriggi, A. & Mazzoni, R., 2004. Dynamics of a reintroduced population of red-legged partridges *Alectoris rufa* in central Italy. Wildlife Biol. 10:1-9.
7. Santilli, F., Dell'Omodarme, A. & Bagliacca, M., 2005. Acclimatisation of farm reared red-legged partridges (*Alectoris rufa* L.) in two protected areas of southern Tuscany. Ann Fac Med Vet Univ Pisa 58:213-218.
8. Santilli, F., Galardi, L. & Bagliacca, M., 2012. First evaluation of different captive rearing techniques for the re-establishment of the red-legged partridge populations. Avian Biol. Res. 5:147-153. http://dx.doi.org/10.3184/175815512X13441821968945
9. Hill, D. & Robertson, P., 1988. The pheasant: ecology, management and conservation. London Blackwell Scientific, London, UK.
10. Tapper, S.C., 1988. Population changes in gamebirds. In: P.J. Hudson and M.R.W. Rands (eds.) Ecology and management of gamebirds. BSP Professional Books, Oxford, UK, pp. 18-47.
11. Warner, R.E., Mankin, P.C., David & L.M., Etter, L.S., 1999. Declining survival of ring-necked pheasant chicks in Illinois during the late 1900s. J. Wildlife Manage. 63:705-710. http://dx.doi.org/10.2307/3802660
12. Csànyi, S., 2000. The effect of hand-reared pheasants on the wild population in Hungary: a modeling approach. Hungarian Small Game Bulletin 5:71-82.
13. Draycott, R.A.H., Bliss, T.H., Carrol, J.P.& Pock, K., 2009. Provision of brood-rearing cover on agricultural land to increase survival of wild ring-necked pheasant *Phasianus colchicus* broods at Seefeld Estate, Lower Austria, Austria. Conservation Evidence 6:6-10.
14. Chamberlain, D.E., Fuller, R.J., Bunce, R.G.H., Duckworth, J.C. & Shrubb, M., 2000. Changes in the abundance of farmland birds in relation to the timing of agricultural intensification in England and Wales. J. Appl. Ecol. 37:771-778. http://dx.doi.org/10.1046/j.1365-2664.2000.00548.x
15. Robinson, R.A. & Sutherland, W.J., 2002. Post-war changes in arable farming and biodiversity in Great Britain. J. Appl. Ecol. 39:157-176. http://dx.doi.org/10.1046/j.1365-2664.2002.00695.x
16. Stoate, C., Boatman, N.D., Borralho, R.J., Carvalho, C., Snoo, G.D. & Eden, P., 2001. Ecological impacts of arable intensification in Europe. J. Environ. Manage. 63:337-365. http://dx.doi.org/10.1006/jema.2001.0473
17. Butler, S.J, Boccaccio, L., Gregory, R.D., Vorisek, P. & Norris, K., 2010. Quantifying the impact of land-use change to European farmland bird populations. Agr. Ecosyst. Environ. 137:348-357. http://dx.doi.org/10.1016/j.agee.2010.03.005
18. ●● Buenestado, F.J., Ferreras, P., Delibes-Mateos, M., Tortosa, F.S., Blanco-Aguiar, J.A. & Villafuerte, R., 2008. Habitat selection and home range size of red-legged partridges in Spain. Agr. Ecosyst. Environ. 126:158-162. http://dx.doi.org/10.1016/j.agee.2008.01.020
19. Pérez, J.A., Alonso, M.E., Gaudioso, V.R., Olmedo, J.A., Díez, C. & Bartolomé, D., 2004. Use of radiotracking techniques to study a summer repopulation with red-legged partridge (*Alectoris rufa*) chicks. Poultry Sci. 83:882-888. http://dx.doi.org/10.1093/ps/83.6.882
20. Blanco-Aguiar, J.A., González-Jara, P., Ferrero, M.E., Sánchez-Barbudo, I., Virgós, E., Villafuerte, R. & Dávila, J.A., 2008. Assessment of game restocking contributions to anthropogenic hybridization: the case of the Iberian red-legged partridge. Anim. Conserv. 11:535-545. http://dx.doi.org/10.1111/j.1469-1795.2008.00212.x

21. Santilli, F.& Bagliacca, M., 2008. Factors influencing pheasant *Phasianus colchicus* harvesting in Tuscany (Italy). Wildlife Biol. 14:281-287.
http://dx.doi.org/10.2981/0909-6396(2008)14[281:FIPPCH]2.0.CO;2

22. Sokos, C.K., Birtsas, P.K. & Tsachalidis, E.P., 2008. The aims of galliforms release and choice of techniques. Wildlife Biol. 14:412-422.
http://dx.doi.org/10.2981/0909-6396-14.4.412

23. Gaudioso, V.R., Pérez, J.A., Sánchez-García, C., Armenteros, J.A., Lomillos, J.M. & Alonso, M.E., 2011. Isolation from predators: a key factor in the failed release of farmed red-legged partridges (*Alectoris rufa*) to the wild? Brit. Poultry Sci. 52:155-162.
http://dx.doi.org/10.1080/00071668.2010.549668

24. Gaudioso, V.R., Sánchez-García, C., Pérez, J.A., Rodríguez, P.L., Armenteros, J.A. & Alonso, M.E., 2011. Does early antipredator training increase the suitability of captive red-legged partridges (*Alectoris rufa*) for releasing? Poultry Sci. 90:1900-1908.
http://dx.doi.org/10.3382/ps.2011-01430

25. Santilli, F. & Bagliacca, M., 2012. Occurrence of eggs and oocysts of intestinal parasites of pheasant (*Phasianus colchicus*) in droppings collected in differently managed protected areas of Tuscany (Italy). Eur. J. Wildlife Res. 58:369-372.
http://dx.doi.org/10.1007/s10344-011-0552-8

26. Boatman, N.D.& Brockless, M.H., 1998. The Allerton project: farmland management for partridges (*Perdix perdix*, *Alectoris rufa*) and pheasants (*Phasianus colchicus*). In: M. Birkan, L.M. Smith, N.J. Aebischer, F.J. Purroy, & P.A. Robertson (eds.) Perdix VII: International Symposium on Partridges, Quails and Pheasants. Gibier Faune Sauvage, Paris, France, pp. 563-574.

27. Stoate, C., 2002. Multifunctional use of a natural resource on farmland: wild pheasant (*Phasianus colchicus*) management and the conservation of farmland passerines. Biodivers. Conserv. 11:561-573.
http://dx.doi.org/10.1023/A:1015564806990

28. ● Buner, F., Jenny, M., Zbinden, N. & Naef-Daenzer, B., 2005. Ecologically enhanced areas–a key habitat structure for re-introduced grey partridges *Perdix perdix*. Biol. Conserv. 124:373-381.
http://dx.doi.org/10.1016/j.biocon.2005.01.043

29. Potts, G.R., 2012. Partridges. Collins, London, UK.

30. Chiverton, P.A., 1999. The benefits of unsprayed cereal crop margins to grey partridges *Perdix perdix* and pheasants *Phasianus colchicus* in Sweden. Wildlife Biol. 5:83-92.

31. Rodgers, R.D., 2002. Effects of wheat-stubble height and weed control on winter pheasant abundance. Wildlife Soc. B. 30:1099-1112.

32. Stoate, C., Henderson, I.G. & Parish, D., 2004. Development of an agrienvironment scheme option: seed bearing crops for farmland birds. Ibis 146(Suppl.2):203-209.
http://dx.doi.org/10.1111/j.1474-919X.2004.00368.x

33. Thomas, S.R., Goulson, D. & Holland, J.M., 2001. Resource provision for farmland gamebirds: the value of beetle banks. Ann. Appl. Biol. 139:111-118.
http://dx.doi.org/10.1111/j.1744-7348.2001.tb00135.x

34. Hoodless, A.N., Draycott, R.A.H., Ludiman, M.N. & Robertson, P.A., 1999. Effects of supplementary feeding on territoriality, breeding success and survival of pheasants. J. Appl. Ecol. 36:147-156.
http://dx.doi.org/10.1046/j.1365-2664.1999.00388.x

35. Draycott, R.A.H., Woodburn, M.I.A., Carroll, J.P. & Sage, R.B., 2005. Effects of spring

supplementary feeding on population density and breeding success of released pheasants *Phasianus colchicus* in Britain. Wildlife Biol. 11:177-182.
http://dx.doi.org/10.2981/0909-6396(2005)11[177:EOSSFO]2.0.CO;2

36. Santilli, F. & Azara, S., 2011. Effect of supplementary feeding on survival and breeding success of wild pheasants. Extended abstract of the International Union of Game Biologist, XXX Congress and Perdix XII, Barcelona, pp. 61-62.

37. Mayot, P., 1999. Aménagements pour la perdrix: résultats d'une enquête régionale. Bull. Mens. ONC 249:28-32. Available from: http://www.oncfs.gouv.fr/IMG/file/oiseaux/galliformes/plaine/pg_amenagement.pdf

38. Peel, M.C., Finlayson, B.L. & McMahon, T.A., 2007. Updated world map of the Köppen-Geiger climate classification. Hydrol. Earth Syst. Sci. 11:1633-1644.
http://dx.doi.org/10.5194/hess-11-1633-2007

39. Pepin, D., 1983. Utilisation et valeur de diverses méthodes d'estimation de la densité de la Perdrix rouge (*Alectoris rufa*) au printemps. In: Actos XV Congr. Int. de Fauna Cinegetica y Silvestre, Trujillo, Peru, pp. 725-735.

40. Bibby, C.J., 2000. Bird census techniques. Elsevier, Amsterdam, The Netherlands.

41. Tizzani, P., Negri, E., Silvano, F., Malacarne, G., & Meneguz, P.G., 2012. Does the use of playback affect the estimated numbers of red-legged partridge *Alectoris rufa*? Anim. Biodivers. Conserv. 35:429-435.

42. Siegel, S., 1956. Non parametric statistics: for the behavioral sciences. Hill Book Company, New York, USA.

43. Neu, C.W., Byers, C.R. & Peek, J.M., 1974. A technique for analysis of utilization-availability data. J. Wildlife Manage. 38:541-545.
http://dx.doi.org/10.2307/3800887

44. Byers, C.R., Steinhorst, R.K. & Kraussman, P.R., 1984. Clarification of a technique for analysis of utilization-availability data. J. Wildlife Manage. 48:1050-1053.
http://dx.doi.org/10.2307/3801467

45. Alldredge, J.R. & Ratti, J.T., 1986. Comparison of some statistical techniques for analysis of resource selection. J. Wildlife Manage. 50:157-165.
http://dx.doi.org/10.2307/3801507

46. ●● Nelli, L., Meriggi, A., & Vidus Rosin, A., 2012. Effects of Habitat Improvement Actions (HIAs) and reforestations on Pheasants (*Phasianus colchicus*) in northern Italy. Wildlife Biol. 8:121-130.
http://dx.doi.org/10.2981/11-022

47. Sánchez-García, C., Armenteros, J.A., Alonso, M.E., Larsen, R.T., Lomillos, J.M. & Gaudioso, V.R., 2012. Water-site selection and behaviour of red-legged partridge *Alectoris rufa* evaluated using camera trapping. Appl. Anim. Behav. Sci. 137:86-95.
http://dx.doi.org/10.1016/j.applanim.2012.01.013

48. Meriggi, A., Montagna, D. & Zacchetti, D., 1991. Habitat use by partridges (*Perdix perdix* and *Alectoris rufa*) in an area of northern apennines, Italy. Boll. Zool. 58:85-89.
http://dx.doi.org/10.1080/11250009109355733

49. Tapper, S.C., Potts, G.R., & Brockless, M.H., 1996. The effect of an experimental reduction in predation pressure on the breeding success and population density of grey partridges *Perdix perdix*. J. Appl. Ecol. 33:965-978.
http://dx.doi.org/10.2307/2404678

50. Borralho, R., Stoate, C. & Araújo, M., 2000. Factors affecting the distribution of Red-legged Partridges *Alectoris rufa* in an agricultural landscape of southern Portugal. Bird Study 47:304-

310.
http://dx.doi.org/10.1080/00063650009461190

51. ● Casas, F. & Vinuela, J., 2010. Agricultural practices or game management: which is the key to improve red-legged partridge nesting success in agricultural landscapes? Environ. Conserv. 37:177-186.
http://dx.doi.org/10.1017/S0376892910000299

52. Rands, M.R.W., 1986. Effects of hedgerow characteristics on partridge breeding densities. J. Appl. Ecol. 23:479-487.
http://dx.doi.org/10.2307/2404030

53. Rueda, M.J., Baragaño, J.R. & Notario, A., 1993. Nidification de la Perdix Rouge (*Alectoris rufa*) dans la région de La Mancha (Espagne). Bull. Mens. ONC 184:2-9.

54. Rands, M.R.W., 1987a. Hedgerow management for the conservation of partridges *Perdix perdix* and *Alectoris rufa*. Biol. Conserv. 40:127-139.
http://dx.doi.org/10.1016/0006-3207(87)90063-2

55. Rands, M.R.W., 1987b. Recruitment of gray and red-legged partridges (*Perdix perdix* and *Alectoris rufa*) in relation to population density and habitat. J. Zool. 212:407-418.
http://dx.doi.org/10.1111/j.1469-7998.1987.tb02912.x

56. Vargas, M., Duarte, J. & Farfan, M.A, 2011. Red legged partridges (*Alectoris rufa*) chick survival in relation to habitat structure in mediteranean farmlands. Book of Abstract XXXth IUGB Congress and Perdix XIII, Barcelona, Spain.

57. Peiró, V. & Blaanc, P.C., 1998. Système d'Information Géographique et gestion de la Perdrix Rouge (*Alectoris rufa*) dans la plaine viticole de l'Hérault (France). Gibier Faune Sauvage 15:355-378.

58. Ponce-Boutin, F., Mathon, J. & Le Brun, T., 2009. Impact of game crops intensification and hunting management on red-legged partridge. In: S.B. Cederbaum, B.C. Faircloth, T.M. Terhune, J.J. Thompson, J.P. Carroll (eds.) Gamebird 2006: Quail VI and Perdix XII, Warnell School of Forestry and Natural Resources, Athens, GA, USA, pp. 267-270.

59. Millán, J., Gortazar, C. & Villafuerte, R., 2003. Does supplementary feeding affect organ and gut size of wild red-legged partridges *Alectoris rufa*? Wildlife Biol. 9:229-233.

60. Meriggi, A., Mazzoni della Stella, R., Brangi, A., Ferloni, M., Masseroni, E., Merli, E. & Pompilio, L., 2007. The reintroduction of Grey and Red-legged partridges (*Perdix perdix* and *Alectoris rufa*) in central Italy: a metapopulation approach. Ital. J. Zool. 74:215-237.
http://dx.doi.org/10.1080/11250000701246484

61. Bro, E., Mayot, P. & Reitz, F., 2012. Effectiveness of habitat management for improving grey partridge populations: a BACI experimental assessment. Anim. Biodivers. Conserv. 35:405-413.

62. Drake, J.F., Kimmel, R.O., Smith, J.D. & Oehlert, G., 2009. Conservation Reserve Program grasslands and ring-necked pheasant abundance in Minnesota. In: S.B. Cederbaum, B.C. Faircloth, T.M. Terhune, J.J. Thompson, J.P. Carroll (eds.) Gamebird 2006: Quail VI and Perdix XII, Warnell School of Forestry and Natural Resources, Athens, GA, USA, pp. 302-314.

63. Matthews, T.W., Taylor, J.S. & Powell, L.A., 2012. Mid-contract management of Conservation Reserve Program grasslands provides benefits for ring-necked pheasant nest and brood survival. J. Wildlife Manage. 76:1643-1652.
http://dx.doi.org/10.1002/jwmg.409

64. Matthews, T.W., Taylor, J.S. & Powell, L.A., 2012. Ring-necked pheasant hens select managed Conservation Reserve Program grasslands for nesting and brood-rearing. J. Wildlife Manage. 76:1653-1660.
http://dx.doi.org/10.1002/jwmg.410

65. Santilli, F., Scarselli, D., Mazzarone, V. & Mazzoni della Stella, R., 2009. Miglioramenti ambientali e popolazioni di fagiano in provincia di Pisa. Alula 16:64-66.

66. Lopez-Antia, A., Ortiz-Santaliestra, M.E., Mougeot, F. & Mateo, R., 2013. Experimental exposure of red-legged partridges (*Alectoris rufa*) to seeds coated with imidacloprid, thiram and difenoconazole. Ecotoxicology 22:125-138.
http://dx.doi.org/10.1007/s10646-012-1009-x

67. ● Rantanen, E.M., Buner, F., Riordan, P., Sotherton, N. & Macdonald, D.W., 2010. Habitat preferences and survival in wildlife reintroductions: an ecological trap in reintroduced grey partridges. J. Appl. Ecol. 47:1357-1364.
http://dx.doi.org/10.1111/j.1365-2664.2010.01867.x

68. Parish, D.M. & Sotherton, N.W., 2004. Game crops as summer habitat for farmland songbirds in Scotland. Agr. Ecosyst. Environ. 104:429-438.
http://dx.doi.org/10.1016/j.agee.2004.01.037

69. Parish, D.M. & Sotherton, N.W., 2004. Game crops and threatened farmland songbirds in Scotland: a step towards halting population declines? Capsule during winter songbirds were far more abundant in game cover crops than conventional agricultural habitats. Bird Study 51:107-112.
http://dx.doi.org/10.1080/00063650409461341

70. Sage, R.B., Parish, D.M., Woodburn, M.I. & Thompson, P.G., 2005. Songbirds using crops planted on farmland as cover for game birds. Eur. J. Wildlife Res. 51:248-253.
http://dx.doi.org/10.1007/s10344-005-0114-z

4

Alternative Methodology for Handling and Marking Meso-Mammals for Short-Term Research

I.D. Parker[1,*], R.R. Lopez[1], N.J. Silvy[1], D.S. Davis[2] & J.C. Cathey[1]

[1] Department of Wildlife and Fisheries Sciences, Texas A&M University, College Station, TX 77801.

[2] Department of Veterinary Pathobiology, Texas A&M University, College Station, TX 77801.

* Corresponding author address: 663 Fellows Ave., Syracuse, NY 13210;
E-mail: iparkergto@yahoo.com.

Keywords

Capture;
Dye;
Ink;
Mark-recapture;
Squeeze cage.

Abstract

We evaluated squeeze cages and water-soluble inks and dyes as methods for handling and marking of meso-mammals in a short-term capture-recapture study. Study animals exhibited no adverse physiological impacts from capture or marking techniques. We also observed no negative impacts on trap success for the duration of the study. We found these methods quick (5 minutes maximum hold time), safe (no observed injuries of animals or researchers) and reliable (effective marking and mark retention for a minimum of 12 days).

Handling and marking techniques may be inappropriate or cost prohibitive for short-term meso-mammal studies. These studies are often characterized by large numbers of simultaneous animal captures that must be documented and marked quickly with limited personnel, but do not require long-term or permanent markings (e.g., ear tags, branding). Meso-mammals are loosely defined as mammals larger than rodents (though sometimes including the largest rodents) up to roughly fox-sized mammals (*Vulpes* spp.) [1]. Handling and marking meso-mammals is a time intensive, expensive, and potentially dangerous undertaking that requires solid justification and methodological scrutiny. Research has linked wildlife capture, handling, handling duration, and marking with increased mortality and behavioral changes in study subjects; thus highlighting the need for safe and simplified methods [2,3,4,5]. Accepted guidelines now mandate a well-structured protocol that places emphases on the welfare of the study species and the safety of the researcher [6,7]. Current and past researchers have followed a variety of methodologies to balance the needs of the project with the obligations of ethical research. Methodologies for handling and marking generally follow similar frameworks: 1) immobilization either through physical restraint or chemical immobilization, 2) internal marking (e.g., Passive Integrated Transponders [PIT]), and 3) external marking (temporary, semi-permanent, or permanent marking [8]). Many of these techniques are unnecessary or impractical for short-term meso-mammal studies. For example, chemical immobilization often requires large time investments (60-120 minutes per animal is typical) as researchers must carefully monitor animals through chemical application and recovery [9,10]. Additionally, research has demonstrated increased mortality and behavioral alterations may result from chemical immobilization [11,12]. These methods often require expensive equipment (e.g., chemicals) and extended preparation time, and cause increased difficulty in attaining institutional approval. These methods also become highly restrictive when researchers expect to capture many individual animals at once, the environment is intemperate (e.g., warm temperatures), or both. Therefore, researchers and managers look for handling and marking methods for meso-mammals that increase wildlife survival and researcher safety at minimum cost and maximum efficiency.

Researchers have often used squeeze cages in controlled environments (e.g., laboratory settings) to immobilize wildlife for injections or data collection, though we found little to indicate that squeeze cages have been used to immobilize wild animals in-situ for marking [13,14,15]. The concept is similar to squeeze chutes (often integrated into corrals) used to safely and temporarily immobilize large domestic animals and many ungulate species for agricultural or research purposes [16]. Although squeeze cages cause stress in captured animals; we believed handling time reduction and safe immobilization would decrease the risk of capture myopathy [17]. Squeeze cages also reduced the risk of injury to the handler by minimizing physical contact between animal and handler (i.e., animal always in trap or cage, squeeze cage manipulated through handles set back from cage surface). Additionally, temporary marking with dyes and inks has proven effective in many previous studies with no recorded adverse physiological effects [8,18,19]. We attempted fur clipping as an alternative temporary marking technique with little success. Raccoon fur is thick and requires powerful electric clippers or manual clippers. These were imprecise in marking and it was difficult to safely access and control the animal for clipping. We did not use manual clippers due to concerns of animal lacerations or punctures.

In 2008, we began researching the role of meso-mammals in water quality problems. Our overall analyses required estimation of meso-mammals population densities. As such, one of the goals of our research was to develop handling and marking methodologies for meso-mammals that fulfilled the prescribed protocols of wildlife handling and the needs of our larger research project (seasonal density estimates of meso-mammals). Since we needed only age, sex, and capture/recapture data, we emphasized short holding times. The challenge was to handle and mark conscious meso-mammals in a quick and safe manner using physical restraint and ephemeral dyes and inks. Additionally, as our data collection occurred in a grassland floodplain and our study species were likely to encounter water either through direct contact with the creek or wet grass, our markings needed to last even when wet. Many meso-mammals are large enough to cause handling difficulty outside of a trap; thereby, often necessitating chemical immobilization. We felt that the risks of chemical immobilization were not justified by the needs of our research (i.e., short-term marking). We sought to apply methods that limited handling time, subjected the meso-mammals to the minimum of stress and discomfort, provided reliable short-term marking (3-week retention), and provided safety to researchers. All capture and handling was conducted in compliance with approved permits issued by Texas Parks and Wildlife Department and Institutional Animal Use and Care Committee at Texas A&M University.

From 2008–2009, we conducted research on 2 private ranches (Property A, 518 ha; Property B, 660 ha) in Brazos County, Texas. Each ranch stocked cattle (Property A, 1 cow/10.4 ha; Property B, 1 cow/2.2 ha) on typical post oak savannah that consisted of a mix of upland/bottomland grasslands with scattered post oak woodlands (*Quercus stellata*). We determined meso-mammal density by analyzing trapping numbers in live-traps (81 cm x 25 cm x 30 cm; Tomahawk Live Trap, Tomahawk, Wisconsin, USA). We trapped on both properties using a grid-design ([20]; 42 traps total for each property) with 250-m spacing between traps that had been shown to adequately sample animals highly attracted to baits (e.g., raccoons (*Procyon lotor*), Virginia opossums (*Didelphis virginiana*)). Trapping grids and arrays were randomly placed on each property and trapped for 12 consecutive days using traps baited with canned dog food, apples, bananas and fish scent.

We immobilized Virginia opossums and raccoons using a Tomahawk squeeze cage (50 cm x 27.5 cm x 30 cm, 5.9 kg; Tomahawk Live Trap, Tomahawk, Wisconsin, USA; Figure 1). Total handling time ranged from 3-5 minutes. We decided against chemical immobilization due to concerns of capture myopathy and increased holding times [8,21]. The squeeze cage door was removed and placed against the opening of the box trap. The animal was blocked from the opening by 2 dividers (1 divider can be breached by the animal) as the door to the trap was locked open. The cage and trap were held together by holding the trap by the handle and wedging a foot behind the cage. The dividers were then removed and most animals moved voluntarily to the squeeze cage seeking exit from the trap. If unwilling, the animal was spurred by gentle prodding from the dividers. At this point, the door to the trap was allowed to close and the door to the squeeze cage was set into place. The animal was immobilized through gentle squeeze pressure from the movable wall. The squeeze cage was only used for marking and was not used for recaptured animals.

Fig. 1: Tomahawk squeeze cage (50 cm x 27.5 cm x 30 cm, 5.9 kg; photo courtesy of Tomahawk Live Trap).

We uniquely marked captured animals that were immobilized in the squeeze cage using a small paint brush to apply semi-permanent water-based hair dye (Hot Topic, San Jose, California, USA) or India ink (Dr. Ph. Martin, Golden, Colorado, USA) and released them 3–5 minutes later. All inks and dyes were bright colors (e.g., pink, magenta, blue) likely to show up on variably colored individuals. The hair dye was artificially scented by the manufacturer though we found through multiple recaptures the scent dissipated within 24 hours. We avoided colors found in the animals' fur (e.g., orange, black, white, yellow). Markings were placed in a variety of combinations and colors unique to each individual. Markings were limited to the back, rump, sides, or hips to reduce loss due to grooming or everyday activities (e.g., walking, swimming). Care was taken to work the colorant deep into the fur. Finally, the animal was released from the squeeze cage by removing the main door.

We grid-trapped 2,520 traps-nights during the study and used the squeeze cage to mark 192 raccoons and 23 Virginia opossums. We later recaptured 136 raccoons (80 separate individuals) and 52 Virginia opossums (15 separate individuals). We found the squeeze cage method was impractical or dangerous for nine-banded armadillos (*Dasypus novemcinctus*) and striped skunks (*Mephitis mephitis*) due to unique morphology (armadillo shell) or chemical defenses (skunk).

Although all marks were strongly recognizable for the course of the mark-recapture sessions, marks faded sooner on raccoons than opossums likely due to behavior (e.g., frequent movement in and out of water). We found an increasing percentage of captures were marked animals over the course of the trapping periods (Figure 2-A). This implied mark resilience over time as individuals marked in the first few days remained marked until the last few days. We also had relatively stable capture numbers over the same time period (summed over 3 day intervals; Figure 2-B). We observed no changes in body condition over the course of the trapping sessions. We also had no injuries to captured wildlife or researchers due to use of squeeze cage and no impacts on fur quality from dye and ink application. No animals escaped during handling and marking activities.

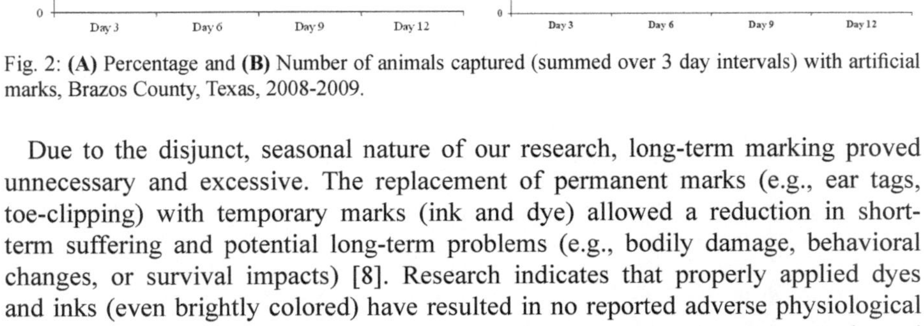

Fig. 2: **(A)** Percentage and **(B)** Number of animals captured (summed over 3 day intervals) with artificial marks, Brazos County, Texas, 2008-2009.

Due to the disjunct, seasonal nature of our research, long-term marking proved unnecessary and excessive. The replacement of permanent marks (e.g., ear tags, toe-clipping) with temporary marks (ink and dye) allowed a reduction in short-term suffering and potential long-term problems (e.g., bodily damage, behavioral changes, or survival impacts) [8]. Research indicates that properly applied dyes and inks (even brightly colored) have resulted in no reported adverse physiological affects [8]. It also reduced concerns from permitting agencies (i.e., Texas Parks and Wildlife Department and Texas A&M University's Institutional Animal Use and Care Committee) related to animal suffering and distress. The nature of marking obviated the need for lengthy restraint and placed the onus on short but productive marking and examination followed by quick release. Many hair dyes used in wildlife management and research have a strong chemical smell, but the artificial scent supplied by the manufacturer was relatively mild and quickly dissipated. Researchers are ethically bound to impact wildlife only for the period necessary for data collection [7]. However, the size and aggression of many meso-mammal species obviates established small-mammal handling techniques (e.g., handling cones). We found that squeeze cages provided a quick, safe, and reliable method for immobilizing raccoons and Virginia opossums. The water-based hair dye and ink provided durable, but short-term marks that minimized impact on these animals. Our methods minimized handling time (3–5 minutes) and impact on animals. Although difficult to make direct comparisons due to different species, circumstances, and research needs; our handling times were less than or comparable to other studies that did not use chemical immobilization to handle wildlife [e.g., 22,23]. Additionally, we required only a single researcher to safely capture and mark meso-mammals. Finally, this methodology was faster and safer for the animal than chemical immobilization (i.e., short handling time, no mortalities or observed declines in body condition).

Our use of a relatively light-weight, commercially available squeeze cage and vibrant, water-based dyes and inks with simplified application allowed individually identifiable marking of animals with a variety of color pelages (including dark colors that generally obviate application of black dye) at a minimum of handling time and animal trauma. This simplified handling of often difficult species to control and provided the handler with the opportunity to individually mark an animal. For instance, raccoons were often very mobile within the trap and aggressive towards the handler; but use of the squeeze cage prevented excessive movement and reduced opportunities for raccoon aggression. This ultimately reduced handling time. Our technique would likely work for other meso-mammals with potential for aggression (e.g., weasels (Mustelids), feral cats (*Felis catus*)). We recommend these methodologies for short-term meso-mammal research studies with high numbers of expected captures and restricted time and monetary budgets.

Acknowledgements

We thank U.S. Department of Agriculture, U.S. Geological Survey, Texas A&M University and private landowners for their support.

References

Five "key references", selected by the authors, are marked below (Three recommended (●) and two highly recommended (●●) papers).

1. Hoffmann, A., J. Decher, F. Rovero, J. Schaer, C. Voigt and G. Wibbelt. 2010. Field Methods and Techniques for Monitoring Mammals. Pages 482-529 in J. Eymann, J. Degreef, C. Häuser, J. C. Monje, Y. Samyn and D. Vanden Spiegel, editors. Manual on field recording techniques and protocols for All Taxa Biodiversity Inventories and Monitoring. Pensoft Publishers, Sofia, Bulgaria.
2. Livezey, K.B. 1990. Toward the reduction of marking-induced abandonment of newborn ungulates. Wildlife Society Bulletin 18: 193-203.
3. Beringer, J., L.P. Hansen, W. Wilding, J. Fischer and S. L. Sheriff. Factors affecting capture myopathy in white-tailed deer. Journal of Wildlife Management 60: 373-380. DOI: 10.2307/3802238

4. McMahon, C., J. van den Hoff and H. Burton. 2005. Handling intensity and the short- and long-term survival of elephant seals: addressing and quantifying research effects on wild animals. Royal Swedish Academy of Sciences 34: 429-429.
5. McCleery, R.A., R.R. Lopez and N.J. Silvy. 2007. An improved method for handling squirrels and similar-sized mammals. Wildlife Biology in Practice 1: 39-42.
6. Animal Care and Use Committee. 1998. Guidelines for the capture, handling, and care of mammals as approved by the American Society of Mammalogists. Journal of Mammalogy 79: 1416-1431.
DOI: 10.2307/1383033
7. ●● Powell, R.A. and G. Proulx. 2003. Trapping and marking terrestrial mammals for research: integrating ethics, performance criteria, techniques, and common sense. Institute for Laboratory Animal Research Journal 44: 259-276.
8. ●● Silvy, N.J., R.R. Lopez and M.J. Peterson. 2012. Techniques for marking wildlife. Pages 230-257 in N.J.Silvy, editor. The Wildlife Techniques Manual. The Johns Hopkins University Press, Baltimore, MD, USA.
9. Belant, J.L. 2004. Field immobilization of raccoons (*Procyon lotor*) with Telazol and Xylazine. Journal of Wildlife Diseases 40:787-790.
10. Baldwin, J.R., J.B. Winstead, L.D. Hayden-Wing, T.J. Kreeger and M.R. Dzialak. 2010. Field sedation of coyotes, red foxes, and raccoons with medetomidine and atipamezole. Journal of Wildlife Management 72: 1267-1271.
DOI: 10.2193/2007-395
11. Peterson, M.N., R.R. Lopez, P.A. Frank, M.J. Peterson and N.J. Silvy. 2003. Evaluating capture methods for urban white-tailed deer. Wildlife Society Bulletin 31: 1176-1187.
12. Arnemo, J.M., P. Ahlqvist, R. Andersen, F. Berntsen, G. Ericsson, J. Odden, S. Brunberg, P. Segerström and J.E. Swenson. Risk of capture-related mortality in large free-ranging mammals: experiences from Scandinavia. Wildlife Biology 12: 109-113.
DOI: 10.2981/0909-6396(2006)12[109:ROCMIL]2.0.CO;2
13. Gregg, D.A. and L.D. Olson. 1975. The use of ketamine hydrochloride as an anesthetic for raccoons. Journal of Wildlife Diseases 11: 335-337.
14. Ratnaswamy, M.J., R.J. Warren, M.T. Kramer and M.D. Adam. 1997. Comparisons of lethal and nonlethal techniques to reduce raccoon depredation of sea turtle nests. Journal of Wildlife Management 61: 368-376.
DOI: 10.2307/3802593
15. Terio, K.A., S.B. Citino, and J.L. Brown. 1999. Fecal cortisol metabolite analysis for noninvasive monitoring of adrenocortical function in the cheetah (*Acinonyx jugatus*). Journal of Zoo and Wildlife Medicine 30: 484-491.
16. ● Schmitt, S.M., T.M. Cooley, L.D. Schrader and M.A. Bradley. 1983. A squeeze chute to restrain captive deer. Wildlife Society Bulletin 11: 387-380.
17. Nicholson, D.S., R.L. Lochmiller, M.D. Stewart, R.E. Masters and D.M. Leslie, Jr. 2000. Risk factors associated with capture-related death in eastern wild turkey hens. Journal of Wildlife Diseases 36: 308-315.
18. ● Ellis, D.H. and C.H. Ellis. 1975. Color marking golden eagles with human hair dyes. Journal of Wildlife Management 39: 445-447.
DOI: 10.2307/3799927
19. Facka, A.N., P.L. Ford and G.W. Roemer. 2008. A novel approach for assessing density and range-wide abundance of prairie dogs. Journal of Mammalogy 89: 356-364.
DOI: 10.1644/06-MAMM-A-450R.1
20. ● Gehrt, S.D., L.L. Hungerford and S. Hatten. 2001. Drug effects on recaptures of raccoons. Wildlife Society Bulletin 29: 833-837.
21. Main, M.B. and L.W. Richardson. 2002. Response of wildlife to prescribed fire in southwest Florida pine flatwoods. Wildlife Society Bulletin 30: 213-221.
22. Lindner, E. and O. Fuelling. 2002. Marking methods in small mammals: ear-tattoo as an alternative to toe-clipping. Journal of Zoology 256: 159-163.
DOI: 10.1017/S0952836902000195
23. Craft, M.E. 2007. Capture and rapid handling of jackals (*Canis mesomelas* and *Canis adustus*) without chemical immobilization. African Journal of Ecology 46: 214-216.
DOI: 10.1111/j.1365-2028.2007.00782.x

5

Home Ranges of the Radio-Collared Bengal Tigers (*Panthera tigris tigris* L.) in Pench Tiger Reserve, Madhya Pradesh, Central India

A. Majumder[1], S. Basu[1], K. Sankar[1,*], Q. Qureshi[1], Y.V. Jhala[1], P. Nigam[1], R. Gopal[2]

[1] Wildlife Institute of India, P.O. Box # 18, Chandrabani, Dehra Dun-248001, Uttarakhand, India.
[2] Member Secretary, National Tiger Conservation Authority, Government of India, New Delhi.
[*] Corresponding author: sankark@wii.gov.in,

Keywords
Ranging Pattern;
Radio-Telemetry;
Monitoring.

Abstract
Home ranges of three radio-collared Bengal tigers [one adult female (AF), one adult male (AM) and one sub-adult male (SAM)] were studied between March 2008 and December 2011 in Pench Tiger Reserve (PTR), Madhya Pradesh. Using 100% Minimum Convex Polygon (MCP), the estimated home ranges of AF (n= 750 locations), AM (n= 136 locations) and SAM (n= 739 locations) were 43 km^2, 55.1 km^2 and 52.2 km^2 respectively. Using 95% Fixed Kernel (FK), the corresponding figures were 32.1 km^2, 64.1 km^2 and 19.1 km^2. The core area of individual activity for each tiger, as determined by 65% FK, was 10.3 km^2 for AF, 20.3 km^2 for AM and 6.6 km^2 for SAM. The estimated overlap area between AM and AF using 95% FK was 19.2 km^2 (65%), whereas it was 15.4 km^2 (48%) between AF and SAM. The AF recruited three, four and five cubs in her 1st (May 2008 and all died during post natal stage), 2nd (October 2008) and 3rd litter (October 2010) respectively. Minimum 44% of the original natal area was used by AF at the time of raising her 2nd litter and 46% in her 3rd litter. Though there was a gradual extension of the annual home range, as observed during the first two years (2008-2010), but the core activity area remained considerably the same for AF over the study period. The study revealed that a minimum of 25 to 30 km^2 undisturbed area was required for a breeding female in PTR where both the wild prey density (348.2/km^2) and biomass (12384.7 kg/ km^2) were found to be high with adequate ground cover for the successful raising of cubs up to the dispersal stage. Pench and its surrounding forested habitat need to be protected for the breeding and dispersing tiger population.

Introduction

Home range is the area normally traversed by an individual animal or group of animals during activities associated with feeding, resting, reproduction, shelter-seeking and other factors important to an animal's survival [1, 2, 3]. The ranging patterns and social behaviour of many mammals show distinct intersexual differences [4, 5]. Female reproductive success is linked with an ability to exploit resources, whereas male reproductive success is governed by an ability to find and mate with females [4, 6]. Radio telemetry provides data on carnivore home range size and social organization, which can be used to derive, estimates of densities [7, 8, 9, 10, 11], habitat use [12, 13] and survival rates [14, 15].

The tiger (*Panthera tigris* L.) is the largest of all felids found in diverse habitat types and show remarkable tolerance to the variation in altitude, temperature and rainfall regimes [16]. Being a large and charismatic carnivore, the tiger has attracted considerable attention in the conservation community the world over [17]. In spite of this overwhelming care, the status of this predator has been deteriorating rapidly [17, 18]. Recently, this has reached more serious proportions than at any other time in the history of tiger conservation in India. A total of eight tiger subspecies are commonly recognised: Bengal tiger (*Panthera tigris tigris*), Sumatran tiger (*P. t. sumatrae*), Amur tiger (*P. t. altaica*), Indo-China tiger (*P. t. corbetti*), South China tiger (*P. t. amoyensis*), Caspian tiger (*P. t. virgata*), Javan tiger (*P. t. sondaica*) and Bali tiger (*P. t. balica*) [19]. Three of these, the Caspian, Javan and Bali tiger, went extinct in the last fifty years. The last Bali tiger was killed in 1937, the Caspian tiger became extinct in the 1950s and the last wild Javan tiger was seen in 1972. The wild South China tiger has not been directly observed since the 1970s and is now believed to be extinct in the wild too [20].

In wild condition tiger can potentially hunt prey varying from the small mammals, medium-sized ungulates to the large bovids [21], the preferred prey weighing between 60 and 250 kg [22]. They are territorial and wide-ranging and the effective size of their territory is a function of the density and biomass of prey species in its habitat [7, 23]. Therefore, studies undertaken to address these issues on tiger are important because it can provide a better insight of tiger ecology and behavior. Radio-telemetry studies on tiger have been effectively used to address these questions [7, 8, 17, 24, 25, 26, 27]. However, the fact those home ranges can vary widely within a species, which makes data transfer between study areas difficult [28]. It is therefore important to estimate site specific ranging patterns of tigers for better understanding, conservation and management. Our study area the Pench Tiger Reserve (PTR), Madhya Pradesh is one of the important source populations of tiger in the Central Indian landscape [21]. No detailed long-term study on ranging, recruitment and dispersal pattern of tiger has been documented earlier in this source population site. The present study was aimed to estimate the home ranges, recruitment and dispersal pattern of three radio-collared tigers in PTR.

Study area

The 758 km^2 area of PTR, Madhya Pradesh (79° 9' E to 79° 22' E and 21° 38 N' to 21°50' N) comprises of Indira Priyadarshini Pench National Park (PNP) (292 km^2), Mowgli Pench Wildlife Sanctuary (PWS) (118 km^2) and Reserved Forest (348 km^2) of Seoni and Chindwara Forest Circles. The intensive study was carried out in the critical tiger habitat or Core Area of PTR (410 km^2) i.e. PNP and PWS. Jhala *et al.* [29] reported a population of 23± 0.77 tiger in this intensive study area. The broadly classified forest types of PTR are tropical dry deciduous ($5A/C_{1b}$ and $5A/C_3$) and tropical moist deciduous forests ($3B/C_{1c}$) [30]. Teak *Tectona grandis* L. and its associated species in the area represent a transition from the tropical dry deciduous to the tropical moist deciduous forests [31]. According to Forest Survey of India [www.fsi.org.in], the forest cover of the study area was classified into 1) Non-Forest (Canopy density < 10%), 2) Open Forest (Canopy density between 10% and 40%), 3) Moderately Dense Forest (Canopy density between 40% and 70%) and 4) Very Dense Forest (Canopy density > 70%), (Figure 4). Being located in the southern lower reaches of the Satpura range of hills, the terrain is undulating (350 m–650 m) in most areas of the tiger reserve [31].

The climate of PTR is sub-tropical and characterized by winter (October-February), summer (March-June) and monsoon (July-Sep¬tember) with a mean annual rainfall of 1400 mm. The temperature ranged from 2°C in winter to 49.5°C in summer during the study period. The River Pench is the main source of water, which flows, north to south, 74 km through the reserve. On its extreme south a dam was constructed. The draw-down area coming under the submergence of Totladoh (Pench) reservoir is 11.7 km^2 and covers 1.55% of PTR, Madhya Pradesh [31]. As summer approaches, these areas, from where the water gradually recedes downstream, become lush green meadows attracting high numbers of wild herbivores. Apart from tiger, the other carnivore species found in PTR are leopard (*Panthera pardus*), dhole (*Cuon alpinus*), jackal (*Canis aureus*) and jungle cat (*Felis chaus*). Palm civet (*Paradoxurus hermaphrodites*), ratel (*Mellivora capensis*), small Indian civet (*Viverricula indica*), two species of mongoose viz. common mongoose (*Herpestes edwardsii*) and ruddy mongoose (*Herpestes smithii*) are the small carnivores. Wolf (*Canis lupus*) and Indian fox (*Vulpes bengalensis*) occur on the fringes and outside the protected area limits. Sloth bear (*Melursus ursinus*) is the only bear species found in this reserve [31].

The wild ungulates found here are chital (*Axis axis*), sambar (*Rusa unicolor*), nilgai (*Boselaphus tragocamelus*), gaur (*Bos gaurus*), barking deer (*Muntiacus muntjac*), chousingha (*Tetraceros quadricornis*) and wild pig (*Sus scrofa*) [21, 31]. There are over 51648 inhabitants, 36143 livestock in 99 villages around the notified buffer zone of PTR [32]. The locals are predominantly tribal (62%) belonging to Gond and Baiga community [32].

Methods

Three tigers, one adult female (AF: T-15), one adult male (AM: T-2) and one sub-adult male (SAM: T-38), were radio-collared and monitored between March 2008 and December 2011 (Table 1) to understand the ranging and dispersal pattern of tigers. The animals were radio-collared following immobilization with "Hellabrunn mixture" (125 mg xylazine + 100 mg ketamine/ml) [33]. Initial dose depended on estimated body weight and sex of the animal and ranged from 2.5 ml for sub-adult to 2.75 ml for adult female and 3 ml for adult male. The estimated body weight of male, female and sub-adult were 200-220 kg, 140-150 kg and 100-120 kg respectively. Adult male however required supplementation and was administered a second injection of 1.2 ml. The drug mixture was remotely injected from the elephant back using Tele-inject projector (Model 4V.310). The induction time ranged between 26-36 minutes and effective handling time was 25-29 minutes. The drug combination induced reliable immobilization with good muscle relaxation and no significant alteration in physiological responses. The sedated animals were measured, weighed, radio-collared, and photographed. Individuals were aged using tooth wear, eruption, and body size, evidence of sexual development and previous births, and overall body condition. Both, Very High Frequency (Telonics TM) and Satellite radio-collars (Telonics™ and Vectronics™ weighed less than 1% of the body weight of the tiger (Table 1). Yohimbine hydrochloride administered intramuscularly at a dose of 5 mg for every 50 mg of xylaxine [34] provided effective reversal for all the tigers and

Table 1: Home range estimates of three radio-collared Bengal tigers (*Panthera tigris tigris*) in Pench Tiger Reserve, Madhya Pradesh (March 2008 - December 2011).

Tiger Id	Age-Sex	Body weight (kg)	Collaring period	Collar type	Collar weight (kg)	Number of location	Home Ranges in km² 100 % MCP	95% FK	65 % FK
T-15	Adult female	145	11/3/2008 to 31/12/2011	Very High Frequency (VHF) (Telonics ™)	0.763	750	44	32.1	10.3
T-2	Adult male	>200	1/1/2010 to 30/1/2010	Satellite & VHF (Telonics ™)	1.346	136	55.1	64.1	20.3
T-38	Sub-adult male	180	11/3/2010 To 30/8/2010	Satellite & VHF (Vectronics™)	1.324	739	52.2	19.1	6.6

MCP= Minimum Convex Polygon, FK= Fixed Kernel

resulted in complete recovery that ranged from 10-28 minutes. The radio-collared tigers were radio-tracked and monitored either from the elephant back or by vehicle at different times of the day and night so as to record location data throughout the 24 hour period. During the study period, attempts were made to locate at least one and when possible two animals on daily basis following their signals. The located animals were 'homed in' from the elephant back at a distance between 50 and 75 m. Care was taken not to disturb tiger's natural behaviour or movement [35]. Once the animal was sighted, a hand-held Global Positioning System (GPS) was used to record the location with its coordinates. The actual sighting of each animal helped the accuracy of the GPS locations obtained for use in home-range analysis and also gave additional data on activity, behaviour and interaction with other tiger especially with cubs. These locations were also used in home range analysis taking only locations separated by a 6 hour interval in order to avoid auto correlation [36].

In addition, recruitment pattern of the AF was assessed with her home range during the cub-rearing period and compared with her own natal area. As AF was not radio-collared during her early growing stage (< 2 years), the locations of AF along with her mother were collected between May 2006 and June 2007 using Intensive Tiger Search Method (ITSM) [37] and camera trapping [18].

In total 80 pairs of camera traps were deployed (Figure 4) in a systematic 2 km X 2 km grid [18] covering an area of > 410 km^2 during the study period. Individual tigers were identified from their unique stripe patterns [38]. As studied by Sharma and Jhala [37] in Kanha Tiger Reserve, Madhya Pradesh, an album of photographs of all the individual tigers operating in the study area was developed with a brief description of location–capture history including AM and AF and also to identify non collared mother of AF to study the natal area of AF. Using ITSM, tigers were located without affecting their natural movement between 6 am to 10 am from elephant back (n=3 elephants). Tiger pugmarks and alarm calls by prey species (chital, sambar, nilgai and common langur) were followed to locate the animal. In addition, we also collected information on tiger locations from forest beat guards, patrolling their individual beats (lowest administrative setup for any forest in India) on foot on a regular basis. Every GPS locations of tigers obtained from ITSM and camera traps were recorded and plotted on the map to estimate minimum home range.

The 100 % MCP method is the most widely used method [39] and is also more useful for comparison with other relevant studies [7, 8, 24, 26, 40, 41]. The FK is rated as one of the most robust and least biased estimator of home range and performs well even with small sample sizes, auto-correlated data and considers the shape of underlying utilization distributions [42, 43, 44]. The smoothing parameter of FK isopleths was chosen using the Least Square Cross Validation (LSCV) procedure [43, 45]. The 95% fixed kernel method also takes care of exploratory movements and outlying fixes [36, 43, 46]. Thus, home ranges of all three tigers were estimated using both 95% FK and 100% MCP for the present study. In case of the AF, the annual home ranges for four consecutive years were also estimated using both MCP and FK method. The MCP (100%) method was used to estimate the minimum home range or natal area of the AF.

FK (%) isopleths versus the area of the home range were plotted and the point of inflexion of the home range curve was considered for the isopleth defining the core area of the home range [40]. Overlap comparisons were calculated using the 95% FK estimator. Percent overlap between adjacent home ranges was computed as the average proportion of overlap between two home ranges [43, 46].

Distance sampling by line transect method [47, 48, 49] was used to asses prey density and biomass around the den sites of collared female. In total 12 line transects were laid and walked early in the morning when animals found to be more active [50]. Potential prey species of large carnivores were recorded during transect walk along with following parameters such as species, cluster size, animal bearing (with compass) and angular sighting distance (using laser range finder). The density of prey species was calculated using programme DISTANCE *Version 4.1* [51].

Results

A total of 750 locations were collected from AF, 136 locations from AM and 739 locations from SAM (Table 1). GPS fixes were plotted on the map and home ranges were estimated using Hawths tool of Arc GIS 9.3 [40]. The actual body weights of all three tigers were presented on Table 1. The 100% MCP ranges versus sample size plot showed an asymptote for the adult male and female confirming adequacy of sampling. The asymptote was reached at different sample sizes for individual animals during the study period (for AM between 115 and 120 from overall 136 locations and for AF between 460 and 465 from overall 750 locations) (Fig. 1). The sub-adult tiger ranges were found expanding and hence did not reach asymptote (Fig. 1). On an average, Fixed Kernel isopleth graphs for all three tigers showed an inflexion at 65% (Fig. 2), and this isopleth value was used as an estimate of the core area of home ranges for all three tigers. The estimated home range (100% MCP and 95% FK) of all three tigers was given in Table 1 and Figure 3a and 3b. The core activity (65% FK) area of AF, as determined by 65% FK, was 7.1 km^2 for the 2nd litter and 5 km^2 for the 3rd litter (Table 1, Fig. 3a and 3b). The estimated overlap area between AM and AF (using 95% FK) was 19.2 km^2 (65%), whereas it was 15.4 km^2 (48%) between AF and SAM. As the collaring-period was not the same for both AM and SAM, their overlapped area was not assessed.

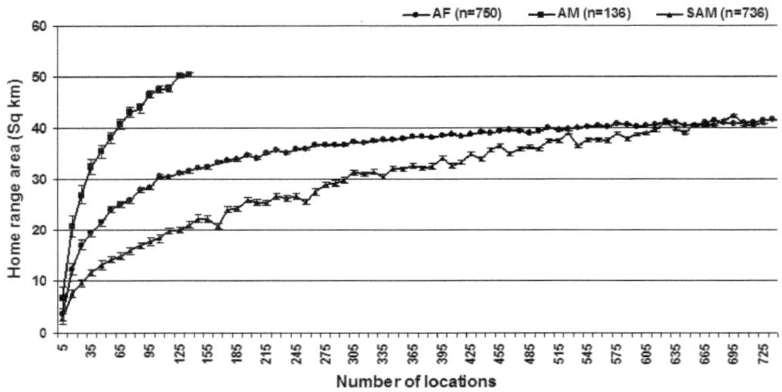

Fig. 1: 100% Minimum Convex Polygon home ranges (± Standard Error) of individual tigers obtained by cumulative sequential samples plotted versus number of cumulative locations showing adequacy of sampling effort for Adult Female (AF, total locations or n= 750), Adult Male (AM, n=136) and her Sub-Adult Male (SAM, n= 739) cub in Pench Tiger Reserve, Madhya Pradesh.

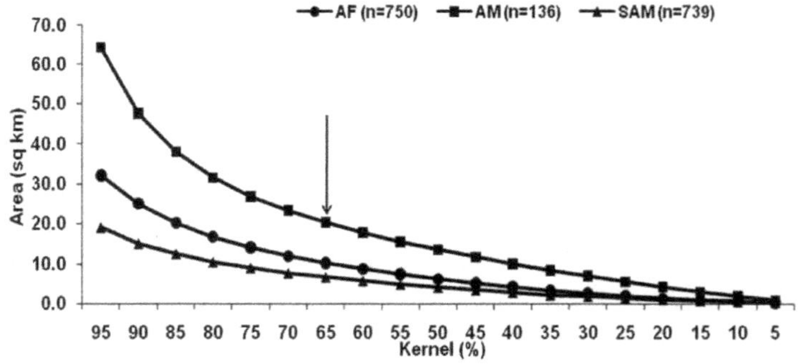

Fig. 2: Fixed Kernel (%) isopleths versus the area of the home range for defining the core area of the home range at point of inflexion (↓) for Adult Female (AF), Adult Male (AM) and Sub-Adult Male (SAM) in Pench Tiger Reserve (March 2008 and December 2011).

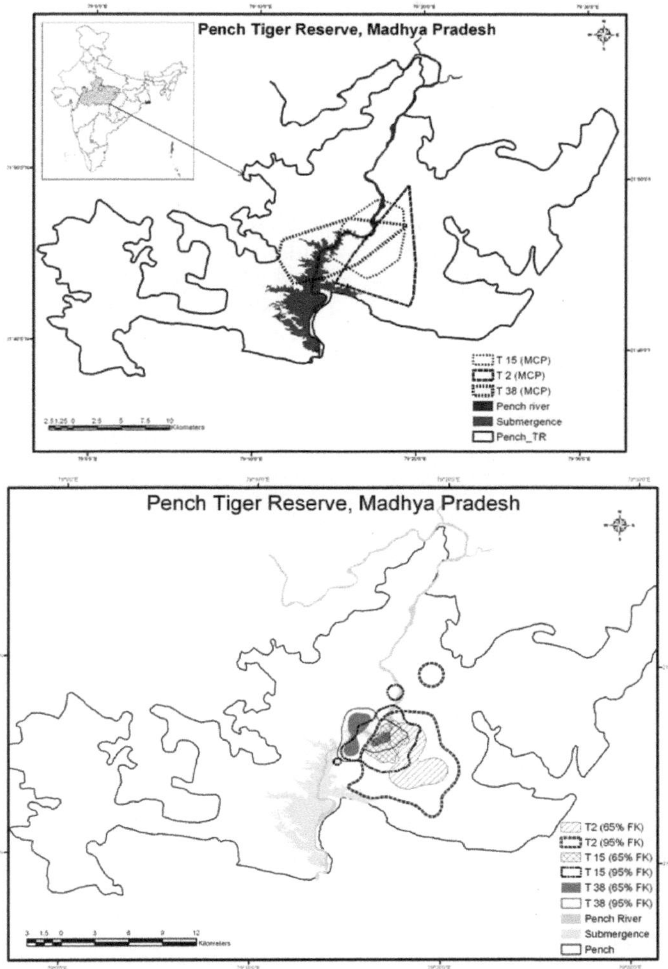

Fig. 3: **(A)** Minimum Convex Polygon (100%) based home ranges of tiger in Pench Tiger Reserve, Madhya Pradesh (March 2008- December 2011) and **(B)** Fixed Kernel (95% & 65% FK) based home ranges of tiger in Pench Tiger Reserve, Madhya Pradesh (March 2008 to December 2011).

The estimated home range of AF using 100% MCP for the 1st year (March 2008 to February 2009, n=120) was 15 km^2 followed by 28 km^2 for the 2nd year (March 2009 to February 2010, n=110), 35 km^2 for the 3rd year (March 2010 to February 2011, n=390) and 23 km^2 for the 4th year (March 2010 to December 2011, n=130 locations).

In total, 12 cubs were recruited by AF in three consecutive litters. She delivered her 1st litter at the age of 31 month (cubs first observed on 21st May 2008). As confirmed by the post-mortem report, all three cubs from her first litter died on August 2008 because of pneumonia (Veterinarian of PTR, Dr. Akhilesh Mishra pers. com.), and subsequently she recruited her second litter (n=4, three males and one female) within three months (cubs first observed on 20th October 2008) and the third litter (n=5; four females and one male) after 24 months of 2nd litter. The estimated gestation period of AF only for her 3rd litter was in between 107 and 108 days; she was observed mating on 6th and 7th July 2010 and cubs were observed first on 23rd October 2010. The 100% MCP home ranges of AF during the 2nd and 3rd litter and the estimated minimum home range of her natal area is shown in Fig. 4. The estimated home range of AF using 100% MCP at the time of raising cubs up to dispersal stage of 2nd litter (n=313 locations) was 29 km^2 and 28 km^2 for 3rd litter (n=291 locations). The home range during rearing of her 1st litter could not be assessed due to low sample size.

In total 2336 man days (Total 480 man days from 3 elephant mahout, 1664 man days from 8 beat guards, 192 man days from one researcher) and 3600 trap nights, resulted a total of 128 locations from ITSM and 12 locations of camera trap photographs of AF and her non collared mother. The estimated minimum home range or natal area of AF was 23.9 km^2. The estimated natal area of SAM was 15.4 km^2.

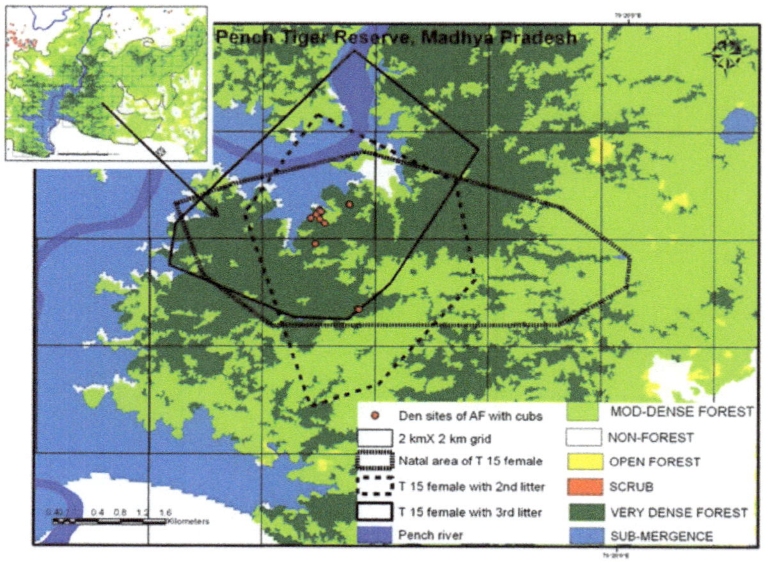

Fig. 4: Home range estimates (100% MCP) of two successive litters of a radio-collared female, their den sites (n=12) and her natal area (100% MCP) in Forest Cover Map of Pench Tiger Reserve, Madhya Pradesh (May 2006 to December 2011).

Discussion

Radio-telemetry is found more reliable to estimate home range of tiger, as they are elusive and nocturnal or move about the landscape using dense cover [7, 8, 17]. However issues regarding sampling size have received less attention. The large number of radio-locations (n=1622) and a reasonable number of radio-collared animals (3 tigers) across all stages from adult male and female to sub adult male provided representative estimates of tiger ranges at different stages. The use of technology such as radio-collars with satellite link (ARGOS and IRIDIUM) also helped us in obtaining precise locations of tiger. As reported by Sharma and Jhala [37] in Kanha Tiger Reserve, ITSM was also found a useful monitoring technique for the routine management activities of PTR. The AM and AF home ranges in this study had reached their asymptote (Fig. 1) and, therefore, provided robust estimates [39].

The estimated home ranges of radio-collared female, male and sub-adult tigers were compared with other available radio-telemetry studies among different sub-species of tigers (Table 2). In most of the radio-telemetry studies on tiger addressed the issues on their home ranges [7, 17, 35], social organizations [7], land tenure system [8] dispersal pattern [25] and survival pattern [17, 27].

Table 2: Estimates of mean home range sizes of different sub-species of tiger (*Panthera tigris*) in Nepal, Bangladesh, India, Indonesia and Russia.

Place (Country)	Method	[1]MCP Home range in km^2			[2]Prey*		[3]Tiger Density/100 km^2	References
		AM (n)	AF (n)	SAM (n)	Density/km^2	Biomass(kg/km^2)		
Chitwan (Nepal) [a]	RT+CT	NA	9.3 (1)	NA	26	2933	3.7	[1,2]Seidesticker [24], [3] Sunquist [7]
Chitwan (Nepal) [a]	RT+CT	54.1(2)	16 (3)	44.7 (1)	59	2600	3.7	[1,2,3]Sunquist [7, 58]
Chitwan (Nepal) [a]	RT+CT	NA	20.7 (7)	NA	59	2600	3.7	[1]Smith *et al.* [8], [2],[3]Sunquist [7, 58]
Panna (India) [a]	RT+CT	188.6 (2)	42.2 (4)	NA	30.9	3100	6.9	[1,3]Chundawat *et al.*[17], [2]Sunquist [58]
Nagarahole (India) [a]	RT+CT	25.7 (1)	16.5 (1)	44.7 (2)	70	15094	11.9	[1,2,3] Karanth & Sunquist [41,59]
Sikhote-Alin (Russia) [b]	RT+SC	1385 (6)	390 (14)	NA	NA	400	0.6	[1]Goodrich *et al.* [27], [2]Sunquist [58], [3] Smirnov & Miquelle [65]
Kanha (India) [a]	RT+CT	102 (2)	10.5 (4)	NA	57.3	3900	13	[1,3] Sharma *et al.* [57], [2]Sunquist [58]
Sariska (India) [a]	RT	168.6 (1)	202.4 (2)	NA	315.7	21618	NA	[1,2] Sankar *et al.* [55]
Sunderban (Bangladesh) [a]	RT	NA	12.3 (2)	NA	41.8	998	4.3	[1] Burlow *et al.* [66],[2,3] Khan [67]
Sunderban (India) [a]	RT	NA	44 (1)	NA	NA	NA	NA	[1]Sharma *et al.* [68]
Indonesia [c]	RT+CT	16 (1)	NA	NA	17.5	400	1.7	[69], [2,3]Obrien *et al.*[70]
Ranthambore (India) [a]	RT+CT	46.1(4)	20.1(2)	NA	69.3	6430	7	[1]Chakravarthi [60],[2] Bagchi *et al* [71], [3]Jhala *et al.* [29]
Indonesia [c]	RT	236 (4)	NA	NA	17.5	400	1.7	Priatna *et al.* [65], [2,3]Obrien *et al.*[70]
Pench (India) [a]	RT+CT	55.1 (1)	44 (1)	52.2 (1)	167.9	6238.1	3.6	[1]Present study, [2]Majumder *et al.*[72], [3]Jhala *et al.*[29]

RT (MCP) = Radio Telemetry, NA= Not Assessed/ Available, RT+CT= Radio-telemetry & Camera trap, TC= Track Count on Snow, * Large & medium body sized prey, [a]= Bengal tiger (*Panthera tigris tigris*), [b]= Amur tiger (*Panthera tigris altaica*), [c] = Sumatran tiger (*Panthera tigris sumatrae*). N.B. -All tiger den sites were estimated using ½ Mean Maximum Distance Moved or ½ MMDM Method (Efford [73]), [1]MCP= Minimum Convex Polygon.

Earlier studies revealed that, female spatial organization of tigers depend on the distribution of resources while male spatial patterns are determined by the distribution of females [52, 53]. In Indian sub-continent [7, 8, 25], relatively small size of adult female home range is attributed to the spatially homogenous and higher abundance of prey species (Table 2). On the contrary, larger home ranges in the female Amur tiger (*Panthera tigris altaica*) of Russia have been associated with lack of adequate prey (Table 2) and poor habitat quality [27, 40, 54]. Although prey biomass/km^2 in Sariska Tiger Reserve was very high as compared to other areas (Table 2), the reported annual home ranges of both the male (n=1) and females (n=2) were considerably larger since the re-introduced tigers explored larger areas during the initial period [55]. The successful raising of all cubs by AF from her 2nd (n=4) and 3rd litter (n=5) to dispersal stage, suggested that all requirements for survival (as abundance of prey, cover for stalking and resting, no anthropogenic pressure and perennial water sources) were fulfilled in the study area. She used 44.3% (10.6 km^2) of her original natal area for raising the 2nd litter and 46.4% (10 km^2) area for raising the 3rd litter. The larger area beyond the original natal area, that was being used year-after-year by the radio-collared female (AF), was attributed to the philopatric nature of female tiger [25]. The sub-merged habitat was largely used by both the radio-collared breeding female and her non-collared mother during the cub-raising period and this may be attributed to the availability of major prey species, *i.e.* chital (> 60% of their diet composed of chital) and *Lantana* sp. as cover in and around the submerged habitat at Totladoh. The core activity (65% FK) area (Fig. 3a and 3b) was found to be smaller for the radio-collared breeding female (AF) during the first three months of recruitment and it was gradually expanded as the cubs grew up to about 12 months. Similar findings were also reported from Chitawan National Park, Nepal [7, 24, 25, 56]. In total, twelve den sites were used by the AF for rearing her three litters (1st litter=3 den sites, 2nd litter=4 den sites and 3rd litter=5 den sites). All the dens were located in areas with no anthropogenic disturbance, Very Dense Forest cover (Fig. 4) and the estimated mean distance to the water source (i.e. back water or submergence of Pench river) was <250 m in all seasons. The estimated overall wild prey density around the den sites was 348.2/km^2 (chital 176.8/km^2, sambar 10.6/km^2, common langur 144.5/km^2, nilgai 2.3/km^2, wild pig 12.2/km^2 and gaur 1.5/km^2) with a total biomass of 12384.7 kg/km^2 (Total effort 476.4 km). The reported overall female home ranges (Table 2) from five tiger reserves in India (Kanha - Sharma *et al.* [57], Sunquist [58];

Nagarahole- Karanth and Sunquist [41, 59], Ranthambore - Chakrabarty [60], Jhala *et al.* [29]; Panna-Chundawat *et al.* [17]; Pench- Present study, Jhala *et al.* [29]) were compared with their respective tiger densities (Table 2). It was observed that the size of female (mean) home ranges are negatively related (Linear regression, $R^2 = 0.9116$) with overall tiger densities (Fig. 5a) *i.e* when the overall tiger density increases, the female home range size decreases.

A wide variation was observed in case of male home range sizes in different study areas (Table 2). Minimum home range (100% MCP) of the resident male tiger in Chitwan varied from 60 to 100 km^2 [7]. McDougal [61] recorded a male home range of 90–105 km^2 (MCP) in Chitwan. Karanth and Sunquist [41] recorded a home range (95% MCP) of 56.7 km^2 for an adult radio-collared male in Nagarhole, although the tiger had a 95% adaptive kernel home range of 98.8 km^2. Schaller [50] recorded a minimum range of 78 km^2 for an adult territorial male and Panwar [62] recorded a minimum range of 30–50 km^2 for the male in Kanha. Matyushkin [63] estimated minimum ranges of 800–1000 km^2 for the male tigers in Russia. Miquelle et al. [40] reported seasonal ranges of 235 to 238 km^2 for a male tiger in Russia. Goodrich et al. [54] reported ranges of 299–614 km^2 for two translocated radio-collared males in Silkote-Alin in Russia, whereas the mean home range of six adult male tigers was 1,385 km^2 (95% MCP).

The home range size of male tiger may be influenced by various ecological factors. According to Chundwat *et al.* [17], larger home ranges of radio collared male tiger in Panna may well be a response to the high spatial and temporal unpredictability of available resources. According to Priatna *et al.* [64], the home range size of male tiger is not only determined by the abundance of prey species but also by the density of local tigers that already occupy the area as well as their social structure. Although there may not be sufficient data to compare (Table 2), we found an indication that size of male tiger (mean) home ranges are positively related with the size of female tiger (mean) home ranges (Fig. 5b).

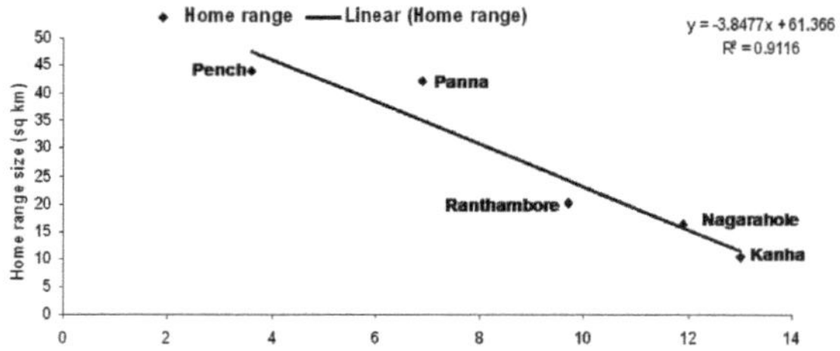

Fig. 5a: Minimum Convex Polygon home range (Mean) sizes (km^2) of female tigers (using radio-telemetry) were compared with available overall tiger densities/100 km^2 (using camera trap) from five different tiger reserves of India, (Kanha - Sharma *et al.* [57], Sunquist [58]; Nagarahole - Karanth and Sunquist [41, 59], Ranthambore - Chakrabarty [60], Jhala *et al.* [29]; Panna - Chundawat *et al.* [17]; Pench - Present study, Jhala *et al.* [29]) (R^2 value obtained from Linear regression).

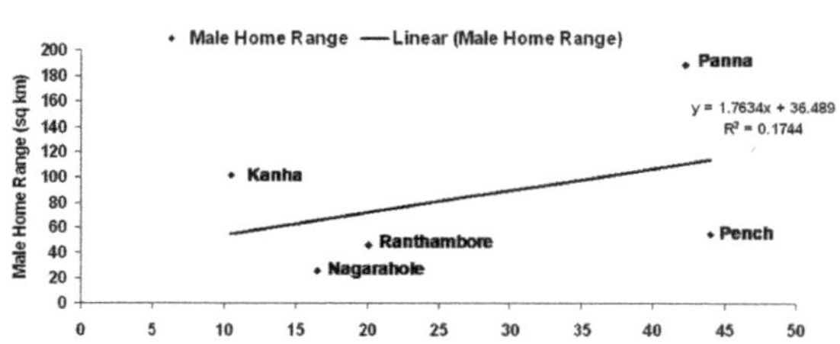

Fig. 5b: Minimum Convex Polygon home range (Mean) sizes (km^2) of male tigers (using radio-telemetry) were compared with mean home range size of female tigers from five different tiger reserves of India, (Kanha - Sharma *et al.* [57], Sunquist [58]; Nagarahole - Karanth and Sunquist [41, 59], Ranthambore - Chakrabarty [60], Jhala *et al.* [29]; Panna - Chundawat *et al.* [17]; Pench - Present study, Jhala *et al.* [29]) (R^2 value obtained from Linear regression).

It was observed that the SAM home range remained smaller in its natal area (MCP home range 15.4 km^2) and his home range expanded (MCP home range 52.2 km^2) gradually during the dispersal stage. Similar findings were reported by Sunquist [7] and Smith in Nepal [25]. Although the SAM's radio-collar ceased functioning after five months of deployment (Table 1), a long-term (January to May 2011) camera trap study, covering a larger area (> 410 km^2, 102 photographs of 23 individual tigers; 16 female and 7 male) confirmed the dispersal of this individual from its natal area.

Conservation Implication

Our study gave an insight on the breeding tiger habitats in this human dominated landscape. The study revealed that a minimum of 25 to 30 km^2 undisturbed area was required for a breeding female in the study area. The intensive study area can potentially hold a maximum of 10 breeding tigress based on 44 km^2 as home range size of a tigress. We have recorded nine breeding tigresses operating in this area (Camera trapping and ITSM). Twenty breeding females are needed for long term survival of tiger population at a site [29]. Thus, it is important to have corridor connectivity with neighboring population of Kanha Tiger Reserve, Satpura Tiger Reserve and Pench Tiger Reserve (Maharastra) to form a meta-population structure for long term viability of tiger population.

According to Smith [25], dispersal plays an important role in the genetic and demographic organization of tiger populations. Our ongoing camera trap study confirmed the dispersal of radio-collared SAM tiger from its natal area (approximately 50 km south-east) to the fringes of Pench Tiger Reserve (Maharastra) and a non-collared sub-adult male tiger from the intensive study area (photo-captured in 2006) to Kanha Tiger Reserve (travelled approximately 150 km north-east and photo-captured in 2010, 2011 and 2012). Therefore, habitat and prey population in Pench- Kanha and Pench-Satpura corridors need to be protected for dispersing tiger individuals.

Though the intensive study area is relatively undisturbed, there is continuous biotic pressure exert from the 99 villages located around the notified buffer zone of PTR, Madhya Pradesh [32, 65]. Poisoning of domestic livestock kills and water holes outside the protected area are the two major threats for tiger population in Pench. The radio-collared AM died due to poisoning (Dr. Akhilesh Mishra Pers. Comm.). Ensured protection is the need of the hour for the long-term survival of tigers in Pench and surrounding areas.

Acknowledgements

We thank Madhya Pradesh Forest Department (MPFD), National Tiger Conservation Authority (NTCA), all officials and staff of PTR (MP and Maharastra), Director and Dean, Wildlife Institute of India (WII) for their support and facilities extended to the project in various ways. Special thanks to Dr. H.S. Pabla (former Chief Wildlife Warden, MPFD), Mr. N.S. Dungariyal (former Field Director, PTR, MP), Dr. Khageshwar Naik (former Field Director, PTR, MP), Dr. Akhilesh Mishra (veterinarian of PTR MP) and Dr. A.B. Srivastav (senior veterinarian, MP) for their support during radio-collaring of tigers. All elephant mahouts and our field assistants are acknowledged for their support in the field. We are also thankful to three anonymous reviewers for their valuable comments on this manuscript.

References

Five "key references", selected by the authors, are marked below (Three recommended (●) and two highly recommended (●●) papers).

1. Burt, W. H. 1943. Territoriality and home range concepts as applied to mammals. J. of Mammal 24:346–352.
DOI:10.2307/1374834
2. Harestad, A. S. & Bunnell, F. L.1979. Home range and body weight – a reevaluation. Ecology 60:389–402.
DOI:10.2307/1937667

3. Sanderson, G. C. 1966. The study of mammal movements – a review. J. Wildl Manage 52:316–320.
4. Wilson, E. O. 1975. Sociobiology: The New Synthesis. Harvard University Press, Cambridge, Mass.
5. Crook, J. H., Ellis, J. E. & Goss–Cus–Tard, J. D. 1976. Mammalian social systems: structure and function. Anim Behav 24: 261–274. DOI:10.1016/S0003-3472(76)80035-8
6. Clutton–Brock, T. H. 1989. Mammalian mating systems. Proceedings of the Royal Society of London. Biol Sci 236:339–372.
7. Sunquist, M. E. 1981. The social organization of tigers (*Panthera tigris*) in Royal Chitwan National Park. Smithsonian Contributions to Zoology 336: 1–98. DOI:10.5479/si.00810282.336
8. Smith, J.L.D., McDougal, C. & Sunquist, M.E. 1987a. Female land tenure system in tigers. In: Tilson, R.L. & Seal, U.S. (eds). Tigers of the world: the biology, biopolitics, management and conservation of an endangered species. Noyes Publications, Park Ridge, New Jersey, USA, pp. 97–109.
9. Smith, J.L.D. Wemmer, C. & Mishra, H. R. 1987b. A tiger geographic information system: the first step in a global conservation strategy. In: Tilson, R.L. & Seal, U.S. (eds). Tigers of the world: the biology, biopolitics, management and conservation of an endangered species. Noyes Publications, Park Ridge, New Jersey, USA, pp. 464–474.
10. Quigley, H.B., Koehler, G.M. & Hornocker, M.G. 1989. Dynamics of a mountain lion population in central Idaho over a 20–year period. In: Smith, R.H. (ed) Proceedings of the 3rd mountain lion workshop. The Wildlife Society, Arizona Chapter, Phoenix, p. 54.
11. Soisalo, M.K. & Cavalcanti, S.M.C. 2006. Estimating the density of a jaguar population in the Brazilian Pantanal using camera–traps and capture–recapture sampling in combination with GPS radio–telemetry. Biol Conserv 129:487–496. DOI:10.1016/j.biocon.2005.11.023
12. Eric, M., Gese, E.M, Rongstad, O.J., Mytton, W.R. 2008. Home range and habitat use of coyotes in south eastern Colorado, J. of Wildl Manage 52 (4): 640-646.
13. Jhala, Y.V., Mukherjee, S., Shah, N, Chauhan, K.S., Dave, C.V., Meena, V. & Banerjee, K. 2009. Home range and habitat preference of female lions (*Panthera leo persica*) in Gir forests, India Biodivers Conserv 18:3383–3394. DOI:10.1007/s10531-009-9648-9
14. Trent, T.T. & Rongstad, O.J. 1974. Home range and survival of cottontail rabbits in South-western Wisconsin. J. of Wildl Manage 38:459–472. DOI:10.2307/3800877
15. Kelly, M.J., Noss, A.J., Bitetti, M.S., Maffei, L., Arispe, R.L., Paviolo, A., De Angelo, C.D. & Di Blanco, Y.E. 2008. Estimating puma density from camera trapping across three study sites: Bolivia, Argentina, and Belize. J. of Mammal 89:408–418 DOI:10.1644/06-MAMM-A-424R.1
16. Sunquist, M.E., Karanth, K.U. & Sunquist, F. 1999. Ecology, behaviour and resilience of the tiger and its conservation needs. In: Seidensticker, J., Christie, S. & Jackson, P. (eds). Riding the tiger; Tiger conservation in human dominated landscapes. Cambridge: Cambridge University Press, pp. 5–18.
17. Chundawat, R.S., Gogate, N. & Malik, P.K. 2002. Understanding tiger ecology in the tropical dry forest of Panna Tiger Reserve. WII (1996–2002) technical report. p.105.
18. Jhala, Y.V., Gopal, R., & Qureshi, Q. 2008. Status of tigers, co–predators and prey in India. National Tiger Conservation Authority, Govt., of India, New Delhi and Wildlife Institute of India, Dehradun, TR 2011/003, pp.302.
19. Luo, S.J, Kim, H.J., Johnson, W., Joelle van der Walt. J., Martenson, J., Yuhki, N.G., Miquelle. D., Uphyrkina, O., Goodrich, J., Quigley, H., Tilson, R., Brady, G., Martelli, P., Subramaniam, V., McDougal, C., Hean, S., Huang, S., Pan, W., Karanth, K.U., Sunquist, M., Smith, J. & O'Brien, S. 2004. Phylogeography and genetic ancestry of tigers (*Panthera tigris*). PLoS Biol 2: 2275–2293. DOI:10.1371/journal.pbio.0020442
20. Ronald, T., Muntifering, H. Defu, J. & Nyhus, P.J. 2004. Dramatic decline of wild South China

tigers *Panthera tigris amoyensis*: field survey of priority tiger reserves. Oryx 38(1), 40–47.

21. Biswas, S. & Sankar, K. 2002. Prey abundance & food habit of tiger (*Panthera tigris tigris*) in Pench National Park, Madhya Pradesh, India. J. of Zool (London) 256: 411–420. DOI:10.1017/S0952836902000456
22. Hayward, M.W., Jedrzejewski, W. & Jedrzejewska, B. 2012. Prey preferences of the tiger *Panthera tigris*. J. of Zool (London) 286: 221–231. DOI:10.1111/j.1469-7998.2011.00871.x
23. Karanth, K.U. 1991. Ecology and management of tigers in tropical Asia. In: Maruyama, N., Bobek, B., Ono, Y., Reglin, W., Bartos, L. & Reglin, P.R. (eds.). Wildlife conservation: Present trends & perspectives for the 21st century. Japan Wildlife Research Center, Yushima Bunkyo–Ku, Tokyo, Japan, pp. 156–159.
24. Seidensticker, J. 1976. On the ecological separation between tigers & leopards. Biotropica 8:225–234. DOI:10.2307/2989714
25. Smith, J.L.D. 1993. The role of dispersal in structuring the Chitwan tiger population. Behaviour 124: 165–195. DOI:10.1163/156853993X00560
26. ● Chundawat, R.S., Gogate, N. & Johnsingh, A.J.T. 1999. Tigers in Panna: preliminary results from an Indian tropical dry forest. In: Seidensticker, J., Christie, S. & Jackson, P. (eds), Riding the tiger: tiger conservation in human–dominated landscapes. Cambridge University Press, pp. 123–129.
27. Goodrich, J.M., Miquelle, D.G., Smirnov, E.N., Kerley, L.L., Quigley, H.B. & Hornocker, M.G. 2010. Spatial structure of Amur (Siberian) tigers (*Panthera tigris altaica*) on Sikhote–Alin Biosphere Zapovednik, Russia, J. of Mamm 91(3):737–748. DOI:10.1644/09-MAMM-A-293.1
28. Herfindal, L.J., Linnell, D.C., Odden, J., Nilsen, E.B. & Andersen, R. 2005. Prey density, environmental productivity and home–range size in Eurasian lynx (*Lynx lynx*). J. of Zool (London) 265:63–71. DOI:10.1017/S0952836904006053
29. Jhala, Y.V., Qureshi, Q, Gopal, R. & Sinha, P.R. 2010. Status of tigers, co–predators and prey in India. National Tiger Conservation Authority, Govt., of India, New Delhi and Wildlife Institute of India, Dehradun. TR 11/003 p.302.
30. Champion, H.G & Seth, S.K. 1968. A revised survey of the forest types of India. Manager of Publications, Govt.of India Press, New Delhi. p.404.
31. Sankar, K., Qureshi, Q., Mathur, V.B., Mukherjee, S.K., Areendran, G., Pasha, M.K.S., Thapa, R. & Lal, P. 2001. Mapping of the Protected Area (PA) and surrounding areas in Pench Tiger Reserve, Madhya Pradesh. A consultancy task assignment report under the India Eco–development project for Pench Tiger Reserve, Madhya Pradesh. Final Report, Wildlife Institute of India, Dehradun. p.130.
32. Dungariyal, N.S. 2008. Management plan of Pench Tiger Reserve, Madhya Pradesh, p.233.
33. Hafner S, Wiesner H, von Hegel G, Halm S. & Erhardt W. (1989). The antagonism of ketamine/xylazine anesthesia ("Hellabrunn mixture") in wild zoo ruminants. Berl Munch Tierarztl Wochenschr. 1;102(6):181–184.
34. Seal, U.S., Armstrong, D.L. & Simmons, L.G. 1987. Yohimbine hydrochloride reversal of ketamine hydrochloride and xylazine hydrochloride immobilization of Bengal tigers and effects on hematology and serum chemistries. J. Wildl Dis 23 (2): 296–300.
35. Vattakaven, J., Jhala, Y.V. & Qureshi, Q. 2010. Radio–telemetry study of tigers in Kanha Tiger Reserve, India. *In*: Sivakumar, K & Habib, B. (eds), Telemetry in Wildlife Science, ENVIS Bulletin: Wildlife & Protected Areas, Wildlife Institute of India, Dehradun, 13(1): 45–47.
36. ●● White, G.C. & Garrott, R.A. 1990. Analysis of wildlife radio–tracking data. Academic Press, San Diego, California, USA, p.383.
37. ●● Sharma R.K. & Jhala Y.V. 2010. Monitoring tiger populations using intensive search in a capture–recapture framework. Pop Ecol 53:373–381. DOI:10.1007/s10144-010-0230-9
38. Karanth, K.U., 1995. Estimating tiger (*Panthera tigris*) populations from cameras trap data using capture–recapture models. Biol Conserv 71: 333–338.

DOI:10.1016/0006-3207(94)00057-W

39. Harris, S., Cresswell, W.J., Forde, P.G., Trewhella, W.J., Wollard, T. & Wray, S. 1990. Home–range analysis using radio–tracking data: a review of problems and techniques applied to the study of mammals. Mamm Rev 20(2/3):97–123. DOI:10.1111/j.1365-2907.1990.tb00106.x
40. Miquelle, D.G., Smirnov, E.N., Merrill, W.T., Myslenkov, A.E., Quigley, H.B., Hornocker, M.G. & Schleyer, B. 1999. A habitat protection plan for the Amur tiger. In Riding the tiger: meeting the needs of people and wildlife in Asia. Seidensticker, J., Christie, S. & Jackson, P (eds). Cambridge: Cambridge University Press, pp. 273– 295.
41. Karanth, K.U. & Sunquist, M.E. 2000. Behavioural correlates of predation by tiger (*Panthera tigris*), leopard (*Panthera pardus*) and dhole (*Cuon alpinus*) in Nagarahole, India. J. of Zool (London) 250: 255–265. DOI:10.1111/j.1469-7998.2000.tb01076.x
42. Renata, V.J. & David, B. C. 1987. Comparison of techniques to determine eastern grey kangaroo home range. J of Wildl Manage 51 (4): 921–930. DOI:10.2307/3801761
43. Kernohan, B.J., Gitzen, R.A. & Millspaugh, J.J. 2001. Analysis of animal space use and movements. In: Millspaugh, J. J. & Marzluff, J. M. (eds), Radio Tracking Animal Populations: Academic Press, pp. 125–166. DOI:10.1016/B978-012497781-5/50006-2
44. Worton, B.J. 1989. Kernel methods for estimating the utilization distribution in home range Studies. Ecology 70:164–168. DOI:10.2307/1938423
45. Seaman, D.E., Millspaugh, J.J., Kernohan, B.J., Brundige, G.C., Raedeke, K.J., & Gitzen, R.A. 1999. Effects of sample size on kernel home range estimates. J. of Wildl Manage 63: 739–747. DOI:10.2307/3802664
46. Mizutani, F. & Jewell, P.A. 1998. Home–range and movements of leopards (*Panthera pardus*) on a live–stock ranch in Kenya. J. Zool (Lond) 244: 269–286. DOI:10.1111/j.1469-7998.1998.tb00031.x
47. Anderson, D.R., Laake, J.L., Crain, B.R. & Burnham, K.P. 1979. Guidelines for transect sampling of biological populations. J. of Wildl Manage 43(1): 71-79. DOI:10.2307/3800636
48. Burnham, K.P., Anderson, D.R. & Laake, J. L. 1980. Estimation of density from line transect sampling of the biological populations. Wildl Mono 72: 202.
49. Buckland, S.T., Anderson, D.R., Burnham, K.P. & Laake, J.L. 1993. Distance sampling: estimating abundance of biological populations. Chapman and Hall, London, pp. 446.
50. Schaller, G.B. 1967. The deer and the tiger. University of Chicago Press, Chicago. p. 384.
51. Laake, J.L., Buckland, S.T., Anderson, D.R. & Burnham, K.P. 1998. DISTANCE user's guide, Version 4.1. Colorado Cooperative Fish and Wildlife Research Unit, Colorado State University, Fort Collins, Colorado, USA.
53. Smith, R.M. 1978. Movement patterns and feeding behavior of the leopard in the Rhodes Matopos National Park, Rhodesia. Carnivora 1:58–69.
52. Sandell, M. 1989. The mating tactics and spacing patterns of solitary carnivores. *In*: Gittleman, J. L. (ed.). Carnivore behavior, ecology and evolution. Cornell University Press, Ithaca, New York, pp. 164–182.
53. Macdonald, D.W. 1983. The ecology of carnivore social behaviour. Nature 301: 379–384. DOI:10.1038/301379a0
54. Goodrich, J.M., Kerley, L.L., Miquelle, D.G., Smirnov, E.N., Quigley, H.B. & Hornocker, M.G. 2005. Social structure of Amur tigers on Sikhote–Alin Biosphere Zapovednik. In: Miquelle, D.G., Smirnov, E.N. & Goodrich, J.M. (eds). Tigers of Sikhote–Alin Zapovednik: Ecology and Conservation, PSP, Vladivostok, Russia (in Russian), pp. 50–60.
55. Sankar, K. , Qureshi, Q., Nigam, P., Malik, P. K., Sinha, P. R., Mehrotra, R. N., Gopal, R., Bhattacharjee, S., Mondal, K. & Gupta, S. 2010. Monitoring of reintroduced tigers in Sariska Tiger Reserve, Western India: preliminary findings on home range, prey selection and food habits. Trop Conserv Sci 3: 301–318.

56. McDougal, C. 1977. The face of the tiger. London, Rivington Books. p.180.

57. Sharma, R.K., Jhala, Y.V., Qureshi, Q. Vattakaven, J., Gopal, R. & Nayak, K. 2010. Evaluating capture–recapture population and density estimation of tigers in a population with known parameters. Anim Conserv 13: 94–103.
DOI:10.1111/j.1469-1795.2009.00305.x

58. ● Sunquist, M.E. 2010. Tigers: Ecology and behavior. In: Tilson, R. & Nyhus, P. J. (eds.). Tigers of the World: The science, politics and conservation of *Panthera tigris*, Elsevier, San Diego, pp. 19–33.

59. Karanth, U.K. & Sunquist, M.E. 1995. Prey selection by tiger, leopard and dhole in tropical forests. J. of Anim Ecol 64 (4): 439–450.
DOI:10.2307/5647

60. Chakravarty, P.P.J. 2009. Ranging patterns of tigers in Ranthambhore Tiger Reserve, Rajasthan. Abstract. 23nd Annual Research Seminar, Wildlife Institute of India, Dehra Dun, India. (http://www.wii.gov.in/research/abstracts/irs_abstracts_2009.pdf).

61. Panwar, H.S. 1979. Population dynamics and land tenure of tigers in Kanha national park. Indian Forester (Special Issue) pp. 18–36.

62. Matyushkin, E.N. 1977. The choice of route and territory used by the Amur tiger (according to winter tracking data). Povedenie mlekopitaiuschikh. Voprosy teriologii M: (in Russian), 146–178.

63. Priatna, D., Santosa, Y., Prasetyo, L.B. & Kartono, A.P. 2012. Home range and movements of male translocated problem tigers in Sumatra. A.J. of Conserv Biol 1(1): 20–30.

64. Qureshi Q., Gopal R., Kyatham, S., Basu S., Mitra A, & Jhala, Y.V. 2006. Evaluating tiger habitat at the Tehsil level. Project Tiger Directorate, Govt. of India, New Delhi, and Wildlife Institute of India, Dehradun TR No.06/001, p.162.

65. Smirnov, E.N. & Miquelle, D.G. 1999. Population dynamics of the Amur tiger in Sikhote–Alin State Biosphere Reserve. *In*: Seidensticker, J., Christie, S. & Jackson, P (eds). Riding the tiger: tiger conservation in human–dominated landscapes. Cambridge University Press, Cambridge, United Kingdom, pp. 61–70.

66. Barlow, A.C.D., Smith, J.L.D., Ahmad, I.U., Hossain, A.N.M., Rahman, M & Howlader, A. 2011. Female tiger *Panthera tigris* home range size in the Bangladesh Sundarbans: the value of this mangrove ecosystem for the species' conservation. Fauna and Flora International, Oryx, 45(1): 125–128.
DOI:10.1017/S0030605310001456

67. Khan, M.M.H. 2004. Ecology and conservation of the Bengal tiger in the Sundarbans Mangrove forest of Bangladesh. Ph.D. Thesis, University of Cambridge, p. 296.

68. Sharma, R.K., Jhala, Y.V. & Qureshi, Q. 2011. Home range size of a tigress in Sundarbans, India: preliminary results. CATNEWS, 54: 13–16.

69. Report of Zoological Society of London, 2003. Pigs, palms, people and tigers–21st Century Tiger Progress, p. 6.

70. O'Brien, T., Kinnaird, M. & Wibisono, H. 2003. Crouching tigers, hidden prey: Sumatran tiger and prey populations in a tropical forest landscape. Anim Conserv 6: 131–139.
DOI:10.1017/S1367943003003172

71. Bagchi, S., Goyal, S. P. & Sankar, K. 2003. Prey abundance and prey selection by Tigers (*Panthera tigris*) in a semi arid, dry deciduous forest in western India. J of Zool (London) 260: 285–290.
DOI:10.1017/S0952836903003765

72. ● Majumder A, Basu S, Sankar K, Qureshi Q, Jhala YV, Gopal R, 2012. Prey Selection, food habits and temporal activity patterns of sympatric carnivores in Pench Tiger Reserve, Madhya Pradesh, Central India. J of Sci Trans in Environ and Tech, 5(3):110-120.

73. Efford, M. 2004. Density estimation in live-trapping studies. Oikos 106, 598–610.
DOI:10.1111/j.0030-1299.2004.13043.x

6

Dietary Variation of the Stone Marten (*Martes foina*): A Meta-Analysis Approach

M. Papakosta[1], K. Kitikidou[2,*] D. Bakaloudis[1] & C. Vlachos[1]

[1] Aristotle University of Thessaloniki, Faculty of Forestry and Natural Environment, University Campus, 54124, Thessaloniki, Greece.

[2] Democritus University of Thrace, Department of Forestry and Management of the Environment and Natural Resources, Pandazidou 193, 68200, Orestiada, Greece.

* Corresponding author phone: email: kkitikid@fmenr.duth.gr.

Keywords
Food Habits;
Stone Marten;
Systematic Review.

Abstract
Goal: Understanding the biogeographical trophic patterns of generalist predators, such as the stone marten (*Martes foina*), is particularly challenging due to their wide distributions and the broad trophic spectrum. In this research, the geographical variation in food habits of the stone marten was reviewed. *Method:* We collected data from 43 studies, using the available literature on food studies from locations covering most of the stone marten's European range. We explored whether or not a pattern exists between the dietary composition of the species and the geographical range. The original data were expressed as weight, percentage of numerical abundance (%N) and frequency of occurrence of the food items measured in stomachs and scats, in north and central Greece, and we compared them using meta-analysis. *Results and Conclusions:* We observed a clear heterogeneity in the main food categories (mammals, birds' eggs, insects, vegetation and fruits. and reptiles and amphibians), with no apparent latitudinal differentiation. Consequently, we concluded that variations in the feeding habits of stone martens can be associated with local climate and, therefore, the availability of alternative potential food. On a general geographical scale, stone martens respond to varying food availability by adjusting their foraging strategy and so should be considered an opportunistic and flexible feeder.

Introduction

The spatial and temporal abundance and distribution of food resources influences foraging strategies [1,2], because energy expenditure for detecting, capturing and handling preys is related to diet quantity and quality [3,4]. Therefore, understanding the biogeographical trophic patterns of a generalist predator is central to understanding the ecology of the species [5]. The study of stone Stone martens' (*Martes foina*) dietary habits is particularly challenging, due to their wide geographical spread (open forested areas and rocky outcropping throughout Europe and in western and central Asia) and broad trophic spectrum.

Members of the genus Martes (family Mustelidae) are medium-sized carnivores that are distributed widely in Europe and Asia. Mustelidae is the largest family within Carnivora and comprises 57 species in 22 genera. The exact number of species is still debated, while up to 15 *Martes foina* subspecies have been proposed. Stone martens are found in open forested areas and rocky out cropping. They can often be found near areas of human habitation where they make dens in buildings; otherwise, natural den sites include tree hollows, rock crevices, and abandoned burrows. They are mainly solitary and their territory can be up to 80 hectares in size. They are omnivores and they feed on smaller mammals, rodents, birds, arthropods, eggs, reptiles/amphibians and fruits [6]. This dietary plasticity makes Martes an exemplary model genus with which to study adaptive foraging.

Feeding habits of the species vary across regions, and the composition and diversity of the marten's diet is not well understood. The abundance and diversity of potential prey (food types) determines a predator's (consumer's) diet composition and diversity [7]; thus, we hypothesize that widely-distributed species are opportunistic and adaptable to variation in prey communities. However, it has not yet been confirmed that the stone marten is indeed an opportunist through the systematic review of the numerous studies concerning its food habits. In this work, a number of studies of the food habits of the stone marten were reviewed. We collected data from 43 studies in the available literature, from locations covering most of the stone marten's European range. We then applied a meta-analysis in order to attempt a systematic review of these independent studies of the stone marten's dietary habits, in order to confirm the adaptive capacity of the species to regional abundance and diversity of food. The novelty of this work is that a conclusion that applies to a regional scale (opportunism of the stone marten concerning its dietary habits) can be generalized to a large geographical scale with the use of sophisticated statistical analysis.

Methods

Data Collection

We For the application of the method of meta-analysis, in exploring the dietary ecology of the stone marten, its European distribution—based on latitude—was divided into two regions: the Mediterranean region and the region of central and northern Europe. From the literature review 43 studies were found; these are those that have been conducted in Europe, 22 of them registered in the Mediterranean region and 21 in central and northern Europe. The techniques used to express data were the barometric method, the frequency of occurrence (frequency method), and the percentage of numerical abundance % (relative frequency method).

The independent variables used were quantitative (sample size: stomachs and scats) and qualitative (area: Mediterranean – central and northern Europe; data type: stomachs – scats; season: spring – summer – autumn – winter; analytical methods: barometric – frequency – relative frequency; food group: mammals – birds and eggs – insects – vegetation and fruits – reptiles and amphibians). The weighting of the results of the three analytical techniques by sample size (number of stomachs and scats) was performed using the *ES* type, as described in section 2.2.

No	Study	Classification	No	Study	Classification
1	[8]	1	23	[30]	2
2	[9]	1	24	[31]	1
3	[10]	1	25	[32]	2
4	[11]	1	26	[33]	1
5	[12]	1	27	[34]	1
6	[13]	1	28	[35]	2
7	[14]	2	29	[36]	1
8	[15]	2	30	[37]	1
9	[16]	2	31	[38]	2
10	[17]	1	32	[39]	2
11	[18]	1	33	[40]	1
12	[19]	2	34	[41]	1
13	[20]	1	35	[42]	2
14	[21]	2	36	[43]	2
15	[22]	2	37	[44]	2
16	[23]	1	38	[45]	2
17	[24]	2	39	[46]	1
18	[25]	1	40	[47]	2
19	[26]	2	41	[48]	2
20	[27]	2	42	[49]	2
21	[28]	1	43	[50]	1
22	[29]	1			

Classification 1: Mediterranean, 2: Central and northern Europe.

Meta-analysis

There are a number of researchers who describe the method of meta-analysis in detail [51,52,53,54,55]. Meta-analysis focuses on the direction and magnitude of the differences and similarities between studies. This method can be applied to research that is empirical, rather than theoretical, produces quantitative results rather than qualitative, examines the same constructs and relationships, and has comparable findings. The effect size (*ES*) is the dependent variable that standardizes findings across studies that can be directly compared. There are many different types of *ES* measures, each suited to different research scenarios. In this study we used a proportion *ES* type, weighting barometric, frequency and relative frequency results by sample size (numbers of stomachs and scats, *w*). The mean *ES* is equal to:

$$\overline{ES} = \frac{å\ (w'\ ES)}{å\ w}$$

The standard error of the mean is:

$$se_{\overline{ES}} = \sqrt{\frac{1}{\sum w}}$$

The 95% confidence interval is calculated as:

$$Lower = \overline{ES} - 1.96(se_{\overline{ES}})$$

$$Upper = \overline{ES} + 1.96(se_{\overline{ES}})$$

We then applied a homogeneity analysis in order to test the assumption that every *ES* estimates the same population mean.

The index *Q* is calculated as (Cochran's *Q* statistic, [56]):

$$Q = \sum (w \times ES^2) - \frac{\left[\sum (w \times ES)\right]^2}{\sum w}$$

If the calculated *Q* is less than the critical value for a chi-square with a known *df* and *p*-value (e.g. 0.05), we fail to reject the null hypothesis of homogeneity. In this case, the variability across effect sizes does not exceed what would be expected based on sampling error.

On the other hand, if *ES*s distribution is heterogeneous, we should analyse the excess variability between studies.

The fixed-effect model assumes that all of the variability between effect sizes is due to sampling error, while the random-effect model assumes that the variability between effect sizes is due to sampling error plus variability in the populations.

Fixed-effect model weights each study by the inverse of the sampling variance:

$$w_i = \frac{1}{se_i^2}$$

The random-effect model weights each study by the inverse of the sampling variance plus a constant that represents the variability across population effects:

$$w_i = \frac{1}{se_i^2 + \hat{v}_\theta}$$

where $\hat{v}_q$ is the random effects variance component.

The random effects variance component is based on *Q*. The formula is:

$$Q_v = \frac{Q - k - 1}{\sum w - \left(\frac{\sum w^2}{\sum w}\right)}$$

If confidence intervals in random-effect model are larger, then populations vary significantly (there are unique differences in the set of true population effect sizes), i.e., not all of the variability between effect sizes is due to sampling error.

If $\overline{ES}$ differs between a fixed- and random-effect model, then the sample size is highly related to effect size.

A graphical method, the so-called forest plot, has been developed for detecting heterogeneity [53]. In this plot, we can examine visually the overlap of confidence intervals of the individual studies in the meta-analysis. In a typical forest plot the results of studies are represented as squares, while the edges of the horizontal line through each square represents the limits of the confidence interval for each study.

The overall assessment of the meta-analysis of both the measure of relationship between studies and the confidence interval is placed in the diagram's basis, with a separate line where boundaries reflect the estimated confidence intervals and a dot of a different reflects shape the overall assessment of the measure.

Results

Meta-analysis

The 43 studies measuring dietary variation of the stone marten are proven to be significantly heterogeneous in all measurement methods (barometric, frequency, relative frequency), since the calculated Q is bigger than the chi-square critical value (Tables 1-5).

Table 1: Meta-analysis statistics for the mammals food group.

Barometric method					
	Fixed effects model	**Random effects model**	**Number of studies (k)**	**Degrees of freedom (df)**	**Chi-square critical value**
Q	3961.386	97.787	10	9	16.919
$\overline{ES}$	36.494	8.6153			
$se_{\overline{ES}}$	0.151	1.681			
Confidence intervals (lower-upper)	36.199-36.790	5.320-11.910			
Frequency method					
	Fixed effects model	**Random effects model**	**Number of studies (k)**	**Degrees of freedom (df)**	**Chi-square critical value**
Q	92824.334	61.302	24	23	35.172
$\overline{ES}$	16.086	13.514			
$se_{\overline{ES}}$	0.043	2.164			
Confidence intervals (lower-upper)	16.001-16.171	9.271-17.755			
Relative frequency method					
	Fixed effects model	**Random effects model**	**Number of studies (k)**	**Degrees of freedom (df)**	**Chi-square critical value**
Q	37654.992	45.189	14	13	22.362
$\overline{ES}$	25.579	12.040			
$se_{\overline{ES}}$	0.093	3.014			
Confidence intervals (lower-upper)	25.396-25.762	6.132-17.949			

Assuming that there are random differences across studies, we calculated a random-effect model, interpreting the variability between effect sizes as a total of sampling error and variability in populations. The random-effect model had higher confidence intervals, confirming the assumption that there is significant variability between populations (Tables 1-5). $\overline{ES}$ differs between the fixed- and random-effect model, confirming that sample size is highly related to effect size (Tables 1-5).

In the forest plots (Figs 1-15) the central axis represents the studies' average. Square dots represent our studies conducted in Greece. Pink dots represent studies in the Mediterranean, while blue dots represent studies in central and northern Europe. The ES

Table 2: Meta-analysis statistics for the birds-eggs food group.

Barometric method					
	Fixed effects model	**Random effects model**	**Number of studies (k)**	**Degrees of freedom (df)**	**Chi-square critical value**
Q	11585.208	39.539	10	9	16.919
$\overline{ES}$	12.916	4.807			
$se_{\overline{ES}}$	0.090	1.702			
Confidence intervals (lower-upper)	12.740-13.092	1.472-8.143			
Frequency method					
	Fixed effects model	**Random effects model**	**Number of studies (k)**	**Degrees of freedom (df)**	**Chi-square critical value**
Q	46441.924	89.686	21	20	31.410
$\overline{ES}$	7.599	7.471			
$se_{\overline{ES}}$	0.032	1.170			
Confidence intervals (lower-upper)	7.535-7.662	5.179-9.764			
Relative frequency method					
	Fixed effects model	**Random effects model**	**Number of studies (k)**	**Degrees of freedom (df)**	**Chi-square critical value**
Q	18572.286	142.587	13	12	21.026
$\overline{ES}$	6.755	6.770			
$se_{\overline{ES}}$	0.049	1.287			
Confidence intervals (lower-upper)	6.659-6.851	4.247-9.293			

Table 3: Meta-analysis statistics for the insects food group.

Barometric method					
	Fixed effects model	**Random effects model**	**Number of studies (k)**	**Degrees of freedom (df)**	**Chi-square critical value**
Q	3686.921	63.946	9	8	15.507
$\overline{ES}$	1.769	0.962			
$se_{\overline{ES}}$	0.035	0.366			
Confidence intervals (lower-upper)	1.702-1.837	0.245-1.680			
Frequency method					
	Fixed effects model	**Random effects model**	**Number of studies (k)**	**Degrees of freedom (df)**	**Chi-square critical value**
Q	63090.403	55.096	23	22	33.924
$\overline{ES}$	15.734	10.658			
$se_{\overline{ES}}$	0.044	1.806			
Confidence intervals (lower-upper)	15.647-15.820	7.119-14.197			
Relative frequency method					
	Fixed effects model	**Random effects model**	**Number of studies (k)**	**Degrees of freedom (df)**	**Chi-square critical value**
Q	64495.695	36.822	13	12	21.026
$\overline{ES}$	20.985	11.798			
$se_{\overline{ES}}$	0.088	3.830			
Confidence intervals (lower-upper)	20.813-21.157	4.291-19.304			

Table 4: Meta-analysis statistics for the vegetation-fruits food group.

Barometric method					
	Fixed effects model	**Random effects model**	**Number of studies (k)**	**Degrees of freedom (df)**	**Chi-square critical value**
Q	34597.254	25.086	10	9	16.919
$\overline{ES}$	10.389	7.063			
$se_{\overline{ES}}$	0.080	3.134			
Confidence intervals (lower-upper)	10.231-10.546	0.921-13.205			
Frequency method					
	Fixed effects model	**Random effects model**	**Number of studies (k)**	**Degrees of freedom (df)**	**Chi-square critical value**
Q	108492.760	85.134	24	23	35.172
$\overline{ES}$	25.157	24.628			
$se_{\overline{ES}}$	0.054	2.960			
Confidence intervals (lower-upper)	25.051-25.263	18.827-30.429			
Relative frequency method					
	Fixed effects model	**Random effects model**	**Number of studies (k)**	**Degrees of freedom (df)**	**Chi-square critical value**
Q	56505.725	40.337	12	11	19.675
$\overline{ES}$	25.022	15.044			
$se_{\overline{ES}}$	0.096	4.357			
Confidence intervals (lower-upper)	24.834-25.210	6.505-23.584			

Table 5: Meta-analysis statistics for the reptiles-amphibians food group.

Barometric method					
	Fixed effects model	**Random effects model**	**Number of studies (k)**	**Degrees of freedom (df)**	**Chi-square critical value**
Q	2550.398	76.582	7	6	12.592
$\overline{ES}$	1.015	0.789			
$se_{\overline{ES}}$	0.038	0.338			
Confidence intervals (lower-upper)	0.941-1.089	0.125-1.452			
Frequency method					
	Fixed effects model	**Random effects model**	**Number of studies (k)**	**Degrees of freedom (df)**	**Chi-square critical value**
Q	4409.940	175.224	9	8	15.507
$\overline{ES}$	0.718	0.650			
$se_{\overline{ES}}$	0.015	0.171			
Confidence intervals (lower-upper)	0.689-0.746	0.315-0.984			
Relative frequency method					
	Fixed effects model	**Random effects model**	**Number of studies (k)**	**Degrees of freedom (df)**	**Chi-square critical value**
Q	3266.196	82.984	7	6	12.512
$\overline{ES}$	2.544	1.095			
$se_{\overline{ES}}$	0.039	0.401			
Confidence intervals (lower-upper)	2.468-2.621	0.309-1.880			

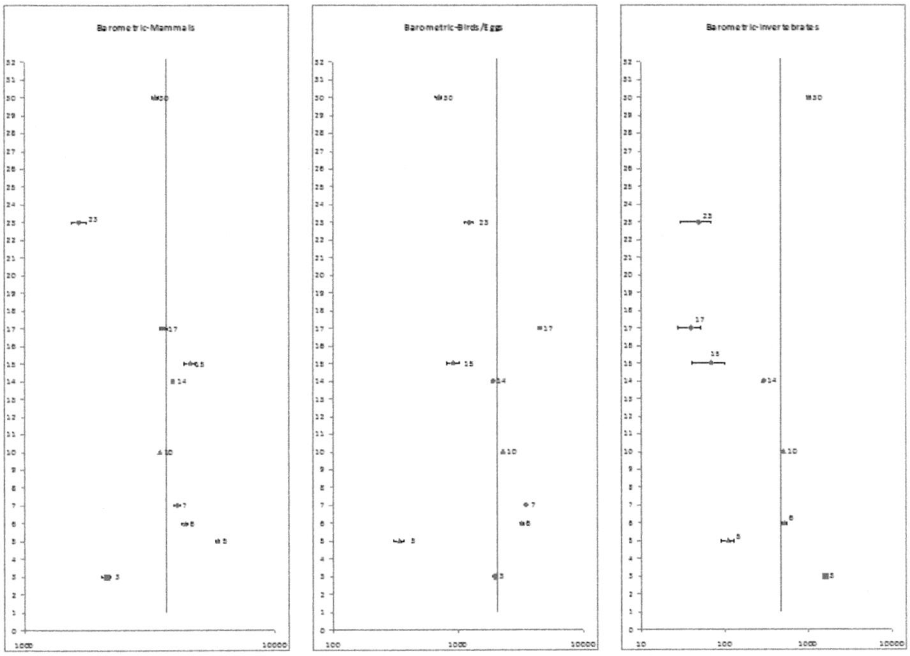

Figure 1-3: Forest plots of each food group and method.

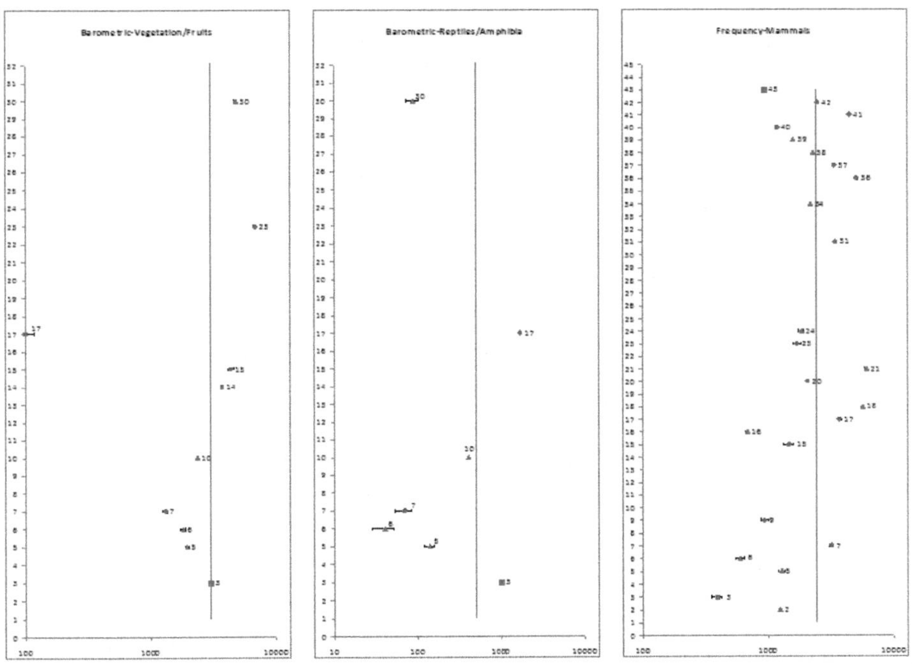

Figure 4-6: Forest plots of each food group and method.

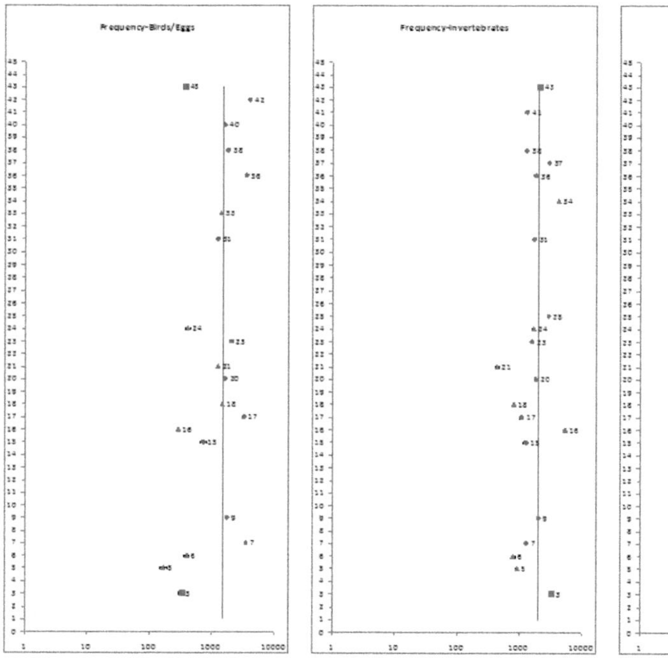

Figure 7-9: Forest plots of each food group and method.

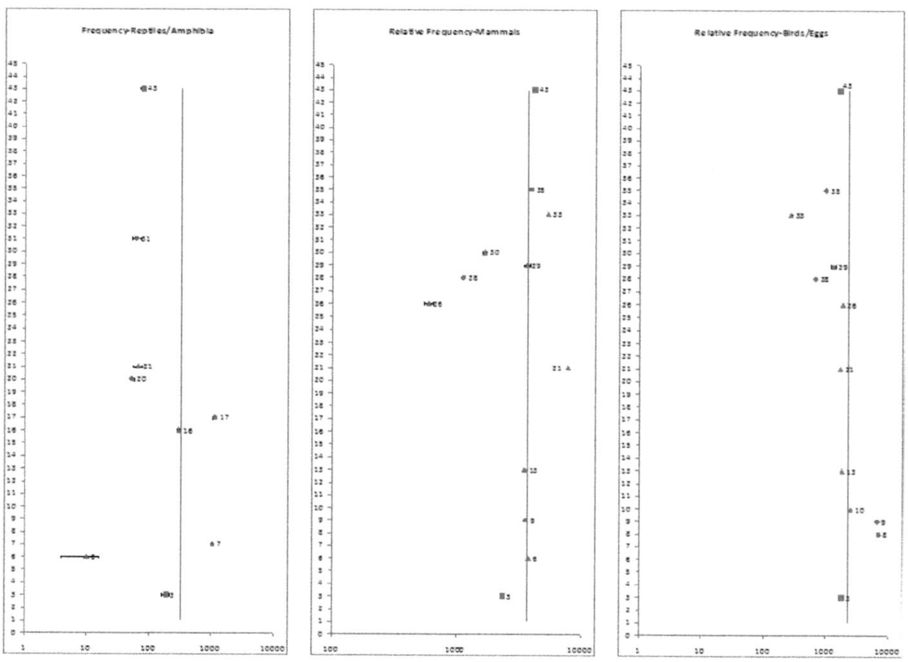

Figure 10-12: Forest plots of each food group and method.

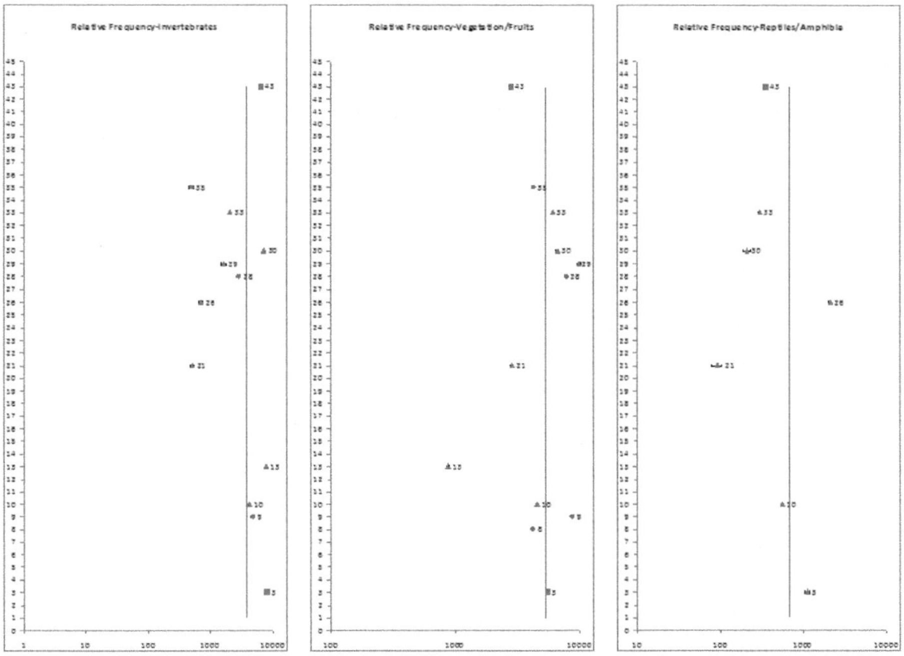

Figure 13-15: Forest plots of each food group and method.

values in the x-axis are multiplied by 100 and transformed on a logarithmic scale, for an easier interpretation through graphical illustration. Study numbers are represented on the vertical axis. No general pattern seems to exist; however, we can recognize a trend of fewer insects in the stone marten's diet in central and northern Europe when using the barometric method (Fig 3), fewer mammals in the Mediterranean with the frequency method (Fig 6), and more birds' eggs in central and northern Europe with the frequency method (Fig 7).

Further analysis with meta-analysis of variance (meta-ANOVA)

In order to support our conclusions, we applied an Analysis of Variance using *ES* as the dependent variable (meta-ANOVA, Table 6). Method of analysis (barometric – frequency – relative frequency), food groups, study classification (Mediterranean – central and northern Europe), and type of data (stomachs – scats) were used as factors, while sample size (number of stomachs and scats) was used as covariates. As was expected, methods of analysis and food groups significantly differ statistically, (p-value<0.05), while type of data and sample size do not, regarding *ES* (p-value >0.05). Geographical variation does not form a clear diet pattern (geographical classification p-value>0.05, i.e., Mediterranean does not differ from central and northern Europe as regards to *ES*); however, they differ statistically significantly for a 10% probability (p-value<0.10), confirming that there is a trend for certain food groups, as we concluded from forest plots (Figs 3, 6, 7)

Table 6: Meta-ANOVA.

Source	Type III Sum of Squares	df	Mean Square	F	Sig.
Corrected Model	50175.595	11	4561.418	14.446	0.000
Intercept	30145.796	1	30145.796	95.474	0.000
Method of analysis	7922.318	4	1980.579	6.273	0.000
Food groups	40743.880	4	10185.970	32.260	0.000
Study classification	982.218	1	982.218	3.111	0.079
Type of data	132.212	1	132.212	0.419	0.518
Sample size	52.140	1	52.140	0.165	0.685
Error	78305.849	248	315.749		
Total	316824.001	260			
Corrected Total	128481.444	259			

Discussion

Stone martens, as demonstrated by the systematic review of 43 studies in Europe, are broad-ranged and opportunistic feeders who take advantage of the seasonal availability of food, occupy and use a wide variety of habitats, and adjust their food requirements depending on environmental conditions.

Meta-analysis application suggests no clear geographical diet pattern; however, we can distinguish a trend for fewer insects in stone martens' diet in central and northern Europe, using the barometric method (Fig 3), fewer mammals in the Mediterranean with the frequency method (Fig 6), and more birds-eggs in central and northern Europe, using with the frequency method (Fig 7). These trends may indicate differentiation of food habits of stone martens depending on their local environmental conditions and the availability of alternative sources of food [11,25,31], and/or competition with other species [37].

On the contrary, [5] observed a clear latitudinal gradient in dietary composition, with mammals as the primary food type in the diet of martens at lower latitudes, and vegetables and insects found more frequently in southerly regions. However, we should note that [5] examined the dietary habits of martens (*Martes spp.*) using data expressed as relative frequency of occurrence from 163 studies, while our study specializes in *Martes foina* and data are expressed as weight, percentage of numerical abundance (%N) and frequency of occurrence (three different methods). The conclusion of [5] is in accordance with the [57] study concerning the 'latitudinal richness gradient'; however, [58] found no evidence supporting the relationship between latitude and dietary diversity for stone martens and martens.

Overall, on a general geographical scale, stone martens respond to varying food availability by adjusting their foraging strategy and should be considered opportunistic and flexible feeders. These findings are similar to those reported on the regional scale in the Chickagof Island in southern Alaska, USA [59]) and in the Hakel Forest in central Germany [39].

In general, species of the Martens genus have a very elastic diet, which includes small mammals, birds, insects, fruits and carrion [5]. On the contrary, other animal species include only certain food types in their diet. The first food tactic shows the particularly adaptive capacity of these animals.

The southern latitudes and low elevations are characterized by warm and mostly stable climates where fruits and insects are abundant and available for a longer period during the year [57,60,61]. Many reports show that, in the southern regions, species of the genus Martes feed on larger percentages of fruit [62,63,64]. In tropical areas there are more fresh fruit and for a long time, while towards the poles there are less fruit and for a short time [61]. In contrast, in northern latitudes it has been observed that some species of the Mustelidae family tend to specialize in rodents, regardless of the availability of alternative sources of food [65].

Similar results are found in the present study for other medium-sized carnivores of the same family (Mustelidae) and different genera, such as the badger (genus Meles), which has been considered a specialized consumer of earthworms, in southern England [66,67]. However, wider scale studies found that the diet of the badger shifts towards plant species when fruits and cereals are abundant, in central Poland [68]. The same trophic pattern was observed on a wider scale in southwestern Portugal [69]. Also, the ferret (family Mustelidae, genus Mustela) has been found to consume lower levels of shrub forest fruits (berries) when snow levels increase, possibly due to non-accessibility [70]. Instead, [71] demonstrated that in mild winters the ferret fed on a greater number of blueberries. These studies support the claim that the species of the genus Martes prefer fruits over rodents, when fruits are available, because these provide greater energy benefit through their easy access and thus lower energy cost of collection.

For the Stone Marten, seasonal rainfall rates were found to be important for determining diet [5]. Seasonality, determined by fluctuations in environmental conditions and productivity (i.e., food production), is an important evolutionary force that influences the lifestyle and behaviour of mammals [72]. If environmental conditions affect the diet of the species of the genus Martes, on a biogeographical scale it can be argued that these species are vulnerable to global climate change, and this, possibly along with other factors, can change their distribution range [73], their phenology [74], and their behaviour [75,76].

In conclusion, all of the species of the genus Martens display an optional food strategy, demonstrating a preference for fruits when they are in abundance and for small mammals when the availability of fruit decreases. There is a general trend for alternative food sources such as insects, birds, reptiles, carrion and fish when these are in great abundance in different habitats. These observations suggest that they behave as an opportunistic species. All biogeographical studies on the food habits of carnivores reinforce the importance of understanding the ecological role of these species on a wider geographical scale, especially when developing their integrated protection programmes and especially when it comes to rare and endangered species.

References

Five "key references", selected by the authors, are marked below (Three recommended (●) and two highly recommended (●●) papers).

1. Gittleman, J. 1986. Carnivore life-history patterns: allometric, phylogenetic, and ecological associations. Am Nat 127(6): 744–771.
http://dx.doi.org/10.1086/284523.

2. Roff, D. 2002. Life History Evolution. Sunderland, MA.: Sinauer Associates.
3. Pyke, G., Pulliam, H. & Charnov, E. 1977. Optimal foraging: a selective review of theory and tests. Quarterly Review of Biology 52: 137–154. http://dx.doi.org/10.1086/409852.
4. Sundell, J., Eccard, J., Tiilikainen, R. & Ylonen, H. 2003. Predation rate, prey preference and predator switching: experiments on voles and weasels. Oikos 101(3): 615–623. http://dx.doi.org/10.1034/j.1600-0706.2003.12264.x.
5. ●● Zhou, Y., Newman, C., Xu, W., Buesching, C., Zalewski, A., Kaneko, Y., MacDonald, D. & Xie, Z. 2011. Biogeographical variation in the diet of Holarctic martens (genus Martes, Mammalia: Carnivora: Mustelidae): adaptive foraging in generalists. J Biogeogr 38(1): 137–147. http://dx.doi.org/10.1111/j.1365-2699.2010.02396.x.
6. Harrison, D., Fuller, A. & Proulx, G. 2004. Martens and Fishers (Martes) in Human-Altered Environments: An International Perspective. New York: Springer.
7. Schoener, T. 1971. Theory of feeding strategies. Annu Rev Ecol Syst 2: 369–404. http://dx.doi.org/10.1146/annurev.es.02.110171.002101.
8. Sanchez, M. 1996. Dieta de la Garduña (Martes foina Erxleben, 1777) en una localidad de las sierras subbéticas de Granada (sureste de España). [Diet of the Marten (Martes foina Erxleben, 1777) in a village in the mountains of Granada subbéticas (southeast Spain)]. Acta Vertebrata 23(1): 83–90.
9. Ruiz-Olmo, J., & Palazon, S. 1993. Diet of the stone marten (*Martes foina* Erxleben, 1777) in the Northeastern Spain. Acta Vertebrata 20(1): 59–67.
10. ●● Papakosta, M., Bakaloudis, D., Kitikidou, K., Vlachos, C. & Chatzinikos, E. 2010. Dietary overlap among seasons and habitats of Red fox and stone marten in Central Greece. European Journal of Scientific Research 45(1): 122–127.
11. ● Genovesi, P., Secchi, M. & Boitani, L. 1996. Diet of stone martens: an example of ecological flexibility. J Zool 238(3): 545–555. http://dx.doi.org/10.1111/j.1469-7998.1996.tb05412.x.
12. Such, A. & Calabuig, G. 2003. Dieta de la Garduna (*Martes foina* Erxleben, 1777) en la Serra de la Solana (Sur del Pais Valencia). [Diet of the Marten (*Martes foina* Erxleben, 1777) in the Sierra de la Solana (south of Valencia)]. Galemys 15 (no especial): 167–180.
13. Ballesteros, T., Degollada, A. & Plaza, V. 2000. Dieta de la fagina al Parc Natural de Sant Llorenç del Munt i l'Obac. [Diet of beech marten in the Natural Park of Sant Llorenç del Munt i l'Obac]. ECOIMA: 119–122.
14. Lanszki, J. & Széles, G. 2007. A Dráva-mellékén élő nyestek (*Martes foina*) tavaszi táplálék-összetétele. [Diet composition of stone martens (*Martes foina*) living in the Drava district, Hungary, in spring]. Somogyi Múzeumok Közleményei – B Természettudomány, 17: 199–206.
15. Rasmussen, A. & Madsen, A. 1985. The diet of the stone marten *Martes foina* in Denmark. Natura Jutlandica 21(8): 141–144.
16. Apathy, M. 1998. Data to the diet of the urban stone marten (*Martes foina* Erxleben) in Budapest. Opusc. Zool. Budapest, XXXI.
17. ● Delibes, M. 1978. Feeding habits of the stone marten, *Martes foina* (Erxleben, 1777), in northern Brugos, Spain. Zeitschrift für Säugetierkunde 43: 282–288.
18. Basto, M., Matos, H., Pinheiro. P. & Santos-Reis, M. 2007. Diet of stone martens (*Martes foina*) in the Iberian Peninsula: a comparative analysis [abstract]. In: 25th Mustelid Colloquium; 2007 October 4–7; Třeboň. Czech Republic: Institute of Vertebrate Biology, v.v.i., Academy of Sciences of the Czech Republic; Ministry of the Environment of the Czech Republic;

Agency for Nature Conservation and Landscape Protection of the Czech Republic; Czech Otter Foundation Fund.

19. Eskreys-Wojcik, M. & Wierzbowska, I. 2007. The food habits of stone marten (*Martes foina*) inhabiting the city of Krakow. In: 25th Mustelid Colloquium; 2007 October 4–7; Třeboň. Czech Republic: Institute of Vertebrate Biology, v.v.i., Academy of Sciences of the Czech Republic; Ministry of the Environment of the Czech Republic; Agency for Nature Conservation and Landscape Protection of the Czech Republic; Czech Otter Foundation Fund.th Mustelid Colloquium, 4-7 October 2007, Trebon.

20. Amores, F. 1980. Feeding habits of the stone martens, *Martes foina* (Erxleben, 1777) in southwestern Spain. Säugetierkundliche Mitteilungen 28: 316–322.

21. Goszczynski, J. 1986. Diet of foxes and martens in central Poland. Acta Theriol 31(36): 491–506.
http://dx.doi.org/10.4098/AT.arch.86-44.

22. Holisova, V. & Obrtel, R. 1982. Scat analytical data on the diet of urban stone martens, *Martes foina* (Mustelidae, Mammalia). Folia Zool 31(1): 21–30.

23. Alegre, J., Hernandez, A., Purroy, F., Salgado, J. & Fuertes, B. 1991. Dieta otono-invernal de la Garduna, *Martes foina* (Erxleben, 1777), en un habitat rural de Leon (España) [Autumn-winter diet of beech marten, Martes foina (Erxleben, 1777) in a rural habitat of Leon (Spain)]. Ecologia o 5: 265–273.

24. Romanowski, J. & Lesinski, G. 1991. A note on the diet of stone marten in southeastern Romania. Acta Theriol 36: 201–204.
http://dx.doi.org/10.4098/AT.arch.91-19.

25. Lode. T. 1994. Feeding habits of the stone marten *Martes foina* and environmental factors in western France. Zeitschrift für Säugetierkunde 59(3): 189–191.

26. Lanszki, J., Sárdi, B. & Széles, G. 2009. Feeding habits of the stone marten (*Martes foina*) in villages and farms in Hungary. Natura Somogyiensis 15: 231–246.

27. Lanszki, J. 2003. Feeding habits of stone martens in a Hungarian village and its surroundings. Folia Zool 52(4): 367–377.

28. Bertolino, S. & Dore, B. 1995. Food habits of the stone marten *Martes foina* in "La Mandria" Regional Park (Piedmont region, north-western Italy). Hystrix 7(1–2): 105–111.
http://dx.doi.org/10.4404/hystrix-7.1-2-4058.

29. Martinoli, A. & Preatoni, D. 1995. Food habits of the stone marten (*Martes foina*) in the upper Aveto Valley (northern Apennines, Italy). Hystrix 7(1–2): 137–142.
http://dx.doi.org/10.4404/hystrix-7.1-2-4062.

30. Heltai, M., Szocs, E., Balogh, V. & Szabo, L. 2005. Adatok a nyest (*Martes foina* Erxleben, 1777) táplálkozásához és területhasználatához ember által zavart környezetben. [Information about the beech marten (*Martes foina* Erxleben, 1777), nutrition and land use of man-made environment]. Állattani Közlemények 90(2): 75–83.

31. Pandolfi, M., De Marinis, A. & Petrov, I. 1996. Fruit as a winter feeding resource in the diet of stone marten (*Martes foina*) in east-central Italy. Zeitschrift für Säugetierkunde 61(4): 215–220.

32. Sklodowski, J. & Posluszny, M. 2005. Beetles in the diet of the beech marten *Martes foina* and the pine marten *Martes martes*. Baltic Journal of Coleopterology 5(1): 31–36.

33. ● Carvalho, J.C. & Gomes, P. 2004. Feeding resource partitioning among four sympatric carnivores in the Peneda-Gerês National Park (Portugal). J Zool 263(3): 275–283.
http://dx.doi.org/10.1017/S0952836904005266.

34. Brangi, A. 1995. Seasonal changes of trophic niche overlap in the stone marten (*Martes foina*) and the red fox (*Vulpes vulpes*) in a mountainous area of the northern Apennines (N-Italy). Hystrix 7(1–2), 113–118.
http://dx.doi.org/10.4404/hystrix-7.1-2-4059.
35. Prigioni, C., Balestrieri, A., Remonti, L. & Cavada, L. 2008. Differential use of food and habitat by sympatric carnivores in the eastern Italian Alps. Ital J Zool 75(2): 173–184.
http://dx.doi.org/10.1080/11250000701885521.
36. Bermejo, T. & Guitian, J. 2000. Fruit consumption by foxes and martens in NW Spain in autumn: a comparison of natural and agricultural areas. Folia Zool 49(2): 89–92.
37. Barrientos, R. & Virgós, E. 2006. Reduction of potential food interference in two sympatric carnivores by sequential use of shared resources. Acta Oecol 30(1): 107–116.
http://dx.doi.org/10.1016/j.actao.2006.02.006.
38. Lanszki, J., Kormendi, S., Hancz, C. & Zalewski, A. 1999. Feeding habits and trophic niche overlap in a Carnivora community of Hungary. Acta Theriol 44(4): 429–442.
http://dx.doi.org/10.4098/AT.arch.99-41.
39. Rödel, H. & Stubbe, M. 2006. Shifts in food availability and associated shifts in space use and diet in stone marten. Lutra 49(1): 67–72.
40. Padial, J., Avila, E. & Sanchez, J. 2002. Feeding habits and overlap among red fox (*Vulpes vulpes*) and stone marten (*Martes foina*) in two Mediterranean mountain habitats. Mammalian Biology 67(3): 137–146.
http://dx.doi.org/10.1078/1616-5047-00021.
41. Serafini, P. & Lovari, S. 1993. Food habits and trophic niche overlap of the red fox and the stone marten in a Mediterranean rural area. Acta Theriol 38(3): 233–244.
http://dx.doi.org/10.4098/AT.arch.93-19.
42. Baghli, A., Engel, E. & Verhagen, R. 2002. Feeding habits and trophic niche overlap of two sympatric Mustelidae, the polecat *Mustela putorius* and the beech marten *Martes foina*. Zeitschrift für Jagdwissenschaft 48(4): 217–225.
http://dx.doi.org/10.1007/BF02189692.
43. Skirnisson, K. 1986. Untersuchungen zum Raum-Zeit-System freilebender Steinmarder (*Martes foina* Erxleben, 1777). [Studies on the space-time system of the wild stone marten (*Martes foina* Erxleben, 1777)]. Beiträge zur Wildbiologie 6: 1–200.
44. Marchesi, P. 1989. Écologie et comportement de la martre (*Martes martes* L.) dans le Jura Suisse. [Ecology and behavior of the marten (*Martes martes* L.) in the Jura Swiss]. D. Sc. Dissertation, Univ. of Neuchatel, Switz. pp. 185.
45. Tester, U. 1986. Vergleichende Nahrungsuntersuchung beim Steinmarder (*Martes foina* Erxleben, 1777) in großstädtischem und ländlichem Habitat. [Comparative food study of stone martens (*Martes foina* Erxleben, 1777) in large urban and rural habitats]. Säugetierkundliche Mitteilungen 33: 37–52.
46. Cheylan, G. & Bayle, P. 1988. Le regime alimentaire de quatre especes de mustelides en Provence: la fouine *Martes foina*, le blaireau *Meles meles*, la belette *Mustela nivalis* et le putois *Putorius putorius*. [The diet of four species of mustelid in Provence: the marten *Martes foina*, badgers *Meles meles*, weasels *Mustela nivalis* and polecats *Putorius putorius*]. Faune de Provence, 9: 14–26.
47. Ansorge, H. 1989. Die Ernährungsökologie des Steinmarders *Martes foina* in den Landschaftstypen der Oberlausitz. [The foraging ecology of the stone marten *Martes foina* in the landscapes of Upper Lusatia. In: Stubbe, M. (ed.), Population Ecology of Marten-Like Mammals.]. Wiss. Beitr. Univ. Halle, pp. 473–493.

48. Waechter, A. 1975. Écologie de la fouine en Alsace. [Ecology of the weasel in Alsace]. La Terre et Vie 29: 399–457.

49. Libois, R.M. 1991. La fouine (*Martes foina* Erxleben 1777). [The beech marten (*Martes foina* Erxleben 1777)]. In: Artois, M. & Delattre, P. (eds.), Encyclopédie des carnivores de France. Soc. Française l'étude protection mammifères, Nort-sur-Erdre. pp. 53 [Leger, F. (1979), cited in Libois 1991 (France)].

50. Panteloudis, A. 2011. Food habits of *Martes foina* in the Dadia - Lefkimi - Soufli Forest National Park (north-eastern Greece). Diploma thesis conducted in the Wildlife Laboratory of the Faculty of Forestry and Natural Environment, Aristotle University, Greece.

51. Weed, D. 1997. Meta-analysis under the microscope. Journal of the National Cancer Institute 89(13): 904–905.
http://dx.doi.org/10.1093/jnci/89.13.904.

52. Lipsey, M. & Wilson, D. 2000. Practical Meta-Analysis (Applied Social Research Methods). London: SAGE Publications, UK.

53. Panagiotakos, D. 2006. Μεθοδολογία της έρευνας και της ανάλυσης δεδομένων, για τις επιστήμες της υγείας. [Methodology of research and data analysis for the health sciences]. Kostakis Publications, Athens.

54. Borenstein, M., Hedges, L., Higgins, J. & Rothstein, H. 2009. Introduction to Meta-Analysis. London: John Wiley and Sons, UK.
http://dx.doi.org/10.1002/9780470743386.

55. Cooper, H. 2010. Research Synthesis and Meta-Analysis: A Step-By-Step Approach. London: SAGE Publications, UK.

56. Hardy, R.J. & Thompson, S.G. 1998. Detecting and describing heterogeneity in meta-analysis. Stat Med 17(8): 841–856.
http://dx.doi.org/10.1002/(SICI)1097-0258(19980430)17:8<841::AID-SIM781>3.0.CO;2-D.

57. Rosenzweig, M. 1995. Species diversity in space and time. Cambridge: Cambridge University Press.
http://dx.doi.org/10.1017/CBO9780511623387.

58. Clevenger, A. 1994. Feeding ecology of Eurasian pine marten *Martes martes* and stone marten *Martes foina* in Europe. In: Buskirk, S., Harestad, A., Raphael, M. & Powell, R. (eds.), Martens, Sables and Fishers: Biology and Conservation. Comstock Publishing Associates Ithaca, New York, pp. 326–340.

59. Ben-David, M., Flynn, R.W.& Schell, D.M. 1997. Annual and seasonal changes in diets of martens: evidence from stable isotope analysis. Oecologia 111(2): 280–291.
http://dx.doi.org/10.1007/s004420050236.

60. Chapman, R. 1998. The Insects: Structure and Function. Cambridge: Cambridge University Press.
http://dx.doi.org/10.1017/CBO9780511818202.

61. Ting, S., Hartley, S. & Burns, K. 2008. Global patterns in fruiting seasons. Global Ecology and Biogeography 17(5): 648–657.
http://dx.doi.org/10.1111/j.1466-8238.2008.00408.x.

62. Buskirk, S., Harestad, A., Raphael, M. & Powell, R. 1994. Martens, Sables, and Fishers: Biology and Conservation. Ithaca, New York: Comstock Publishing Associates.

63. Lopez-Martin, J. 2006. Comparison of feeding behavior between stone marten and common genet: living in coexistence. In: Santos-Reis, M., Birks, J., O'Doherty, E. & Proulx, G. (eds.), Martes in Carnivore Communities: Proceedings of the Fourth International Martes Symposium. Sherwood Park, Alta: Alpha Wildlife Publications. pp, 137–155.

64. Zhou, Y., Newman, C., Wang, X. & Zhang, S. 2008. Frugivory and seed dispersal by the yellow-throated marten, *Martes flavigula*, in a subtropical forest of China. J Trop Ecol 24(2): 219–223.
http://dx.doi.org/10.1017/S0266467408004793.

65. Glasser, J.W. 1982. A theory of trophic strategies: the evolution of facultative specialists. Am Nat 119: 250–262.

66. Kruuk, H. & Parish, T. 1981. Feeding specialization of the European badger *Meles meles* in Scotland. J Anim Ecol 50: 773–788.
http://dx.doi.org/10.2307/4136.

67. Da Silva, J., Woodroffe, R. & Macdonald, D.W. 1993. Habitat, food availability and group territoriality in the European badger, *Meles meles*. Oecologia 95(4): 558–564.
http://dx.doi.org/10.1007/BF00317441.

68. Goszczynski, J., Jedrzejewska, B. & Jedrzejewski, W. 2000. Diet composition of badgers (*Meles meles*) in a pristine forest and rural habitats of Poland compared to other European populations. J Zool (London) 250(4): 495–505.
http://dx.doi.org/10.1111/j.1469-7998.2000.tb00792.x.

69. Rosalino, M., Loureiro, F., Macdonald, D. & Santos-Reis, M. 2005. Dietary shifts of the badger *Meles meles* in Mediterranean woodlands: an opportunistic forager with seasonal specialisms. Mammalian Biology 70(1): 12–23.
http://dx.doi.org/10.1078/1616-5047-00172.

70. Pulliainen, E. & Ollinmaki, P. 1996. A long-term study of the winter food niche of the pine marten *Martes martes* in northern boreal Finland. Acta Theriol 41: 337–352.
http://dx.doi.org/10.4098/AT.arch.96-33.

71. Helldin, J. 1999. Diet, body condition, and reproduction of Eurasian pine martens *Martes martes* during cycles in microtine density. Ecography 22(3): 324–336.
http://dx.doi.org/10.1111/j.1600-0587.1999.tb00508.x.

72. Boyce, M. 1979. Seasonality and patterns of natural selection for life histories. Am Nat 114: 569–583.
http://dx.doi.org/10.1086/283503.

73. Parmesan, C. 2006. Ecological and evolutionary responses to recent climate change. Annual Review of Ecology, Evolution, and Systematics 37: 637–669.
http://dx.doi.org/10.1146/annurev.ecolsys.37.091305.110100.

74. Visser, M. & Both, C. 2005. Shifts in phenology due to global climate change: the need for a yardstick. P Roy Soc Lond B Bio 272: 2561–2569.
http://dx.doi.org/10.1098/rspb.2005.3356.

75. Nussey, D., Postma, E., Gienapp, P. & Visser, M.E. 2005. Selection on heritable phenotypic plasticity in a wild bird population. Science 310(5746): 304–306.
http://dx.doi.org/10.1126/science.1117004.

76. Chamaille-James, S., Massot, M., Aragon, P. & Clobert, J. 2006. Global warming and positive fitness response in mountain populations of common lizards *Lacerta vivipara*. Glob Change Biol 12: 392–402.
http://dx.doi.org/10.1111/j.1365-2486.2005.01088.x

7

Field Anesthesia of Least Weasels (*Mustela nivalis nivalis*) with Isoflurane

L. Gorini[1,2,*], S.J. Wedul[1], J.M. Arnemo[1,3], J.D.C. Linnell[2], L. Boitani[4], E.B. Nilsen[2]

[1] Department of Forestry and Wildlife Management, Faculty of Applied Ecology and Agricultural Sciences, Hedmark University College, Campus Evenstad, NO-2418 Elverum.
[2] Norwegian Institute for Nature Research, NO-7485 Trondheim, Norway.
[3] Department of Wildlife, Fish and Environmental Studies, Faculty of Forest Sciences, Swedish University of Agricultural Sciences, SE-901 83 Umeå, Sweden.
[4] University of Rome "La Sapienza", Department of Environmental Biology, V.le dell`Università 32, 00185 Rome, Italy.
* Corresponding author address: Department of Forestry and Wildlife Management, Faculty of Applied Ecology and Agricultural Sciences, Hedmark University College, Campus Evenstad, NO-2418 Elverum; e-mail: lucrezia.gorini@hihm.no.

Keywords
Anesthesia;
Least Weasel;
Nest Box;
Portable Anesthesia Unit.

Abstract
The aggressive and agile nature of weasels makes the use of inhalant anesthetics preferable to injectable agents because inhalant anesthetics generally reduce handling time, cause less stress and present no risk of delivering the drug into an unintended area. Here we report on the use of the inhalant anesthetic isoflurane in free ranging least weasels. We used a battery powered portable anesthesia unit (Uninventor 400 Anesthesia Unit, AgnTho's AB, Sweden) to immobilize the animals in the field. The isoflurane liquid was filled in a syringe connected to a vaporizer where it mixed with air coming from an air pump. The animals inhaled the anesthetic first in a wooden chamber and then through a modified facial mask. Mean induction time in the anesthetic chamber was 4.8 minutes (± 0.3 SE) with a setting of 4% isoflurane. The anesthesia was maintained with the facial mask and a setting of 2% isoflurane. Mean recovery time was 16.4 minutes (± 2.1 SE) and mean recumbency time 40.7 minutes (± 2.9 SE). We anesthesized 9 animals, had no fatalities during anesthesia and all animals recovered quickly. One animal was found dead 11 days post-capture of undetermined cause. Isoflurane is safe and easy to use with the tested setting. However we were unsuccessful in monitoring the animals following their release and we therefore recommend further evaluation of the anesthetic. In addition, we recommend the use of a wooden anesthetic chamber with bedding to avoid the risk of hypothermia.

Introduction

Predation by stoats (*Mustela erminea*) and weasels (*M. nivalis*) may have a strong impact on the population dynamics of their prey [1, 2]. Habitat use and hunting behavior of these species have however, mostly been studied using capture-mark-recapture methods, hair-traps, tracking tunnels (two-openings tunnels with paper and ink on the bottom where weasels´ tracks are left after the passage of the animals) or snow-tracking [3, 4]. Few studies have been based on radio-telemetry [but see for example 5], a research method that is often more effective [3] but which requires the anesthesia of captured individuals. A large number of anesthetics have been used in studies on small mustelids: e.g. ether [6, 7], phencyclidine hydrochloride and promazine [8] and ketamine [9]. Some of them are no longer in use or no longer produced (e.g. ether and phencyclidine). Others, such as ketamine, have the disadvantage of causing muscle rigidity and should therefore be used together with sedatives, for example, xylazine or medetomidine [5, 10-12]. This however may prolong the handling time and increase the number of necessary agents. In addition, the use of injectable agents requires a great amount of handling and potential stress which can be reduced by using inhalant anesthetics instead [13]. In small and very agile animals the additional risk of injecting the drug into an unintended area, e.g. close to a nerve, is particularly high. Therefore, the use of inhalant anesthetic is recommended [3] and isoflurane in particular has been successfully used in small and medium sized mammals [e.g. 13, 14, 15, 16]. Kreeger *et al.* [17] found that black-footed ferrets (*Mustela nigripes*) were more "physiologically normal" when immobilized with isoflurane compared to ketamine-medetomidine. Indeed, isoflurane provides general anesthesia with the help of a vaporizer and supply gas with a minimal amount of metabolism within the animal's body, thus contributing to the relatively high degree of safety [14, 18, 19]. Here we report on the use of isoflurane in the least weasel (*Mustela nivalis nivalis*) which, to the best of our knowledge, has not yet been tested in this species in the field [see reference 1 for use of isoflurane on captive least weasels]. We have used a light, battery powered, portable anesthesia unit which we found to be very practical and which has not been used in the field before on this species.

Material & Methods

Study area

The study area was located at Evenstad, Hedmark county, south-eastern Norway near latitude 61°25' N and longitude 11°06' E. The landscape consisted of patches of harvested forest stands, dominated by Scots pine (*Pinus sylvestris*) and Norway spruce (*Picea abies*), interspersed among clear-cuts. Other common tree species were birch (*Betula* spp.), willow (*Salix* spp.) and rowan (*Sorbus aucuparia*).

The study area in a cool-moderate continental climatic zone, was semi-humid with an annual average temperature of 4°C (average temperature over 2001-2010 measured at Evenstad weather station, altitude 255m), and mean precipitation 789mm year^{-1} (www.met.no). Snow cover normally lasts from November to April with large between year variation ranging from 20 to 120 cm.

Methods

The trapping was conducted in 2007 and 2008 in June, July and August. We used wooden Edgar type trap boxes [20] provided with different types of bait (e.g., raw meat mixed with eggs, dry fish or dead voles, etc.). All Edgar traps were connected to a wooden nest box (ca. 36x25x25 cm) that we could use as an anesthetic chamber in the field. The box had a transparent hard-plastic cover on top which made it possible to observe the animal inside. On top of it, a removable steel cover created a dark environment designed to reduce the stress experienced by the captured individual [3]. Weasel trapping was part of a project on habitat use of small mustelids and their main prey, small rodents. Hence, the mustelid traps (Edgar type) were mostly located within or in the vicinity of two trapping grids (1.4 and 1.8 ha respectively) for small rodents containing a total of 291 "multi-capture" live-traps for voles (Ugglan Special N°3, Grahnab AB, Hillerstorp, Sweden) [see 21 for details on the trapping of small rodents]. Weasels could easily enter the vole traps, in which case they were led to the nest box by carefully opening the sliding bottom of the vole trap on top of the entrance to the nest box (the two openings were the same width). In 2007 trapping was preceded by snow-tracking and the use of 16 tracking tunnels to obtain presence indices of small mustelids within the study area. The vole traps were set for a total of 90 days between 2007 and 2009. This effort, added to the 60 trapping days with 20 wooden mustelid traps in 2007 and 2008, gave a total of 150 trapping days and 45150 trap-nights. All traps were checked twice a day. All necessary permits for capture and handling animals were provided by the Norwegian Directorate for Nature Management and the Norwegian Animal Research Authority.

We used an Uninventor 400 Anesthesia Unit, size 23 x 13 x 10 cm, weight 1.8kg (AgnTho`s AB, Sweden, http://www.agnthos.se) with a built-in air flow meter and a vaporizer calibrated for isoflurane. The concentration of isoflurane and volume of the air flow could be constantly adjusted as needed. By connecting the unit to a portable battery, it could be easily used in the field (Fig. 1).

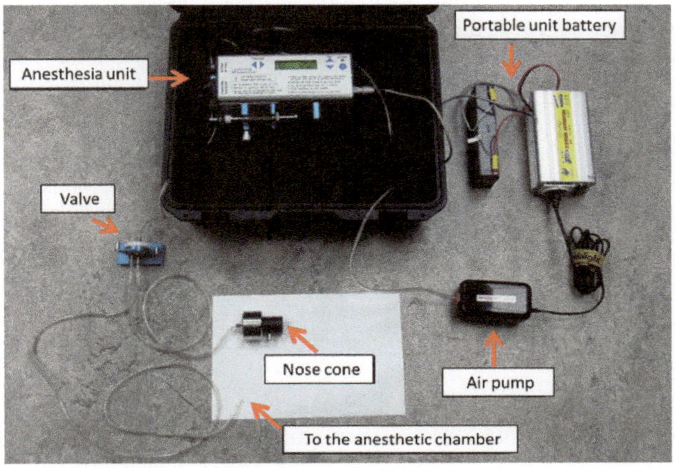

Fig. 1: Battery powered portable anesthesia unit for use in the field.

Isoflurane from a 10 ml gas-tight glass syringe was mixed with air in the vaporizer and was delivered in useful, safe concentrations for the anesthesia via the wooden chamber or a facial mask. The air came from an air pump connected to the unit (Fig. 1). From the vaporizer, isoflurane mixed with air flowed through a tube where an incorporated valve allowed the inhalant anesthetic to be directed either towards the anesthetic chamber or the facial mask. After induction in the anesthetic chamber, by turning a switch in the valve to reroute the anesthetic towards the facial mask placed around the animal`s head, the anesthesia could be maintained outside the anesthetic chamber (Fig. 2).

Fig. 2: Least weasel receiving the anesthetic through the facial mask. The animal was ear-tagged and radio-collared.

In addition, the unit had a built-in alarm system in case the air-flow or the anesthesia was blocked. Induction dose in the nest box was set to 4% on the vaporizer which was reduced to 2%, and 200 ml/min air once the anesthetic was delivered through the facial mask. We chose these doses based on the manufacturer's recommended maintenance dose of 2.2% isoflurane at a flow of 200 ml of air per min for mice, which we were ready to adjust in accordance with the observed effect on the animal. The animal was placed on a hot water bottle covered with a thin blanket to keep it warm during anesthesia. Standard measurements (body weight and length) were taken and a radio-collar (Biotrack, collar cable tie, weight 1.2gr) and ear tag (Hauptner & Herberholz) were fitted. Radio-tracking of the collared individuals started shortly after release and was scheduled in three hours periods throughout day and night, with 15 minutes between each location. Once the collar was fit and the measurements taken, the animal was placed into the wooden nest box provided with bedding to avoid loss of body heat and the nest box placed open next to the original capture site. We waited until the animal was able to leave the box on its own. Induction time was calculated from the moment isoflurane was inhaled in the anesthetic chamber to the moment the animal was removed from the anesthetic chamber and placed on the hot water bottle covered with a blanket. Maintenance time was calculated from the moment the facial mask was placed over the animal`s head until it was removed. Recumbency time was calculated from the moment the induction started until the animal was fully recovered. Recovery time was calculated from the moment the animal stopped receiving isoflurane until they were alert and effortlessly moving.

Results

We captured a total of 9 least weasels in 2008 (Table 1). All were caught in small rodent traps. Weasels were transferred to the wooden nest box where they first inhaled the anesthetic and we did not observe any panic reaction. We fitted collars on 6 animals. For the remaining 3: one individual was considered too small (weighing less than 30 gr) for the collar and the two others were a by-catch in small rodent traps within an experimental exclosure connected to a different project and were thus not suitable for our study on the species` movements. The mean induction time with a setting of 4% isoflurane in the anesthetic chamber was 4.8 (± 0.3 SE) minutes. The maintenance anesthesia time with a setting of 2% isoflurane, was 19.5 minutes (± 1.9 SE). After removal of the facial mask it took 4.1 (± 0.8 SE) minutes before we detected the first movements. The animals were alert and effortlessly moving after 16.4 (± 2.1 SE) minutes. Recumbency time was 40.73 (± 2.9 SE) minutes. We had no fatalities during the anesthesia but we found one dead individual 11 days after anesthesia. Based on our interpretation of the VHF radio-telemetry data it had died a week earlier. We were unable to successfully radio-track the other animals and recovered only two additional collars: one was recovered at a resting site three days after immobilization and one was found the day after

Table 1: Individual least weasels captured in 2008 in the vicinity of Evenstad (61°25′ N 11°06′ E), south-eastern Norway.

Date	Sex/ Age	Weight (gr)	Induction time (min.)	Maintenance time (min.)	Recovery time (min.)	Recumbency time (min.)	Respiration rate		
							10 min	15 min	25 min
19.7.2008	F/A	42	5.0	23.5	14.3	42.8	-	-	-
22.7.2008	F/A	40	4.8	21.2	15.0	41.0	-	-	-
22.7.2008	M/A	50	4.6	21.0	15.0	40.6	-	-	-
22.7.2008	M/*	40	3.0	22.5	25.0	50.5	-	-	-
24.7.2008	M/A	50	5.7	22.3	18.0	46.0	112	-	92
6.8.2008	F/A	40	4.5	22.5	18.0	45.0	92	100	68
6.8.2008	M/*	40	4.8	11.8	3.0	19.7	100	96	-
7.8.2008	F/J	36	6.5	7.5	24.0	38.0	68	-	-
2.9.2008	F/J	30	4.5	22.8	15.7	43.0	88	101	95

* - Uncertain

immobilization. Due to the particular shape of weasels, fitting collars on them has a high risk of the animal losing them [3]. The remaining three collars were never found and we did not succeed in tracking the animals.

Discussion

Despite our efforts in trapping least weasels, we were only able to catch nine. This was probably related to the extremely low density of voles in 2008 and 2009 [21], which most likely was accompanied by a crash in the least weasels population. In 2007, vole density was on the other hand extremely high (ca. 100 voles per hectare) and weasels may have been reluctant to enter a trap when prey was so readily available in most habitat types.

All animals recovered from the anesthesia, none had self-inflicted injuries and in spite of the mild pungency of isoflurane, none of them reacted or panicked when the isoflurane was introduced. We did not observe any emesis effect during anesthesia in accordance with a study by Horn *et al.* [22] showing no sign of isoflurane-induced emesis on a closely related species, i.e. ferrets (*Mustela putorius furo*). Weasels are known for biting at the trap when captured, to the point of breaking their teeth or skinning their noses [3, 23], we therefore believe that the dark environment and adequate bedding provided in the wooden nest box was particularly suitable as an anesthetic chamber for the species [3]. During anesthesia, the animals were monitored by counting the respiratory rate. This provided adequate safety for the animals considering the minimally invasive procedures that were performed and the short time the animals were receiving isoflurane. We prioritized keeping the animal under control continuously and maintaining the anesthesia as short as possible. We did not record systematically the respiratory rate except for 5 individuals (table 1) and we strongly recommend a follow-up study with monitoring of vital signs and blood gases to be carried out. An attempt to record the heart rate was made but due to pulse rates well over 200 it was considered less reliable than counting the respiratory rate. The small size of least weasels poses a challenge for pulse oximeter probe placement as pulse oximeters generally work well in animals weighing more than 200 g [14]. It was impossible to monitor eyelid reflexes as the facial mask covered most of the animals' head (Fig. 2).

The mean recovery time of 16 minutes was short compared to other studies on small mustelids using injectable anesthetics, such as the study from Gehring & Swihart [5] where the authors used ketamine-xylazine on long-tailed weasels (*Mustela frenata*) and recorded a mean recovery time of 26 minutes. In a study on the American marten (*Martes americana*) using isoflurane [16] recovery time was even shorter than in our case with a mean of 6 minutes.

The anesthesia unit was light and easy to operate in the field which allowed us to minimize the handling time. The isoflurane concentration and the air flow could be quickly adjusted and the unit could operate on air as well as on medical grade oxygen as recommended [24]. The anesthesia unit's calibration is such that whether using air or oxygen, the concentration of isoflurane shown on the display will be correct. According to the manufacturer the unit functions without any problems between 5°C and 35°C and our study was conducted at temperatures well within this range.

Unfortunately we did not succeed in monitoring the animals after the handling and this limits our interpretation of the safety of this technique. We therefore recommend further evaluation of the isoflurane on least weasels with consistent adequate telemetry monitoring of the animals after immobilization to assess post capture behavior and survival.

Acknowledgments

We thank S.P. for invaluable help in adjusting the anesthesia unit into a battery powered one. We thank all students who have helped us in the field. The study was funded by the Norwegian Directorate for Nature Management.

References

Five "key references", selected by the authors, are marked below (Three recommended (●) and two highly recommended (●●) papers).

1. Koivisto, E., Huitu, O., Sundell, J., Korpimäki, E. 2008. Species-specific limitation of vole population growth by least weasel predation: facilitation of coexistence? Oikos 117:6-12. http://dx.doi.org/10.1111/j.2007.0030-1299.15938.x

2. Dowding, J. E. & Murphy, E. C. 2001. The impact of predation by introduced mammals on endemic shorebirds in New Zealand: a conservation perspective. Biological Conservation 99:47-64. http://dx.doi.org/10.1016/S0006-3207(00)00187-7

3. ● King, M. C. & Powell, R. A. 2006. The natural history of weasels and stoats: ecology, behavior, and management. Oxford University Press.

4. Hellstedt, P., Sundell, J., Helle, P., Henttonen, H. 2006. Large-scale spatial and temporal patterns in population dynamics of the stoat, *Mustela erminea*, and the least weasel, *M. nivalis*, in Finland. Oikos 115:286-298. http://dx.doi.org/10.1111/j.2006.0030-1299.14330.x

5. Gehring, T. M. & Swihart, R. K. 2000. Field immobilization and use of radiocollars on long-tailed weasels. Wildlife Society Bulletin 28:579-585.

6. King, M. C. 1975. The home range of the weasel (*Mustela nivalis*) in an english woodland. The Journal of Animal Ecology 44:639-668. http://dx.doi.org/10.2307/3616

7. Jedrzejewski, W., Jedrzejewska, B., Szymura, L. 1995. Weasel population response, home-range and predation on rodents in a deciduous forest in Poland. Ecology 76:179-195. http://dx.doi.org/10.2307/1940640

8. Seal, U. S., Erickson, A. W., Mayo, J. G. 1970. Drug immobilization of the Carnivora. International Zoo Yearbook 10:157-170. http://dx.doi.org/10.1111/j.1748-1090.1970.tb01324.x

9. Sidorovich, V. E., Polozov, A. G., Solovej, I. A. 2008. Niche separation between the weasel *Mustela nivalis* and the stoat *M.erminea* in Belarus. Wildlife Biology 14:199-210. http://dx.doi.org/10.2981/0909-6396(2008)14[199:NSBTWM]2.0.CO;2

10. Fahlman, Å., Arnemo, J., Persson, J., Segerström, P., Nyman, G. 2008. Capture and Medetomidine-Ketamine anesthesia of free-ranging wolverines (*Gulo gulo*). Journal of Wildlife Diseases 44:133-142.

11. Belant, J. L. 1992. Field immobilization of American martens (*Martes americana*) and Short-tailed weasels (*Mustela erminea*). Journal of Wildlife Diseases 28:662-665.

12. Arnemo, J. & Søli, N. 1992. Immobilization of mink (*Mustela vison*) with Medetomidine-Ketamine and remobilization with Atipamezole. Veterinary Research Communication 16:281-292. http://dx.doi.org/10.1007/BF01839327

13. ● Mathews, F., Honess, P., Wolfensohn, S. 2002. Use of inhalation anesthesia for wild mammals in the field. Veterinary Record 150:785-787. http://dx.doi.org/10.1136/vr.150.25.785

14. Flecknell, P. 1999 Rabbits, rodents and ferrets. In. Edited by Seymour C, Gleed R. Manual of small animals Anaesthesia and analgesia. BSAVA, U.K., 295-304

15. Kocer, C. J. & Powell, L. A. 2009. A field system for isoflurane anesthesia of multiple species of mesopredators. American Midland Naturalist 161:406-412. http://dx.doi.org/10.1674/0003-0031-161.2.406

16. ●● Desmarchelier, M., Cheveau, M., Imbeau, L., Lair, S. 2007. Field use of isoflurane as an inhalant anesthetic in the american marten (*Martes americana*). Journal of Wildlife Diseases 43:719-725.

17. ●● Kreeger, T. J., Vargas, A., Plumb, G. E., Thorne, E. T. 1998. Ketamine-medetomidine or isoflurane immobilization of black-footed ferrets. Journal of Wildlife Management 62:654-662.
http://dx.doi.org/10.2307/3802341

18. Belant, J. L. 1995. Isoflurane as an inhalation anaesthetic for muskrats (*Ondatra zibethicus*). Journal of Wildlife Diseases 31:573-575.

19. Seal, U. S. & Kreeger, T. J. 1987. Chemical immobilization of furbearers. In. Edited by Novak M, Baker JA, Obbard ME, Malloch B. Wild furbearer management and conservation in North America. Ontario, Canada.

20. King, M. C. & Edgar, R. L. 1977. Techniques for trapping and tracking stoats (*Mustela erminea*): a review, and a new system. New Zealand Journal of Zoology 4:193-212.
http://dx.doi.org/10.1080/03014223.1977.9517953

21. Gorini, L., Linnell, J. D. C., Boitani, L., Hauptmann, U., Odden, M., Wegge, P., Nilsen, E. B. 2011. Guild composition and habitat use of voles in 2 forest landscapes in south-eastern Norway. Integrative Zoology 6:299-310.
http://dx.doi.org/10.1111/j.1749-4877.2011.00258.x

22. Horn, C. C., Meyers, K., Pak, D., Nagy, A., Apfel, C. C., Williams, B. A. 2012. Post-anesthesia vomiting: Impact of isoflurane and morphine on ferrets and musk shrews. Physiology & Behavior 106:562-568.
http://dx.doi.org/10.1016/j.physbeh.2012.03.031

23. Belant, J. L. 1992. Efficacy of three types of live traps for capturing weasels, *Mustela* spp. The Canadian Field Naturalists 106:394-397.

24. ● Lewis, J. C. M. 2004. Field use of isoflurane and air anaesthetic equipment in wildlife. Journal of Zoo and Wildlife Medicine 35:303-311.
http://dx.doi.org/10.1638/03-072

8

Alpine Mountain Hare *Lepus timidus varronis* Defecation Rate: A First Step Toward Faecal Pellet Count Monitoring

P. Tizzani[1], M. Bessone[1,*], S. Lioy[1] & P. Meneguz[1]

[1] Department of Veterinary Science, University of Turin, V. Leonardo da Vinci 44, 10095 Grugliasco (TO), Italy. Authors emails: paolo.tizzani@unito.it; mattia.bessone@gmail.com; simonelioy@yahoo.it; piergiuseppe.meneguz@unito.it.

* Corresponding author e-mail: mattia.bessone@gmail.com.

Keywords
Defecation rate;
Lepus timidus varronis;
Pellet count;
Density estimation method.

Abstract

Lepus timidus varronis, is the alpine sub-species of *Lepus timidus* and an endemic glacial relict of the Alps. Despite the great conservational and biological relevance of the species, the number of studies concerning it's biology and ecology are limited. An essential parameter for species management is to correctly estimate population density. To date there are no standardized nor validated survey techniques for surveying *L. t. varronis*. The faecal pellet count is considered a suitable census method but its use is limited due to the absence of data concerning the daily mean defecation rate of the species. The aim of this work is therefore to estimate the daily mean defecation rate of *L. t. varronis* in a controlled environment in order to allow the implementation of subsequent studies on population density and dynamics. A mean defecation rate of 411.7 ± 41.3 pellets/day per hare was estimated on the 14 breeding Alpine hares that were included in the study.

Introduction

The first step toward the conservation of a species, is to know its population density and demographic trend [1]. When cryptic species have to be monitored indirect estimate methods are the best choice [2]. Among these, the faecal pellet count (FPC) is a highly feasible census method that estimates density on the basis of the number of faecal pellets present in a sample area [3]. FPC has been used to estimate population density of mountain hare *Lepus timidus*, but also to monitor population trends and to compare different presence areas [4].

This method seems to be particularly appropriate for mountain hare *Lepus timidus* (Linnaeus 1758) in the Alps, where a suitable monitoring method has not yet been standardized. The *L. timidus* population is present in the Alps as an endemic sub–species (*L. t. varronis*; Miller 1901) [5].

L. t. varronis is a species of great biological value because of its endemic distribution within the high-altitude Alpine environment and its importance in the local trophic chain [6]. To the best of our knowledge, only few scientific studies attempted to investigate the biology of this species [5-8]. Further effort is needed to improve management and conservation plans of *L. t. varronis* especially because in some areas of the Italian Alps its population seems to be declining [6]. Despite, this potential decreasing trend, presently this species is classified as a game species.

Scandinavian and Scottish authors used different methods for estimation of abundance of other *L. timidus* subspecies such as capture – mark – recapture [9], direct counts [10], snow tracks [11] and game bags [12] but a suitable monitoring method for *L. t. varronis* has not yet been standardized.

In order to correctly use FPC, it is a prerequisite to know i) decay rate and ii) defecation rate [13] of the studied species. Data about decay rate are available for closely related species such as *L. americanus*. The reliability of such data is limited since decay rate is highly influenced by local conditions (weather, land use, etc).

FPC implemented with Bennet's formula or with the more recent distance sampling approach has been successfully used to estimate density from pellets density [14, 15]. Clearing the sample area of all faeces before starting the survey allows to bypass the use of decay rate (Clearance count – [13]).

Nevertheless no data about *L. t. varronis* defecation rate is available; and the only existing datum (250 pellet/day – [16]) refers to *L. timidus scoticus* (sub-species phenotipically different from *L. t. varronis*). The lack of such information makes impossible the estimation of *L. t. varronis* density through FPC method.

The aim of this work is to determine the *L. t. varronis* daily defecation rate, in order to obtain an instrument to monitor mountain hare population, based on the FPC method.

Methods

The study was carried out in the Asiago plateau (45° 52' 34" N, 11° 30' 32" E; Vicenza's province, Northeastern Italy), at 1,000 m a.s.l., where breeding captive mountain hares (eight females and six males) were maintained in grilled metallic cages (Table 1). The animals were not maintained in captivity for this study, but they were detained as ornamental animals in accordance with the Italian Law (National Law no. 157/92). During the essays there was no need to manipulate the animals, because pellets were collected outside the grilled cage.

Animals were fed *ad libitum* with a variety of hayed Alpine meadow grasses, common laburnum *Laburnum anagyroides* branches and rabbits industrial fodder (proteins 16%, lipids 3%, fiber 17.5%, ashes 10%); water was administered *ad libitum*. Moreover, in order to recreate natural conditions, no food was given the first day after a snowfall.

On April 2008 faeces were collected removing the litter from the cages at 24 hour intervals; the number of pellets was counted. Each cage was controlled during three consecutive days. Data were processed and statistically analysed to establish the mean defecation rate value.

The essays were not made with isolated animals, in order to avoid stress factors, associated with the separation of breeders already used to be together in the same cage. There was also no usefulness in the estimation of defecation rates between sexes, because in field trials using faecal pellet counts, it is not possible to separate scats according to the sex,

A non-parametric Mann-Whitney test was used to compare defecation rate values with regard to a different food intake.

Results

The number of pellets collected is reported in table 1. Daily defecation mean rate was 372.5 ± 69.6 SD pellets/day with a coefficient of variation of 18.7%.

Table 1: Number of hares/cage and number of collected pellets for each day: in "Day 1" animals were not fed (fast day implication on defecation rate are observed in "Day 2").

Group	N° hares	Day 1 (mean)	Day 2 (mean)	Day 3 (mean)
A	4	1916 (479)	1047 (261.7)	1712 (428)
B	2	789 (394.5)	520 (260)	800 (400)
C	3	1298 (432.6)	988 (329.3)	1007 (335.6)
D	5	1918 (396.2)	1627 (325.2)	2137 (427.4)

As for food consumption, the first collection day was characterized by a heavy snowfall. During snowfalls, the animals significantly reduce food consumption so we considered separately the daily defecation rate of "normal" and "snow" days. The mean defecation rate value for snow days was 294.1 ± 38.4 SD, and it was statistically different from the samples recorded during the "normal feeding days" (411.7 ± 41.3 SD; $W = 32$; $p < 0.01$) implying that a day without access to food can reduce the defecation rate by 28.6%.

On the basis of these observations and in order to obtain a standardized daily defecation rate, we decided to exclude the snow day defecation rate.

The standardized mean defecation rate (SMDR) was 411.7 ± 41.3 SD, ranging between 336 and 479 pellets/day, with higher frequency values between 380 and 440 pellets/day. The coefficient of variation was 10%.

Discussion

This is the first report concerning *L. t. varronis* daily defecation rate. Despite the fact that these essays were done with captive animals and so in environmental conditions that differ from the natural ones, this datum is very important for species management as it will allow a more accurate monitoring of *L. t. varronis* population status and trend using faecal pellet count (FPC) methodologies.

The actual lack of reference data is explained by difficulties in finding *L. t. varronis* in controlled environment, when compared to other species such as cervids [13]. In fact, breeding individuals are the only ones available for which it is possible to record defecation rates under constant monitoring and full pellet's detection [2, 16].

The mean defecation rate estimated in our study (411.7 ± 41.3 SD pellets/day) is higher (about 39%) than the one reported for *L. t. scoticus* (250 pellets/day – [16]) but very similar to the one of *L. europaeus* (430 pellets/day – [16]). This result shows that the application of Flux's *L. t. scoticus* defecation rate to *L. t. varronis*, will consequently produce a large overestimation of the population.

Population density estimation can be affected by several environmental variables that influence the defecation rate; among these factors we can consider the higher or lower food consumption and the sex and age of the population [17].

Our work highlights that during one "snow" day, when hares generally stay under cover [18], lead to a defecation rate reduction of 28.6%. As previously reported, a statistically significant difference between defecation rate in "snow" and normal days was obtained, resulting in the exclusion of "snow" day samples from the calculation of SMDR value. This allowed to increase the precision of our data and to reduce the coefficient of variation.

The composition of the diet (above all its fiber content) is another factor that could influence the defecation rate [17], especially for captive herbivore species. Even if it could be a source of bias, at the same time it is a common problem in all defecation rate studies because the study of this parameter is only possible with captive individuals [2, 16].

The value of SMRD obtained in our study was similar to a personal observation made in another study area (Introd's Parc Animalier 45° 41' 34" N, 7° 10' 55" E in Aosta's Valley, 869 m a.s.l). In this breeding site, only one male and one female *L. t. varronis* were present and the observed defecation rate value was 403.71 ± 65.92 SD in "normal feeding days" and 272.29 ± 69.39 SD in "snow" days (unpublished data).

We therefore believe that the adoption of our results to FPC can contribute to establish the basis for starting an appropriate system of *L. t. varronis* management-monitoring.

F urther studies are needed to improve the evaluation of the defecation rate specially in what concerns the need to increase the number of sampled animals and extent of the sampling period.

In the same way it will be necessary to evaluate the faecal decay rate with the use of decay station in different environmental and climatic condition.

References

Five "key references", selected by the authors, are marked below (Three recommended (●) and two highly recommended (●●) papers).

1. Sutherland, W. J. (1996). Ecological census techniques, a handbook. Cambridge University Press, 336 pp.
2. Dinerstein, E., Dublin, H. T. (1982). Daily defecation rate of captive Axis Deer. Journal of Wildlife Management, 46(3): 833-835. http://dx.doi.org/10.2307/3808586
3. ●● Newey, S., Bell, M., Enthoven, S., Thirgood, S. (2003). Can distance sampling and dung plots be used to assess the density of mountain hares *Lepus timidus*?. Wildlife Biology, 9(3): 185-192.
4. Preatoni, D. G., Tattoni, C., Bisi, F., Masseroni, E., D'Acunto, D., Lunardi, S., Grimod, I.,

Martinoli, A., Tosi, G. (2012). Open Source evaluation of kilometric indexes of abundance. Ecological Informatics, 7(1): 35-40.
http://dx.doi.org/10.1016/j.ecoinf.2011.07.002

5. Couturier, M. A. J. (1964). Le gibier des montagnes françaises. [French mountains game]. Grenoble (Isère): Arthaud, 463 pp.

6. Bisi, F., Nodari, M., Dos Santos Oliveira, N. M., Ossi, F., Masseroni, E., Preatoni, D. G., Wauters, L. A., Martinoli, A. (2011). Space use patterns of mountain hare (*Lepus timidus*) on the Alps. European Journal of Wildlife Research, 57(2): 305-312.
http://dx.doi.org/10.1007/s10344-010-0429-2

7. ●● Bisi F., Nodari, M., Dos Santos Oliveira, N. M., Ossi, F., Masseroni, E., Preatoni, D. G., Wauters, L. A., Martinoli, A. (2013). Habitat selection and activity patterns in Alpine mountain hare (*Lepus timidus varronis*). Mammalian Biology. Vol. 78: 28-33.
http://dx.doi.org/10.1016/j.mambio.2012.05.004

8. Bouche, M. (1989). Le lièvre variable dans le massif des Écrins. [The mouintain hare in Écrins massif]. Documents scientifiques du Parc National des Écrins, 111 pp.

9. Hewson, R. (1965). Population change in mountain hare *Lepus timidus scoticus*. Journal of Animal Ecology, 34: 587-600.
http://dx.doi.org/10.2307/2452

10. Hakkinen I., Jokinen, M. (1981). Population dynamics of the mountain hare on an island in the Outer Archipelago. Proc. 1st Lagomorphs World Conf. Guelph Ontario, 469-477.

11. Kauhala, K., Helle, P., Helle, E., Korhonen, J. (1999). Impact of predator removal on predator and mountain hare populations in Finland. Annales Zoologici Fennici 36: 139–148.

12. Stoddart, D. M. (1984). Mountain hare *Lepus timidus* – bags and moor management. Journal Zoology London, 204: 563-590.
http://dx.doi.org/10.1111/j.1469-7998.1984.tb02388.x

13. ● Mayle, B. A., Peace, A. J., Gill, R. M. A. (1999). How many deer? A field guide to estimating deer population size. Forestry Commission, Edinburgh, 96 pp.

14. ● Bennet, L. J., English P. F., McCaine, R. (1940). Study of deer population by use of pellet group count. Journal of Wildlife Management, 4: 398-403.
http://dx.doi.org/10.2307/3796010

15. Thomas, L., Buckland, S. T., Rexstad, E. A., Laake, J. L., Strindberg, S., Hedley, S. L., Burnham, K. P. (2010). Distance software: design and analysis of distance sampling surveys for estimating population size. Journal of Applied Ecology, 47(1): 5-14.
http://dx.doi.org/10.1111/j.1365-2664.2009.01737.x

16. ● Flux, J. E. C. (1970). The life history of mountain hare (*Lepus timidus scoticus*) in north-east Scotland. Journal of Zoology, 161: 75-123.
http://dx.doi.org/10.1111/j.1469-7998.1970.tb02171.x

17. Neff, D. J. (1968). The pellet group count technique for big game trend, census and distribution: a review. Journal of Wildlife Management, 32(3): 597-614.
http://dx.doi.org/10.2307/3798941

18. Hewson, R., Hinge, M. D. C. (1990). Characteristics of the Home Range of Mountain Hares *Lepus timidus*. Journal of Applied Ecology, Vol. 27, No. 2 (Aug., 1990), 651-666.
http://dx.doi.org/10.2307/2404309

9

Alternative Financing Schemes for Tiger Conservation in Nepal

H. Kafley[1, 2, *], M. E. Gompper[1], F. Spinelli[3], K. L. Poudel[4], B. P. Thapaliya[5]

1 Department of Fisheries and Wildlife Sciences, University of Missouri, Columbia, MO 65211, USA.
2 Nepal Nature Foundation, Koteshwor, Kathmandu-34, Nepal.
3 221 W, Grant St., Bozeman, MT 59715, USA.
4 Department of Agricultural Economics, University of Missouri, Columbia, MO 65211, USA.
5 Chitwan National Park, Kasara, Chitwan, Nepal.
* Corresponding author email: hkc5b@mail.missouri.edu.

Keywords
Conservation finance;
Conservation economics;
Panthera tigris;
Conservation trust fund.

Abstract
Many scientists and conservation workers agree that global tiger numbers in key areas could double by 2022 if efforts are taken immediately. Countries facing declining tiger numbers and having the ability to meet this goal have produced national tiger recovery plans that outline necessary program activities and their associated costs. To assess the financial feasibility of the tiger conservation program in Nepal and to recommend viable alternatives to secure funds to cover long-term tiger conservation costs, we conducted financial analyses of tiger conservation programs in Nepal. Our results show that the present funding level fails to cover the long-term costs of taking the recommended steps for tiger conservation. Thus there is a need to identify and secure alternative funding sources to supply approximately a 100% increase in revenues currently generated from tiger-bearing protected areas assuming a continuance of the current level of funding by the government. This finding is troublesome given the magnitude of the financial burden associated with necessary steps to increase the tiger population, plus the fact that no policy instrument exists that can target the revenue generated by the protected areas specifically for tiger conservation. To achieve financial sustainability of the tiger conservation program, alternative financial mechanisms warrant serious consideration. One alternative institutional mechanism could be a tiger conservation trust fund that would be entrusted to secure a wide range of financing from domestic and international sources to ensure financial sustainability of the Nepal's tiger conservation program.

Introduction

The successful conservation of ecosystems depends on a mix of environmental, social and economic factors. For nations that struggle financially to provide basic public services, the funding of ecosystem maintenance or conservation agendas often are not the highest priority for limited government dollars. In many of these situations, the opportunity costs of maintaining wilderness is perceived as higher than the benefits to the local community or the government – especially in a short-term context [1]. Because short-term solutions yield poor long-term outcomes, global communities recognize that funding remedies are needed to address this short-term investment decision so that conservation of the world's remaining natural systems can be achieved.

In this context, Nepal is striving to identify and secure a sustainable financing mechanism to support tiger conservation programs. A tiger conservation program has been identified and designed for Nepal to meet its' global commitment of doubling the tiger numbers by 2022 [2, 3].This paper analyses the current funding of Nepal's tiger conservation program and the potential of Nepal to fund tiger conservation into the future. Current and future gross costs of tiger conservation in Nepal are estimated in this analysis, something that is rarely conducted for any species and has never been conducted for tigers. Because this analysis finds a serious short-fall in current funding with funding levels needed to achieve tiger conservation goals, alternative potential funding sources are presented and discussed in the context of ensuring the sustainability of tiger conservation programs in Nepal into the future.

Tigers are globally revered and tiger conservation is probably the highest funded conservation program in Nepal. As tigers are recognized as the umbrella species in the regional ecosystem, conservation efforts targeted for tigers have cascading positive impacts on the conservation of biodiversity system-wide [2, 4, 5]. Therefore, a higher conservation cost for tigers is more easily justified when taking into account its positive, synergistic impact on other species and ecosystem services [5]. However, the ability of a developing country such as Nepal to meet the required costs for conservation and sustain efforts to meet the goal of doubling Nepal's tiger population poses a major obstacle for Nepal and the global conservation community.

Funding for conservation, especially when dealing with a goal of improving overall habitat and biodiversity, is always limited [6, 7]. One concrete example of the umbrella species effect brought about by improved tiger habitat is the enhancement of other native wildlife species, such as elephant, rhino, and deer. Although not quantified in this analysis, such general and widespread benefits are apparent in Nepal and should be convincing even with a limited focus on tiger conservation. At a minimum, the estimation of associated expected costs of tiger conservation can provide crucial information if efforts are implemented successfully to halt extinctions of other species [8]. However for most of the terrestrial sites of global conservation significance, the costs of reducing extinction risk by effective and sustained best management practices are largely unknown [8]. Through a National Tiger Recovery Plan (NTRP) Nepal has stated its commitment to develop a comprehensive financial plan that will ensure that full costs of the tiger program are met and maintained in perpetuity [3, 9]. In Nepal, the government has estimated the annual cost for tiger conservation programs as currently roughly $6.5 million (GTRP 2010-22). This estimate is purely a financial calculation and cannot be considered as a total economic cost. Although this amount represents only 0.125 percent of total government spending, any additional money to increase tiger conservation efforts would need all stakeholders including the Nepalese government authority to fully understand all the benefits and costs of action or inaction.

This holds true whether benefits are measured in monetary values or in physical values (e.g. species or ecosystem attributes wherein carnivores provide a useful purpose of keeping populations of other species in check). In addition, results of economic analyses may not always be the driving force determining the goals of biodiversity conservation investment [10] especially in the case of carnivore conservation where the direct costs at a local level may lead to a short-term solution that does not consider long-term ramifications – a classic case of a "market failure" in the natural resource literature [11]. Large carnivores such as tigers are valued at global or national scale but the costs are usually born at the local level, creating "market failure" that escalates human-carnivore conflict [12] and might be an underlying cause of carnivore declines [11].

Financial analysis establishes the magnitude of costs of capital investments and weighs these against financial benefits of a program. In the context of tiger conservation programs in Nepal, investment costs include financial resources required to maintain and protect the tiger-bearing protected areas (hereafter 'TBPAs') and the broader Terai Arc Landscape (hereafter 'TAL') encompassing approximately 12.3 million acres in Nepal and neighboring India, and conducting other specific tasks pertaining to tiger conservation (see the GTRP 2013-14 implementation plan for a breakdown of program costs) beyond the immediate habitats occupied by tigers [13].

The benefit stream referred to in this case is not the typical set of benefits in most environmental economic studies. It does not include economic values generated by overall ecosystem services and tourist spending generated from tiger conservation, but rather simply refers to the existence of a positive gap between expected program funding and program costs. In a sense, this analysis can be thought of as a budget analysis, weighing the possible annual funding with program costs. Program funding includes income from donations obtained from national and international funding agencies for tiger conservation activities, regular cash flow set-aside by the government, and other finances committed by international development organizations such as the World Bank and the United Nations Development Program. The financial analysis also integrates other possible future financing sources as benefits, modified to account for the terms and conditions of financing. Wildlife premiums, either linked to, or independent of, carbon and non-carbon payments for ecosystem services (PES) [4], could also potentially be included in the benefit stream.

Methods

We used a capital budgeting [14] approach for our analyses and estimated Net Present Value (NPV) and Benefit Cost (B/C) ratio [15]. NPV and B/C ratio are widely used for assessing sustainability of conservation projects [16]. NPV is defined as the present value (PV) of the net income stream accruing to the tiger program [17]. NPV calculation requires a discount rate, the rate reflecting the public's preference for money to be received in the future. The market cost of borrowing money from banks and other financial institutions is the general interest rate charged to entities and is widely interpreted as the public's discount rate for various financial analysis purposes [16, 18, 19]. If the tiger conservation program cannot generate an income stream sufficient to pay back the loan plus the interest, then the program lacks financial sustainability. The B/C ratio is the ratio of the PV of the benefit stream and the PV of the cost stream. Thus, in this paper the B/C ratio can be interpreted as the difference in the stream of expected available funding compared to the stream of expected costs. A ratio greater than 1, at the current discount rate, indicates that the funding levels exceeds the costs (i.e., the program is financially sustainable). If the ratio falls below 1, the costs outweigh the funding level and the financial sustainability of the program cannot be assured [17].

Costs and funding levels for tiger conservation program were estimated based on information from government authorities and other stakeholders, such as international or non-governmental organizations (I/NGOs). We consider two types of costs: basic cost and optimal cost [20] of tiger conservation programs based on the data available and the existing practice of cost estimation [9]. These costs are determined based on two assumptions: 1) Basic cost for tiger conservation is the cost required to maintain TBPAs in Nepal (costs of programs implemented by the government in the TBPAs and the costs of government endorsed conservation projects conducted by partner organizations in or outside the protected areas); 2) Optimal cost is defined as the cost that the government deems necessary to improve tiger conservation outcomes reflected in the proposed required costs associated with the Global Tiger Recovery Program 2010-2022. Because the economic costs of tiger conservation are beyond the scope of this study, any opportunity costs of tiger conservation, or the positive/negative externalities of conservation programs, are not considered. Hence, direct expected financial funding needed for the tiger conservation program is considered as cost. Similarly direct financial inflows in the TBPAs including government funding and TBPAs' income constitute the revenues for further analytic purpose.

The NPV criterion of the Capital Budgeting Approach [17, 21], which uses the discounting formulas for a non-uniform or uniform series of payments to value the projected cash flows for each investment alternative at one point in time, was used to conduct sensitivity analysis of sustainability under different revenue and cost alternatives. The formulae used to obtain the discount factor, NPV and the B/C ratio are:

Discount factor, $$\mathrm{df} = \frac{1}{(1+r)^t} \quad (1)$$

Present value of cashflow, $$\mathrm{PV} = \mathrm{FV} \times \frac{1}{(1+r)^t} \quad (2)$$

Net Present Value, $$\mathrm{NPV} = \sum_{t=1}^{n} \frac{C_t}{(1+r)^t} - C_0 \quad (3)$$

Where

t = the time of the cash flow

n = the total time of the project

r = the discount rate

C_t = the net cash flow (the amount of cash) at time t.

C_0 = the capital outlay at the beginning of the investment time ($t = 0$)

The decision making framework of these financial indicators [21] was adopted to identify the sustainability of the tiger conservation program (Table 1).

Table 1: Decision making framework.

	Decision Criteria		
Financial tool	Sustainable	Indifference	Unsustainable
B/C ratio	If, >1	Equal to 1	< 1
NPV	> 0	Equal to 0	< 0

Sensitivity analyses were conducted based on the same criteria given above by creating alterative scenarios based on different levels of support from existing funding sources and assumed expected higher incoming cash flows from higher revenues generated by TBPAs. Potential finances from any sources that could not be quantified in terms of amount of support are discussed qualitatively and not included in the analyses.

Results and Discussion

Tiger conservation costs and funding

The average annual basic cost of tiger conservation in Nepal from 2014 through 2023 was estimated to be US$5.2 million. The average annual funding level secured from internal government sources for covering these costs was approximately US$3.2 million. Thus there was an annual short-fall of US$2.0 million for meeting the basic cost during the 10-year period from 2014 to 2023 (Table 2). This gap increases when one considers the additional funding needed to finance the 'optimal' level of effort to meet the goal of a doubling tiger population.

Table 2: Average annual costs (in US$) of tiger conservation program. Past (2009-2013) costs are based on known expenditures. Future costs (2014-2023) are cost projections from past expenditures and additional cost estimations. Basic cost is based on government budget that reflects costs for maintaining TBPAs plus regular program budget of currently operating INGOs that implement tiger conservation programs based on government approved programs and activities inside or outside of TBPAs. Government budget represents past annual fiscal support for TBPAs and the projected support by the government of Nepal. Optimum cost represents the costs estimated by the government for implementing tiger conservation program (NTRP, 2010) across the tiger range in Nepal. (See Appendices A.1 and A.2 for cost composition).

Years	Optimum cost	Basic cost	Government budget
2009 - 2013	3710980	2843813	1220512
2014 - 2023	6738893	5244803	3178893

The optimal cost of tiger conservation in Nepal, as proposed by the government to achieve the targets mentioned in NTRP, would add approximately US$1.5 million [9] to the basic 2014-2023 cost level, increasing average annual total costs to US$6.7 million (See appendix for the cost and revenue composition). Compared to the static secured funding level of US$3.2 million, the annual deficit for sustaining the program would increase to US$3.6 million or US$36.0 million over the 10-year period from 2014 to 2023. Hence, the results show that funding must increase by 65% over current levels to meet basic costs and more than double over current funding to match the proposed optimum funding. Thus, our analysis shows that the government funding cannot meet the growing conservation costs, even if income generated by TBPAs is included (Fig. 1). A 20% increase in TBPA income in addition to the government budget would meet the total cost of the basic and optimum conservation efforts by the year 2023. In the best-case scenario, if TBPA income is doubled and combined with the government funding, the total funding would meet the optimum cost of tiger conservation in the year 2016 and beyond.

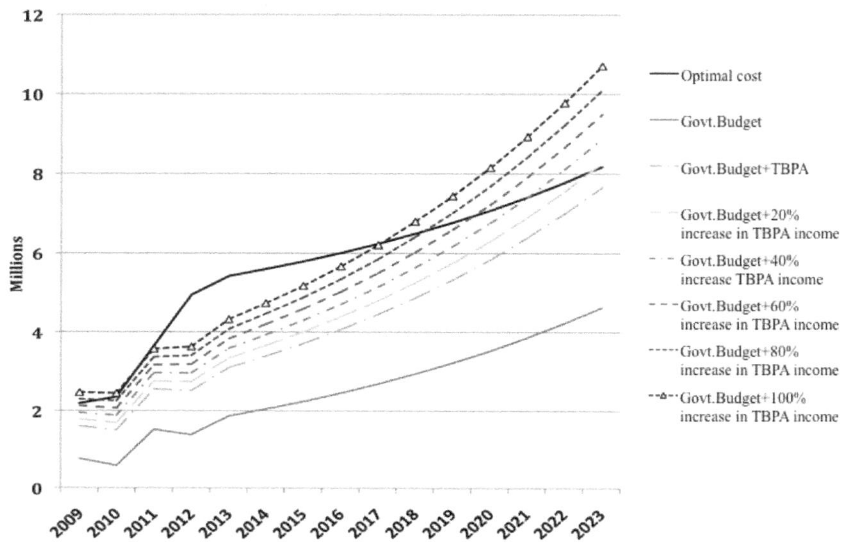

Figure 1: Optimal cost of tiger conservation and income generated under different scenarios that include the government budget, the government budget plus the income of tiger bearing protected area (TBPAs), and the government budget plus TBPA income that is allowed to increase by 20-100%.

In a recent study [22] conducted in the areas within the TAL of Nepal, the proposed cost for managing that landscape for achieving the integrated biodiversity conservation with a special focus on tiger conservation is estimated as >US$20 million per year. Against these annual costs, the study identified only US$4 million per year in secured committed funding and US$2 million per year coming from prospective investors in TAL [22]. Using this scenario, the cumulative total of this annual US$14 million funding short-fall produced a 10-year gap of US$140 million over the 2014 – 2023 projection period. These estimates, reflecting both the basic and optimal cost scenarios, reinforce our findings that Nepal's tiger conservation program is not financially sustainable given current costs and committed funding.

Sensitivity analyses and I/NGO support

Financial analyses found that Nepal's tiger conservation program is not financially sustainable through current government funding alone for both the basic cost scenario and optimal cost scenario (both producing negative NPVs and B/C ratios of < 1). The addition of increased income from TBPAs in the benefit stream did not change these results. To meet the optimum cost of tiger conservation, TBPAs would need to double their gross revenues. Only at this income scenario, would the NPV be positive and the B/C ratio approach 1 (Fig. 2). However, anticipating a sudden increase in park revenue of this magnitude is not presently realistic.

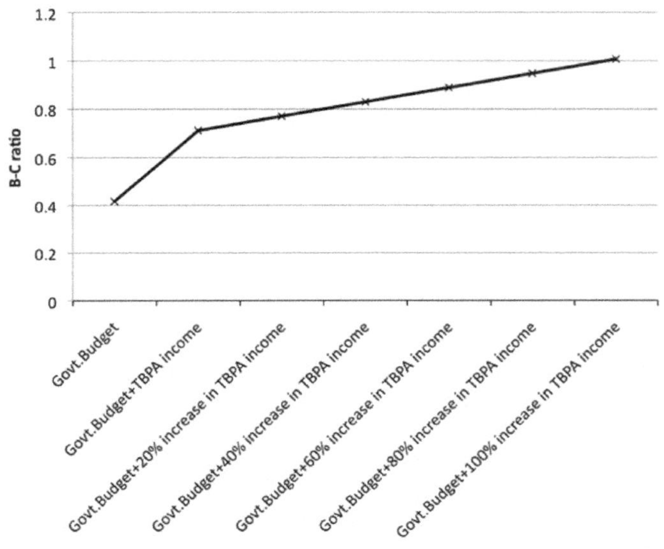

Figure 2: Benefit-cost ratio of the Nepalese tiger conservation program under scenarios that include solely government funding, government funding plus the funds generated by tiger-bearing protected areas (TBPAs), and government funding plus increased income from TBPAs.

One means to remedy this situation is to identify alternative revenue streams. Given that funds pledged to Nepalese conservation programs by international non-governmental organizations (I/NGOs), and in particular, by World Wildlife Fund Nepal (WWF Nepal), has a long history, including these contributions as a sustainable source of revenue deserves attention. Sensitivity analysis including the revenue through WWF Nepal, but excluding revenue from TBPAs, produced a negative NPV and a B/C ratio of less than one. However, revenues from WWF Nepal supplementing constant revenues for TBPAs (adjusting for inflation) produced a positive NPV and a B/C ratio greater than one. Thus, the funding secured through I/NGOs such as the WWF Nepal program is vital for the sustainability of Nepal's tiger conservation program.

Despite the positive result that we identified by increasing TBPA income and by including revenues from I/NGOs, several concerns exist. First, support from I/NGOs may not be forthcoming in any given year. Second, and perhaps more importantly, present Nepal policy does not allow the revenue generated by TBPAs to be directly applied to offset the costs of tiger conservation programs. For example, 30-50% of current total revenues are channeled to buffer zone communities that reside outside the TBPAs (National Parks and Wildlife Conservation Act 1973 Fourth Amendment) and the remaining funds are retained by the central government.

Carbon Payment (REDD+), Wildlife Premiums, and Debt-for-Nature Swaps

The Reduced Emissions from Deforestation and Forest Degradation (REDD) program, and the REDD+ program (which extended the scope of REDD and includes the role of conservation, sustainable forest management, enhancement of forest carbon stock and climate related aspects) are United Nations-based efforts to create financial value for the carbon stored in forests. The program offers incentives to developing countries for reducing emissions from forestlands and adopting low-carbon paths to sustainable development. This international financing mechanism of carbon payments can be leveraged to make payments for forest biodiversity as well [23, 24]. The carbon market potentially provides great opportunity to secure forest resources in developing countries by offering financial incentives to protect forests. In Nepal, this would also protect tigers and generate fiscal resources. The mechanism for REDD implementation in Nepal is still in a preliminary stage. There is no set mechanism nationally and internationally for the appropriate payment system [25]. However, projects examining how forest carbon credits and payments for other ecosystem services (biodiversity, water/food provisioning, soil conservation, etc.) can be pursued are underway [26, 27]. In the current context, carbon payments can be considered as potential financial sources in the long-term, but cannot be viewed as a source of sustainable finances for on-going tiger conservation programs in Nepal. Moreover, no data exists on carbon stock availability in the tiger inhabited landscape or on the price of per unit area carbon sequestrated. Hence, projecting the value of carbon that could be sequestered in Nepal's forests within the range of tiger conservation landscape is analytically unfeasible and unrealistic. We suggest the need to design a pilot project to explore the potential finances that could be generated through carbon payments for supporting tiger conservation program. If appropriately designed and successfully implemented, carbon payment through REDD+ could be a reliable and sustainable source of funding for tiger conservation.

Dinerstein et. al. (2010) suggest that a World Bank-catalyzed wildlife carbon fund that creates a premium carbon-related market for wildlife conservation would be a complementary mechanism to REDD+ [28]. This mechanism would ensure co-benefits for wildlife through the carbon payments. The benefit of this approach for the priority conservation programs such as the tiger conservation program is that the funds will be available from the outset and focused, creating incentives to develop projects targeting wildlife. We recommend the REDD implementing agencies and conservation institutions collaboratively work to identify mechanisms that incorporate wildlife conservation as one of the major targets within REDD+ projects. Indeed, conservation finance in Nepal needs to go beyond traditional government or donor funding by exploring innovative market-based approaches [4, 28-30] such as debt-for-nature swaps, environmental funds, payment for ecosystem services (PES) and any other performance-based payment strategies.

Debt-for-nature swaps are financial transactions that involve renegotiation of the terms of a developing country's debt between an outside agency (I/NGOs or credibly established trust funds) or creditor (typically a developed country) and debtor country. Under this mechanism, the public debt of the developing country is purchased at a discount by a third-party and retired in exchange for government commitments to fund conservation programs by the developing country [31]. Debt-for-nature-swaps have been a major source of funding for securing start-up capital for conservation trust funds in many countries in Latin America and a few countries in Asia and Africa [29, 31]. A study in Nepal [32] emphasized that Nepal has high potential for these swaps because Nepal has a high level of foreign debt, the country's financial inability to otherwise allocate adequate resources to meet the increasing cost of maintenance of the natural habitats, and the country's demonstrated commitment towards the environment conservation. However, a more recent feasibility assessment would be needed to establish Nepal's current position to undertake this venture and to project potential success. If applicable, this mechanism could potentially secure the capital for establishing a conservation trust fund, perhaps one exclusively targeted for tiger conservation, in Nepal.

Increased domestic finances

The TBPAs and community forests in the Nepalese TAL have high potential to generate revenues through tourism. In recent years, the number of tourists visiting protected areas in Nepal has grown (Department of National Parks and Wildlife Conservation, unpublished data), and this tourism is a potential source of employment and foreign currency, thereby creating a mechanism to achieve sustainable economic development [33]. Tourists visiting Chitwan National Park were willing to pay almost three times more than the current park entrance fee for their visit to the park [34]. Periodic adjustment of entrance fees could ensure increased income to the community without the risk of cutting tourist numbers. The increased income that goes to the buffer zone community might not directly contribute for offsetting the costs for tiger conservation but it can tremendously help in lowering the cost of conservation. If the buffer zone, or community forests, can support their costs of conservation through increased income, the overall gap in tiger financing could be substantially reduced.

Securing cash flow from environmentally responsible citizens and corporate businesses in the country can further reduce this funding gap. Locally, Nepal should aim at raising funds from within the country by creating awareness of local entities to accept the responsibility for maintaining the natural system they operate in. Finding a mechanism to generate a sustainable source of funding via local citizenry, regardless of its size, through any form of contributions or charges, would contribute to a narrower funding gap. These contributions and charges could be obtained from any direct or indirect beneficiaries such as Nepali tourists visiting wilderness areas, local or large-scale businesses, and pledges from individuals and organizations. This, as yet untapped, internal financing source could potentially emerge as a sustainable source of finance for Nepal's tiger conservation program.

Tiger Conservation Trust Fund

In many developing countries, environmental trust funds have played a pivotal role in ensuring the long-term financial sustainability of conservation programs [29]. We envision a Tiger Conservation Trust Fund (TCTF) in Nepal that would not only be a financial mechanism, but also an institutional arrangement to convene different actors in tiger conservation for enabling collaborative program planning and implementation. Such a framework could even cross international borders, as India has also pledged to double their tiger population. A well-funded, focused trust fund could avoid duplication of programs, provide a long-term perspective, and help channel funds more appropriately to priority projects. Conservation trust funds are usually registered as autonomous legal entities, although they can be established under a variety of different legal regimes depending on a country's legal provision. Although the legality of the establishment of a Nepali conservation trust fund is beyond the scope of this paper, we recommend identifying an appropriate model that encourages active participation of civil society and non-governmental bodies. Considering the importance placed on tigers by the government, we propose a mixed board including members from relevant government authorities and local stakeholders, as well as natural resource specialists. This would facilitate coordination among stakeholders and avoid the hindrances that might be posed in the absence of government representation. External audits, on a regular basis, would assure accountability of how funds are secured and dispersed. The major capital investments for the trust fund would come from bilateral (bilateral grants and debt-for nature-swaps) and multilateral (e.g. UNDP-Global Environment Facility) donors, international foundations and I/NGOs. Hence, the trust fund would need to be established so that it retains the mandate to negotiate with these funding sources and receive the funds. We also emphasize the potential role of the trust fund in seeking domestic funding sources. Though there are numerous I/NGOs that strive to provide funding for conservation in Nepal, few organizations that we are aware of have been able to secure funds from individual or corporate donors within the country. A TCTF could emerge as the credible institutional mechanism that can tap funding from these sources.

Conclusions

Conservation funding is always in short-supply. The problem is particularly acute in developing countries like Nepal. Despite the will of the government and support from international agencies, adequate funding is difficult to secure for conserving even iconic species like the tiger. Past aid, in combination with government support, has allowed Nepal to make significant progress towards fulfilling its commitment to the global tiger conservation agenda, and there is an active tiger recovery plan in place that objectively aims at contributing to the global goal of doubling the tiger population by 2022 by doubling Nepal's own tiger population. However, most of the funds allocated by the government towards tiger conservation are spent in salaries and wages of staff, expenditure on capital assets and other operating costs. Relatively few funds are left for program expenses and especially for targeted activities such as anti-poaching activities, trade-control and habitat maintenance and management. Therefore, the proposed recommendation to create the TCTF that could aggregate funding from various sources, such as from international and indirect investment made by the income generated by the local communities, deserves serious consideration.

In the context of overall biodiversity conservation as well as for tiger conservation, international funds have played an important role in enabling conservation initiatives. Funds from the WWF have been dedicated to maintaining wilderness characteristics of the TAL, thereby conserving the biodiversity of the system while improving the livelihood of local community. Similarly, the National Trust for Nature Conservation, a prominent Nepalese NGO secures funding from various international funding agencies to conduct conservation programs through the Biodiversity Conservation Center and the Bardia Conservation Program (in the TAL). One of the caveats about internationally secured funding through aid and donations as well as NGO-derived funds, however, is that such funds cannot be guaranteed for long-term project operation and maintenance. The establishment of the TCTF would provide such certainty by creating a national financing mechanism for financing long-term tiger conservation programs.

Acknowledgements

We thank the World Wildlife Fund (WWF) Nepal Program for the financial support to conduct this study. We are also grateful to WWF Nepal Program and the Department of National Parks and Wildlife Conservation Nepal for making data available for this study. Thanks to R. Maharjan, B.R. Gautam, M. Khadka, D. Chapagain and S. Nepal for their advice and support to undertake this study. Our special thanks to J.J. Millspaugh, J. Jones, H. He, and two anonymous reviewers for their critical comments on the manuscript. Additional thanks goes to S. Sharma, J. Thapa, N. Adhikari, M. Sharma and P. Pandey of the Nepal Nature Foundation (NNF) for their constant help and support throughout the study and the manuscript preparation.

References

Five "key references", selected by the authors, are marked below (Three recommended (●) and two highly recommended (●●) papers).

1. Kremen, C., Niles, J., Dalton, M., Daily, G., Ehrlich, P., Fay, J., Grewal, D., & Guillery, R., 2000. Economic incentives for rain forest conservation across scales. Science, 288(5472): 1828-1832.
http://dx.doi.org/10.1126/science.288.5472.1828
2. Wikramanayake, E., Dinerstein, E., Seidensticker, J., Lumpkin, S., Pandav, B., Shrestha, M., Mishra, H., Ballou, J., Johnsingh, A., & Chestin, I., 2011. A landscape based conservation strategy to double the wild tiger population. Conservation Letters, 4(3): 219-227.
http://dx.doi.org/10.1111/j.1755-263X.2010.00162.x
3. GTRP. 2010. Global Tiger Recovery Program 2010-2022. St. petersburg Declaration on Tiger Conservation, International Tiger Summit.
4. ●● Dinerstein, E., Varma, K., Wikramanayake, E., Powell, G., Lumpkin, S., Naidoo, R., Korchinsky, M., Del Valle, C., Lohani, S., & Seidensticker, J., 2013. Enhancing Conservation, Ecosystem Services, and Local Livelihoods through a Wildlife Premium Mechanism. Conservation Biology, 27(1): 14-23.
http://dx.doi.org/10.1111/j.1523-1739.2012.01959.x
5. Ripple, W.J., Estes, J.A., Beschta, R.L., Wilmers, C.C., Ritchie, E.G., Hebblewhite, M., Berger, J., Elmhagen, B., Letnic, M., & Nelson, M.P., 2014. Status and Ecological Effects of the World's Largest Carnivores. Science, 343(6167): 1241484.
http://dx.doi.org/10.1126/science.1241484
6. Tisdell, C., 1999. Biodiversity, conservation and sustainable development: principles and practices with Asian examples. Edward Elgar Publishing Inc. 263.
7. ● Balmford, A. & Whitten, T., 2003. Who should pay for tropical conservation, and how could the costs be met? Oryx, 37(02): 238-250.
http://dx.doi.org/10.1017/S0030605303000413

8. ●● McCarthy, D.P., Donald, P.F., Scharlemann, J.P., Buchanan, G.M., Balmford, A., Green, J.M., Bennun, L.A., Burgess, N.D., Fishpool, L.D., & Garnett, S.T., 2012. Financial costs of meeting global biodiversity conservation targets: current spending and unmet needs. Science, 338(6109): 946-949.
http://dx.doi.org/10.1126/science.1229803
9. NTRP. 2010. National Tiger Recovery Plan, Government of Nepal, Kathmandu, Nepal
10. Tacconi, L. & Bennett, J., 1995. Economic implications of intergenerational equity for biodiversity conservation. Ecological Economics, 12(3): 209-223.
http://dx.doi.org/10.1016/0921-8009(94)00045-W
11. Nelson, F., 2009. Developing payments for ecosystem services approaches to carnivore conservation. Human Dimensions of Wildlife, 14(6): 381-392.
http://dx.doi.org/10.1080/10871200903045228
12. Dickman, A.J., Macdonald, E.A., & Macdonald, D.W., 2011. A review of financial instruments to pay for predator conservation and encourage human-carnivore coexistence. Proceedings of the National Academy of Sciences, 108(34): 13937-13944.
http://dx.doi.org/10.1073/pnas.1012972108
13. GTI. 2013. Global Tiger Recovery Program Implementation Plan 2013-14. The World Bank.
14. Norgaard, R.B. & Howarth, R.B., 1991. Sustainability and discounting the future. Ecological economics: the science and management of sustainability: 88-101.
15. Gitman, L.J. & Forrester Jr, J.R., 1977. A survey of capital budgeting techniques used by major US firms. Financial Management: 66-71.
16. Naidoo, R., Balmford, A., Ferraro, P.J., Polasky, S., Ricketts, T.H., & Rouget, M., 2006. Integrating economic costs into conservation planning. Trends in Ecology & Evolution, 21(12): 681-687.
http://dx.doi.org/10.1016/j.tree.2006.10.003
17. Ross, S.A., 1995. Uses, abuses, and alternatives to the net-present-value rule. Financial Management, 24(3): 96-102.
http://dx.doi.org/10.2307/3665561
18. Abel, A.B. & Eberly, J.C., 1996. Optimal investment with costly reversibility. The Review of Economic Studies, 63(4): 581-593.
http://dx.doi.org/10.2307/2297794
19. Boyle, G.W. & Guthrie, G.A., 1997. Payback and the value of waiting to invest. Department of Economics, University of Canterbury.
20. CBD. Undated. Sustainable Finance. Convention on Biological Diversity. http://www.cbdint/protected-old/sustainable.shtml (accessed 08/22/2013).
21. Lee, A., Lee, J., & Lee, C., 2009. Financial Analysis, Planning and Forecasting: Theory and Application. 2nd ed: World Scientific Publishing Company.
http://dx.doi.org/10.1142/6416
22. Lohani, S. 2010. Cost of getting to TX2: A case study from terai arc landscape (TAL) Nepal. World Wildlife Fund US.
23. ● Busch, J., 2013. Supplementing REDD+ with Biodiversity Payments: The Paradox of Paying for Multiple Ecosystem Services. Land Economics, 89(4): 655-675.
24. CBD. 2011. REDD-plus and Biodiversity. Secretariat of Convention on Biological Diversity.
25. GoN. 2011. Study on REDD Plus Piloting in Nepal. Government of Nepal, Ministry of Forests and Soil Conservation, REDD-Forestry and Climate Change Cell.
26. De Gryze, S. & Durschinger, L. 2009. Payments for Ecosystem Services: Developing Forest Carbon Projects in Nepal. EnterpriseWorks/VITA, Terra Global Capital, USAID and ANSAB.

27. ● GoN. 2014. People and Forests- An SMF-Based Emission Reduction Program in Nepal's Terai Arc Landscape. Forest Carbon Partnership Facility (FCPF) Carbon Fund, The World Bank, MoFSC. https://www.forestcarbonpartnership.org/sites/fcp/files/2014/MArch/March/Nepal%20ER-PIN%20Annexes%20CF9%20(Final).pdf (accessed 06/21/2014).

28. Dinerstein, E., Varma, K., Wikramanayake, E., Lumpkin, S., Seidensticker, J., Shrestha, M., Powell, G., Poor, E., Lovejoy, T., & Kushlin, A. 2010. Wildlife Premium Market+ REDD: Creating a financial incentive for conservation and recovery of endangered species and habitats. World Wildlife Fund Inc.

29. Moye, M. 2002. Innovative Mechanisms to manage Environmental Expenditures in Africa, Asia and Latin America and the Caribbean (LAC). in WWF Centre for Conservation Finance. Conference Paper OECD, Paris. 2002.

30. Moye, M. 2007. Conservation Finance. World Wildlife Fund.

31. Emerton, L., Bishop, J., & Thomas, L., 2006. Sustainable Financing of Protected Areas: A global Review of Challenges and Options. IUCN, Gland, Switzerland and Cambridge: x + 97 pp.

32. Thapa, S. & Thapa, B., 2002. Debt-for-Nature Swaps: potential applications in Nepal. The International Journal of Sustainable Development & World Ecology, 9(3): 239-255. http://dx.doi.org/10.1080/13504500209470120

33. Benavides, D.D. & Perez-Ducy, E. 2001. Tourism in the least developed countries. World Tourism Organization.

34. Cook, J.M., 2011. Valuing protected areas through contingent valuation: a case study of Chitwan National Park, Nepal, in Environmental Applied Science and Management, Ryerson University: Toronto, Ontario, Canada.

10

STATUS OF LONG TAILED MACAQUE (*Macacafascicularis umbrosa*) AND CONSERVATION OF THE RECOVERY POPULATION IN GREAT NICOBAR ISLAND, INDIA

K. Narasimmarajan[1,*] & C. Raghunathan[1]

[1] Zoological Survey of India (ANRC), National Coral Reef Research Institute, Port Blair, Andaman -744102, India.

* Corresponding author address: Door No# 283, Veeravadi, Poonthottam (P.O), Thiruvarur (DT), Tamil Nadu, Southern India -609503; e-mail: wildlife9protect@gmail.com.

Keywords

Macaca fascicularis umbrosa; Recovery; Group size; Sex ratio; Great Nicobar Island.

Abstract

We assessed the present status and population abundance of Long Tailed Macaque (*Macaca fascicularis umbrosa*) in Great Nicobar Island, (India) from November 2011 to February 2012. A total of 34 groups, comprising 1,133 monkeys were sighted, with the group sizes varying from 8 to 127 individuals (mean ± SE = 33.32 ± 4.82). We compared them with previously published reports from 2003 and 2006. However, there were no significant changes between the groups sighted in forest areas located in the interior regions of the island (N=13; 0.12 groups/km), while a significant increase were noticed in the number of groups sighted in coastal forest regions of the Island (N=21; 0.19 groups/km). The survey results revealed that, after the tsunami, the coastal population is slowly recovering. The adult sex ratio (male/female) and adult/immature ratio was 1: 2.14 and 1: 1 indicating a healthy population. Therefore several monkey groups were seen in the abundant coconut farms near the coastal region, which were planted by Islanders who lived there before the tsunami. Comprehensive human–monkey conflict was observed in the interior regions of the Island where monkeys destroyed the coconut and cultivation farms while dogs were used to control the monkey menace. Improving farm management and providing adequate compensation schemes are critical for successfully implementing such conservation efforts.

Introduction

Long Tailed Macaque, commonly known as Nicobar Crab-eating Macaques (*Macaca fascicularis umbrosa*; Miller, 1902) [1] are the only non-human primates found in Nicobar Islands [2-4]. This subspecies is listed as Schedule-I, Part-1, of the Indian Wildlife (Protection) Act 1972 [Anonymous, 1972] and categorized as Vulnerable on the Red List of the IUCN [www.iucnredlist.com]. Furthermore, studies on this species are very scarce [4, 5]. As isolated populations of rare species may be disproportionately affected by both stochastic and deterministic factors such as natural disasters [6], ensuring their long-term conservation requires studies that assess the impacts of such calamities on their population dynamics [1, 5]. As this subspecies occurs in a high-risk seismic zone, conservation programs would have to insure against disasters such as the 2004 Indian Ocean earthquake and tsunami.

On December 26, 2004, a massive earthquake (measuring 9.0 on the Richter scale) occurred in the Indian Ocean, just off the coast of Indonesia (3.244°N; 95.825°E). Known as the Sumatra-Andaman earthquake, it triggered a series of tsunamis, primarily devastating the coasts of Indonesia, Sri Lanka, Thailand and India. Over a million people have been confirmed dead, making this the fourth largest death toll due to an earthquake in recorded history. Several years after the quake and tsunami events, the entire region is still in the process of recovering and rebuilding basic infrastructures. Not only were there losses in terms of human lives and infrastructures, entire coastal forests and wildlife were destroyed [USGS/earthquake.com]. The coastal forests of the Andaman-Nicobar group of Islands were particularly affected [5]. Because of their high endemic diversity, this single natural calamity resulted in the endangering and re-evaluation of the conservation status of many floral and faunal species (e.g., the entire Megapode Island was submerged). Nicobar Crab-eating macaques were particularly affected, since coastal populations recorded a massive decline following the earthquakes and tsunamis. While Sivakumar [5] conducted a post-tsunami assessment of the populations, there have been no follow-up population assessments to examine the present status of this species. It is in this regard that we, in this study, carried out a short-term survey to assess the present status and abundance of the species in the Great Nicobar Island.

Materials and methods

Study Area

Nicobar Islands lies in the Bay of Bengal archipelago [latitudes 6° 45' and 13° 41'; longitudes 92° 12' and 93° 57'] from Arcon, Myanmar in the north to Achin, Sumatra (Indonesia) in the south [4]. The Islands can be subdivided into three sub-groups [7, 8]. To the south lies the Great Nicobar group consisting of two major Islands (>100 km^2) and nine islets (>5 km^2). Among them, Great Nicobar, Little Nicobar, Kondul and Pilo Milo are inhabited [9]. Prior to the tsunami there were 30,000 people inhabited the Nicobar Islands. However, following the tsunami the island population is estimated between 8,000 to 10,000 people [5]. The *Macaca fascicularis umbrosa* subspecies is argued to be of important biological significance as the estimated total population is about 4,800 individuals [1, 4- 5]. The Great Nicobar Island area is about 1045.1 km^2. The habitat is typical tropical rainforest and subtropical moist broadleaf forests and dominant plants in Nicobar include *Cyathea*, *Otanthera*, *Astronia*, *Bentinckia* and *Rhopaloblaste* species [10, 11; Fig. 1]. The Island has several rivers, including the Alexandra, Amrit Kaur, Dogmar and Galathea rivers [8]. The island receives both northeast and southwest monsoons, and rainfall from May to November averages ca. 3800 mm annually. Temperature variation is low from a minimum of 10°C to a maximum of 32°C [4]. Presently, the macaques are protected only in the Great Nicobar Island under two National Parks namely 'Campbell Bay National Park' and 'Galathea National Park' [9, 10]. Historical records, suggest that the species was always restricted to the Nicobar and Nancowry group of islands (Great Nicobar, Little Nicobar and Katchal Island) [4].

Fig. 1: Photo shows the lagoon and its typical Rainforest habitats of Great Nicobar Island.

Data Collection Procedure

From November 2011 to February 2012, the Nicobar Crab-eating Macaques were surveyed in the Great Nicobar Island using a standardized method outlined by Umapathy [4]. Surveys were carried out between 7.00 AM and 12.00 PM by three researchers walking on existing trails, streams, paths and roads [5]. Whenever a monkey group was sighted, its group size, geographical co-ordinates (Fig. 2), distance from the coast and altitude were recorded in the field using a Global Positioning System (Garmin-72). The age and sex composition of 11 groups were also estimated using the morphological features known from monkeys in captivity at Port Blair [5]. A total of 39 transects were laid covering an overall distance of 210.05 km, including 22 transects (110.02 km) in coastal areas and 17 transects (100.03 km) in the interior of the Island. The mean length of transects was 5.0 km (min-max, 1.5 - 9 km). Population, abundance and group composition of monkeys were compared with those obtained from the earlier surveys [4, 5].

Fig. 2: Location of Long Tailed Macaque (*Macaca fascicularis umbrosa*) groups was recorded in the Great Nicobar Island during November 2011 and February 2012.

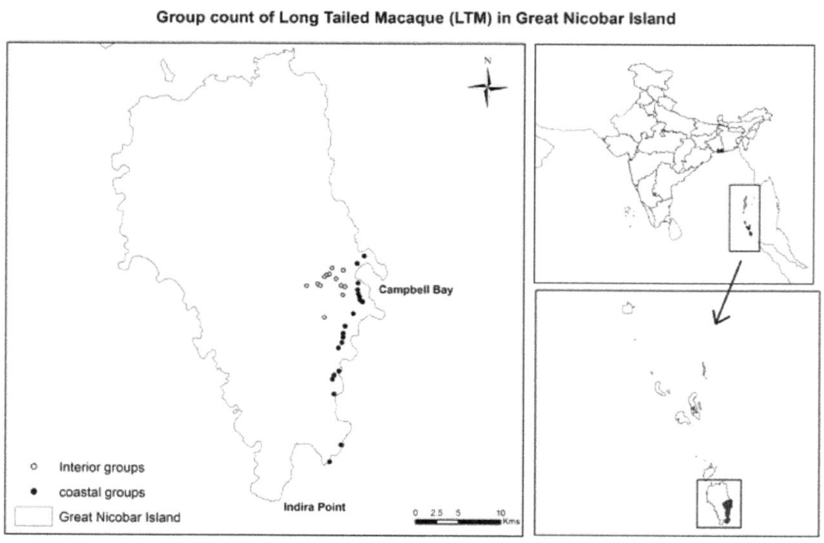

Data Analysis

We analyzed data on demographic parameters such as sex ratio, abundance and group size using nonparametric statistics because the primary level of measurement was often nominal. We used Kruskal-Wallis (K-W) tests to examine for significant changes between interior and coastal forest populations (compared with previous data) [12]. The Welch F test was used to assess significant differences between previous and present total groups sighted in the study area [13]. We used SPSS (v.11.1) software for statistical analysis [14].

Results

A total of 34 groups, including 1,133 individual monkeys were sighted and the group size varied from 8 to 127 (mean ± SD = 33.32 ± 4.82) individuals. The demography of 11 groups of Nicobar Crab-eating Macaques is presented in Table 1. The comparative analysis of previous (in 2006) and present data revealed that there were no significant changes in the number of groups sighted (mean elevation 54m) in interior parts of the Island (K-W test, $x^2 = 0.87$, d.f. = 3, $p<0.85$ N.S.) after 2006, while the mean number of groups sighted in the coastal forest (mean elevation 36m) increased significantly (K-W test, $x^2 = 2.91$, d.f. = 1, $P<0.05$) after 2006. The relative abundance of Nicobar Crab-eating Macaque increased in the coastal region (N=787; 7.13 monkeys/km) after 2006. There were no significant differences in the number of groups sighted in the interior part of the Island (N=346; 3.44 monkeys/km; Table 2). The largest group, comprising 127 individuals, was observed in the coconut farm, besides cropland area near Govind Nagar. The Welch F test indicates that the total number of groups sighted in the coastal region increased significantly ($F= 1.41$, d.f. = 1, $P<0.05$) after 2006. The percentage of adults and immature individuals didn't showing any differences. The number of adult males in a group ranged from 2 to 8 individuals (mean ± SD = 4.05 ± 1.63), whilst that of adult females ranged from 3 to 15 monkeys (mean ± SD = 8.73 ± 3.80). In 2006, the ratio of adults/juveniles was about 1: 0.4 whilst in 2011 it was 1: 1. The adult female/male ratio was stable, ranging from 1:0.48 in 2006 to 1:0.46 in 2011. During the survey period, 22 groups were observed including between 8 to 35 individuals (64.7%), followed by 11 groups ranging between 37- 69 monkeys (32.4%) and the largest group of 127 individuals (2.9%) was recorded for the first time on Nicobar Island. Over 90% of the observed groups present multi-male and multi-female social systems.

Table 1: Demography of Nicobar Crab-eating Macaques (*Macaca fascicularis umbrosa*) in Great Nicobar Island, India.

Groups	Locality	Total	AM	AM%	AF	AF%	IM	IM%	IN	IN%
GNBR 1	Govind Nagar	27	3	11.1	11	37.0	10	37.0	3	11.1
GNBR 2	30 KM, road	15	5	33.3	6	40.0	3	20.0	1	6.7
GNBR 3	35 KM, road	35	8	22.9	15	37.1	8	22.9	4	11.4
GNBR 4	20 KM, road	26	3	11.5	10	38.5	12	46.2	1	3.8
GNBR 5	Bhagam beat	18	4	22.2	10	55.6	3	16.7	1	5.6
GNBR 6	25 KM, road	51	6	11.8	11	21.6	31	60.8	3	5.9
GNBR 7	West road	40	4	10.0	14	35.0	18	45.0	4	5.0
GNBR 8	Nature trail	37	5	13.5	11	29.7	21	54.1	0	0.0
GNBR 9	9 KM, road	31	3	9.7	5	16.1	20	64.5	3	9.7
GNBR 10	Shompen hut road	14	4	28.6	5	50.0	5	35.7	0	0.0
GNBR 11	Galathea road	17	2	11.8	3	17.6	11	64.7	1	5.9
Total		**311**	**47**	**16.9**	**101**	**34.4**	**142**	**42.5**	**21**	**5.9**

AM-Adult Males; AF-Adult Females; IM-Immatures; IN-Infants

Table 2: Comparison with earlier data on group count, abundance and mean group size of the Long Tailed Macaques (*Macaca fascicularis umbrosa*) in Great Nicobar Island, India.

Great Nicobar	Area (km^2)	Number of groups IF	CF	TG	TE (km)	NT	Abundance (Groups/km)	Mean group size
2000[a]		16	37	53	227.1	16	0.23	36.01
2006[b]	1045.1	14	8	22	211.8	41	0.10	20.35
2011*		13	21	34	210.5	39	0.17	33.32

IF = Interior Forest; CF = Coastal Forest; TG = Total Groups; TE = Total Effort; NT = Number of Transects. ([a]Umapathy *et al.*, 2003); ([b]Sivakumar, 2010); *present study.

Discussion

Our study clearly indicates that the coastal population of the Nicobar Crab-eating macaque is recovering following a reduction in population size owing to the tsunami, while the interior population remained fairly stable. The number of groups in the coastal area increased from 8 groups in 2006 to 21 groups in 2011 and the adult/ immature ratio (1: 1) in a group fairly indicated that the coastal population may have achieved constant growth and the population structure may also have been restored (Fig. 3). Besides, after 2006, several human settlements vacated from the coastal region, heading to interior parts of the island, resulting in the recovery of native vegetation along the coast. Despite the positive trends noted in this study, earthquakes are a serious threat to the recovery of the population. If high magnitude earthquakes create tsunamis such as the one in 2004, the future of these populations may be at risk. Since 2010 there have been 11 earthquakes recorded in proximity to the Andaman & Nicobar Islands ranging in magnitude from 4.5 to 5.2 (USGS/earthquake.com). Given this looming threat to the population, it is important to elevate the conservation status of this fragile macaque respectively [15] and conservation programs should include strategies to insure these populations against such devastating calamities.

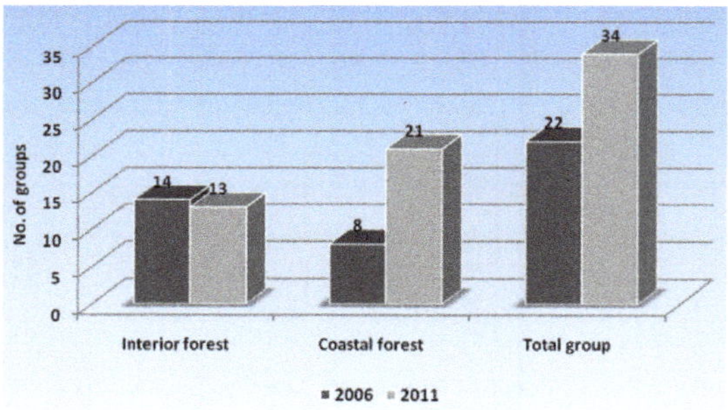

Fig. 3: Comparison of population growth (IF-Interior Forest; CF-Coastal Forest) of the Long-tailed Macaques in Great Nicobar Island.

In addition, abundant coconut and palmnut farms may provide them with enough food sources [5, 11]. Our casual observations indicate that there have been rising incidents of human–macaque conflict in interior parts of the Island with macaques destroying coconut (*Cocos nucifera*) farms. Coconut palms are the most important commercial plants of the Island, being frequently used as food items by the Nicobar Crab-eating Macaque (Fig. 4). With landowners using dogs to ward away macaques, there is the additional threat of disease transmission and potential loss of individual macaques owing to conflict.

Fig. 4: The Nicobar Crab-eating Macaques often raiding/feeding on the cropland and coconut farms.

The most immediate threat in Nicobar is the ongoing free port construction, aimed at creating a dry dock to refuel international shipping near Shastri Nagar beach (36 Km; near Galathea river mouth; Fig. 5). The three kilometer project may have impacts on the southeast coast of Nicobar forests and it may provoke harsh human-wildlife conflict. Since no behavioral and ecological data are available on this subspecies, a long-term ecological and behavioral study must be initiated for the recovery of this population and periodical assessments are mandatory to ensure the long term persistence of these populations.

Threats of Nicobar

Crab-eating macaques often raid crops and coconut trees. In order to chase the monkeys away, farmers almost always keep domestic dogs, which possibly transmit disease and potential loss of individual monkeys [4]. During the survey, we recorded dogs chasing monkeys from croplands (interior area) in Great Nicobar Island, while monkeys were often seen in the ground, feeding on cultivated farms (Fig. 4). Our casual observations and previous report clearly stated that the settlers from Ranchi and some other communities often hunt the Nicobar wild pig (*Sus scrofa nicobaricus*), Megapode (*Megapodius nicobariensis*) and Crocodiles (*Crocodylus porosus*), which are ultimately sold for meat in restaurants [16, 17]. However, seven nylon traps were seen in the interior forest. Indigenous people and settlers subsist on coconut and agriculture in coastal areas. Therefore, after reaching an agreement with the local communities, the entire Great Nicobar Island should be declared as a high sensitive zone in order to enhance the protection of the Long-tailed macaque and other threatened species of the region.

Fig. 5: Immediate threat of east coast is undergoing a new port work at Galathea river mouth in Great Nicobar Island.

Acknowledgements

The author would like to thank the Ministry of Environment and Forests, Government of India, for the financial assistance under the project 'Studies on the faunal diversity of Great Nicobar Biosphere Reserve'. Thanks to Dr. K. Venkataraman, Director Zoological Survey of India, for providing the necessary facilities. KN gratefully acknowledges Mr. Kausik Banerjee, & Mr. Abishek Harihar, WII for their valuable comments and for editing this manuscript before submission. KN sincere thanks to the editors and the anonymous referees for their valuable comments. We thank the Assistant Commissioner (Nicobar Island), field assistants and colleagues.

References

Five "key references", selected by the authors, are marked below (Three recommended (●) and two highly recommended (●●) papers).

1. ● Molur S., Jones B.D., Dittus W., Eudey A., Kumar A., Singh M., Feeroz M.M.., Chalise M., Priya P., Walker S. 2003. Status of South Asian Primates: Conservation Assessment and Management Plan (C.A.M.P.) Workshop Report. Zoo Outreach Organisation / CBSG-South Asia, Coimbatore, India, viii+432pp.
2. Elmqvist T., Folke C., Nyström M., Peterson G., Bengtsson J., Walker B., and Norberg J. 2003. Response diversity, ecosystem change, and resilience. Frontiers in Ecology and the Environment, 1: 488-494.
DOI:10.1890/1540-9295(2003)001[0488:RDECAR]2.0.CO;2
3. Karen V.R. 1998. Evaluating the effects of habitat quality, connectivity, and catastrophes on a threatened species. Ecological Applications, 8: 854–865.
DOI:10.1890/1051-0761(1998)008[0854:ETEOHQ]2.0.CO;2
4. ●● Umapathy G., Singh M., Mohnot M. 2003. Status and Distribution of *Macaca fascicularis umbrosa* in the Nicobar Islands, India. International Journal of Primatology, 24(2): 281- 293.
DOI:10.1023/A:1023045132009
5. ●● Sivakumar K. 2010. Impact of the tsunami (December, 2004) on the long tailed macaque of Nicobar Islands, India. Hystrix It. J. Mamm. (n.s.) 21(1): 35-42
6. Burgman M.A., Ferson S., Akcakaya H.R. 1993. Risk Assessment in Conservation Biology, Chapman and Hall, London.
7. Saldanha C.J. 1989. Andaman, Nicobar and Lakshadweep: An Environmental Impact Assessment, Oxford and IBH Publication, New Delhi.
8. ● Sankaran R. 1997. Developing a protected area network in the Nicobar Islands: The perspective of endemic avifauna. Biodiversity and Conservation, 6: 797- 815.
DOI:10.1023/B:BIOC.0000010403.03662.9a
9. Sivakumar K., Sankaran R. 2003. Incubation mound and hatching success of the Nicobar megapode *Megapodius nicobariensis*. Journal of the Bombay Natural History Society, 100: 375-387.
10. Sivakumar K. 2000. A study on breeding biology of the Nicobar megapode *Megapodius nicobariensis*. Unpublished Doctoral Thesis, Bharathiyar University, Coimbatore.
11. ● Balakrishnan N.P. 1989. Andaman Islands - vegetation and floristics. In: Saldanha C.J. (Eds). Andaman, Nicobar and Lakshadweep: An Environmental Impact Assessment, Oxford and IBH Publication, New Delhi. Pp- 55-61.
12. Zar J.H. 1996. Biostatistical Analysis. Third edition. Prentice Hall. Upper Saddle River, NJ, USA, 662p.

13. Gibbons J.D. 1971. Nonparametric Statistical Inference, McGraw Hill, New York.

14. SPSS 2007. SPSS for Windows. Version 11.1 SPSS, Chicago, IL, USA.

15. Miller G. (2005). The tsunami's psychological aftermath. Science, 309, 1030. DOI:10.1126/science.309.5737.1030

16. Tikader B.K., Das A.K. 1982. Glimpses of Animal Life of Andaman and Nicobar Islands, Zoological Survey of India, Calcutta.

17. Sankaran R. 1995. The distribution, status and conservation of the Nicobar megapode *Megapodius nicobariensis*. Biological Conservation, 72: 17-25. DOI:10.1016/0006-3207(94)00056-V

11

THE EFFECT OF GENDER ON SEVERAL BIOCHEMICAL PARAMETERS IN ROE DEER (*CAPREOLUS CAPREOLUS*)

I. de la Casa-Resino[1,*], D. Hernández-Moreno[1,2], A. López-Beceiro[3], L. Rigueira[3], M. P. Míguez[1], M. Pérez-López[1] & L. E. Fidalgo[3]

[1] Toxicology Area, Fac. of Veterinary Medicine (UEX). 10003 Cáceres, Spain.

[2] Universidad Autónoma de Chile.

[3] Veterinary Clinical Sciences, Dpt. Fac. of Veterinary Medicine (USC), Campus Universitario s/n. 27003 Lugo, Spain.

* Corresponding author email: delacasa@unex.es.

Keywords

Biomarker; *Capreolus capreolus*; Ethoxyresorufin-O-deethylase; Glutathione S-transferase; Malondialdehyde; Roe deer; Uridine diphospho-glucuronosyltransferase.

Abstract

Living beings are constantly exposed to a wide variety of environmental contaminants and their biochemical response can be employed to assess the possible biological impact of such pollution. Sex can play a major role in the uptake, fate, and effects of contaminants in organisms. The present study evaluated endogenous levels of different biomarkers in liver, kidney and lung in both male and female roe deer in order to determine normal levels of these biomarkers and sex-related differences in this species. Post-mitochondrial cytochrome P450 subfamily 1A1, as well as phase II enzymes glutathione S-transferase and uridine diphospho-glucuronosyltransferase, and lipid peroxidation in terms of malondialdehyde levels were determined. The highest values of all the considered parameters were quantified in liver. Results showed no differences by gender in levels of enzymatic activities in the three considered tissues. However, a statistical difference ($P<0.05$) in lipid peroxidation was observed in liver, finding higher values in males. Nevertheless, the differences related to sex in this organ should be taken into account when the effect of pollutants has to be studied on specific ecosystems.

Introduction

The environment is continuously loaded with foreign chemicals (xenobiotics) released by urban communities and industries. In the 20th century, thousands of organic trace pollutants, such as polychlorinated biphenyls (PCBs), organochlorine pesticides (OCPs), polycyclic aromatic hydrocarbons (PAHs), polychlorinated dibenzofurans (PCDFs), and dibenzo-p-dioxins (PCDDs) have been produced and, in part, released into the environment [1]. Wildlife species, particularly mammals, can be employed as bioindicators, as key factors in revealing environmental damage and tracing the paths of specific contaminants [2]. In order to do that, some essential criteria must be fulfilled by a suitable bioindicator: it must have a wide geographical range with an ecological optimum in the ecosystems under investigation, and the ecological and physiological characteristics of the species must be well known [3]. Moreover, standardized sampling and analytical methods must be easy to develop and animals must show an appropriate accumulation rate of different xenobiotics [4-5]. At last, a good correlation between accumulation and input of toxic substances to the ecosystem must be observed [6].

Taking into account several of the above reasons, the roe deer could be considered a good indicator of environmental pollution. First of all, it is among the most abundant deer species in Europe since, throughout the last century, both its density and geographical range have considerably increased. Moreover, it is very adaptable occupying a wide variety of habitats [3]. Also, its territories are remarkably stable from year to year in both extent and boundary limits, thus allowing the analysis of the effects of emissions from single sources on local populations [4, 7].

Biomarkers can be defined as biochemical and/or physiological changes in living organisms, which can be related either to exposure to an environmental chemical or to its toxic effect [8]. According to the World Health Organisation, biomarkers can be subdivided into three classes: biomarkers of exposure, effect, and susceptibility [9].

The most sensitive biomarker of effect is usually related to alteration in levels and activities of biotransformation enzymes [1]. Biotransformation reactions are generally divided into two phases, namely phase I (functionalization reactions including oxidation, reduction, and hydrolysis), and phase II (conjugative reactions) [10]. The system of the cytochrome P450 (CYP450) is known to play a major role in the oxidative biotransformation (phase I) of a wide range of both endogenous and exogenous compounds. The mechanism of enzyme induction has been studied most extensively for CYP450 isozymes, like ethoxyresorufin-O-deethylase (EROD) activity which is a biomarker of the alteration of hepatic CYP450 subfamily 1A1 (CYP1A1). The xenobiotic inducer binds to the cytosolic aryl hydrocarbon receptor (AhR) protein. This inducer-receptor complex migrates to the nucleus of the cell and binds to a DNA recognition sequence upstream of the CYP450 (e.g. CYPlA1) genes. Binding to this transcriptional enhancer leads to an increased transcription of the CYPlA1 gene. Production of mRNA is thus elevated, resulting in an amplified synthesis of the CYP450 proteins [11-12].

Phase II biotransformation enzymes, such as glutathione S-transferase (GST) or uridine diphospho-glucuronosyltransferase (UDPGT), are involved in the conjugation of the xenobiotic parent compound or its metabolites with an endogenous substrate [13]. This facilitates the excretion of these compounds by the addition of more polar groups to the molecule [1]. As for CYPlA1, the mechanism of induction for most forms of phase II enzymes (including GSTs and UDPGTs) in mammals is regulated via the AhR as well [14-15].

Lipid peroxidation (LP) is a marker of effect which informs about the damage caused to biological lipids by reactive oxygen species (ROS), normally formed during all metabolic processes but also enhanced under exposure to xenobiotics [16-17]. LP occurs when ROS damage cannot be counteracted by natural oxidative stress defences and is responsible for membrane and membrane-bound enzymes destabilization [18]. Changes in the levels of certain endogenous metabolites, such as malondialdehyde

(MDA), have been described as biomarkers of oxidative stress caused by different nutritional conditions or by exposure to xenobiotics [19-20]. Since MDA is a product of lipid peroxidation due to overproduction of oxy-radicals in cells and is considered an important feature in cellular injury, it has been extensively used to assess the effects of various pollutants on different ecosystems [21-22].

The use of biomarkers such as CYP1A1, GST, UDPGT activities and MDA levels for pollution monitoring is based on the assumption that an increased/decreased activity or concentration is primarily caused by exposure to contaminants. Nevertheless, the response of biomarkers may be affected by factors such as season, temperature, sex, nutritional status or size [1]. For that reason, the aim of this work was to determine the EROD, GST, and UDPGT activities in liver, kidney, and lung of wild roe deer, from a slightly polluted area, in order to establish normal levels of these enzymatic activities. Levels of lipid peroxidation, in terms of MDA, were also evaluated in the same tissues. Moreover, the effect of sex in these biotransformation enzymes and MDA levels was determined in order to establish some reference values, which could be useful in future biomonitoring studies.

Methods

During the 2008-2009 hunting season (autumn) in Galicia (NW Spain), samples of liver, kidney and lung, from 30 roe deer (21 males and 9 females) were collected. Fieldwork was conducted in a natural area far from apparent sources of pollution (urban or industrialized areas). From each carcass, a sample of liver, kidney, and lung was taken, placed individually in plastic bags, and stored at −80°C until analysis in less than 2 h after dead.

Tissues were homogenized in a 0.1 M buffer phosphate pH 7.4 (KH2PO4/K2HPO4, Panreac). The homogenate was centrifuged at 825xg for 5 min at 4°C, and the supernatant was distributed in three aliquots. Two of them were centrifuged at 13200xg for 20 min at 4°C in order to obtain the supernatant corresponding to the post-mitochondrial fraction. Levels of protein, EROD, GST, and UDPGT enzymatic activities were measured in this post-mitochondrial fraction. The other aliquot was added with 0.1 mL of perchloric acid 70% (VWR International Eurolab S.L.) and centrifuged at 1467xg for 15 min at 4°C. The supernatant was used to determine MDA levels. Protein content was measured in the post-mitochondrial fraction by the method of Bradford using bovine serum albumin (Sigma-Aldrich) as standard [50].

Hepatic CYP1A-mediated ethoxyresorufin-O-deethylase (EROD) activity was determined described in Burke and Mayer [23] later adapted by Sole et al. [24] at 537/583 nm excitation/emission wavelengths. Blood GST activity was determined using the method of Cohen et al. [25] modified by Habig et al. [26] with 1-chloro-2,4-dinitrobenzene (CDNB) as substrate. The reaction was measured spectrophotometrically at 340 nm each 20 s during 5 min. UDPGT activity was measured using a modified version of Clarke et al. [27] with p-nitrophenol (pNP) as substrate. The remaining pNP was measured spectrophotometrically at 405 nm.

Accumulation of MDA was used as an index of lipid peroxidation and evaluated spectrophotometrically at 550 nm by measuring the presence of thiobarbituric acid reactive substances (TBARS), according to the method proposed by Recknagel et al.

[28] using a microplate reader.

In all cases, reproducibility was based on the relative standard deviation of triplicate samples.

Data were expressed as mean ± standard error. Comparisons between groups (males and females) were realized with the non-parametric Mann-Whitney U-test. Non-parametric Spearman test was applied to examine the correlation between the different analyzed enzymes. $P<0.05$ was taken as to indicate statistical significance.

Results

Figures 1 and 2 show the total EROD, GST and UDPGT activity in roe deer in the three tissues considered. The mean hepatic EROD activity was 37.37 ± 5.48 pmol/min/mg protein, whereas no EROD activity was detected for kidney and lung. With respect to sex, the mean hepatic EROD activity was similar in both males and females, with no statistical significance associated to sex (Fig. 1).

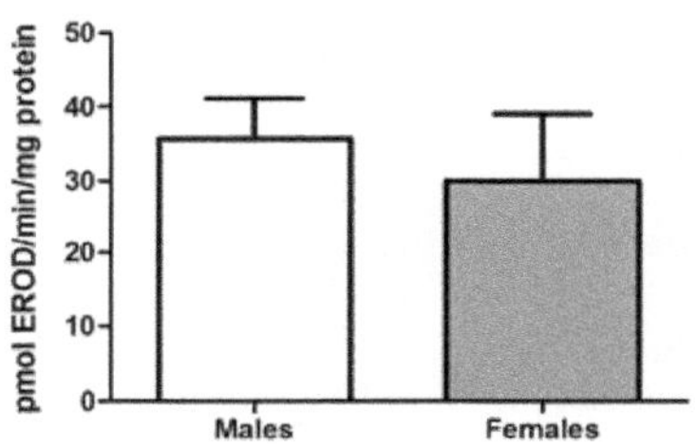

Figure 1: Hepatic EROD activity in male and female roe deer liver.

At the present study the highest levels of GST activity were quantified in liver (1799 ± 159.30 nmol/min/mg protein) when compared with kidney and lung. As previously indicated to EROD activity, no sex-related differences ($P>0.05$) were found in the levels of GST activity in any of the three considered tissues (Fig. 2).

As observed with these previous enzymes, the results obtained for UDPGT activity were higher in liver (0.36 ± 0.05 nmol/min/mg protein) than in the other tissues. Kidney and lung UDPGT levels were very low when compared to liver, and even in some samples were situated below the detection limit. When the influence of sex on the UDPGT activity was studied, the results showed higher levels in males than in females in the three considered tissues, being no statistically significant.

The results of MDA accumulation in liver, kidney, and lung are shown in figure 2. Similar to that observed with the other studied biochemical parameters, liver showed the higher levels, with a mean value of 0.25 ± 0.03 nmol/mg protein. When the MDA results were analyzed based on sex, males showed significant higher levels than females in liver ($P<0.05$). No important sex-related differences were found in kidney and lung.

It was found a positive correlation between both phase II biotransformation enzymes (GST/UDPGT) and between GST/MDA in liver and lung. It was not detected any correlation among the biochemical parameters considered in kidney (Table 1).

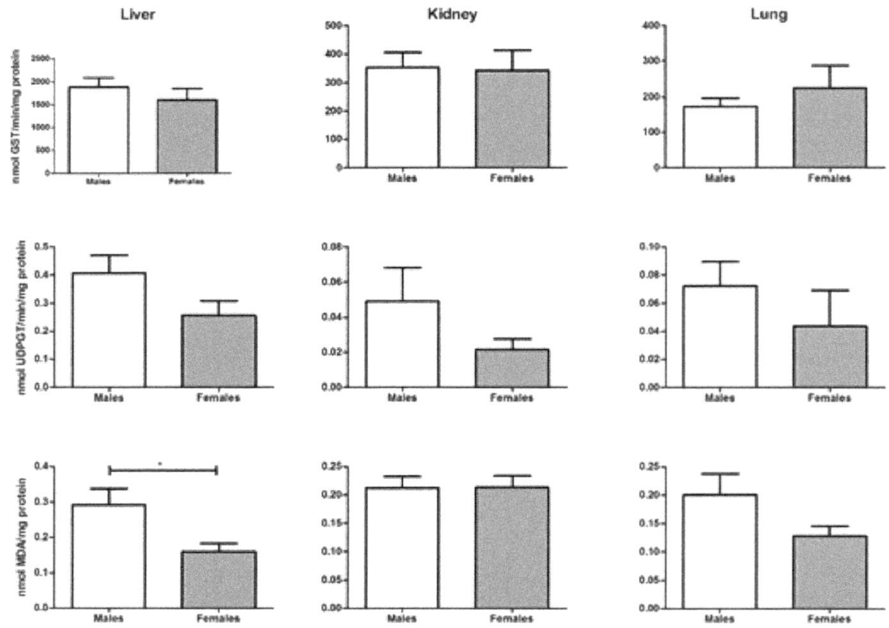

Figure 2: Hepatic, renal and pulmonary (a) GST and (b) UDPGT activities and (c) MDA levels in male and female roe deer.

Table 1: Correlation study of the different considered biomarkers (EROD, GST, UDPGT and MDA) in roe deer liver, kidney and lung. Correlation is significant at *P<0.05 and **P<0.01.

Liver				
		GST	UDPGT	MDA
EROD	Correlation coefficient	0.140	0.075	0.075
	Sig. (2-tailed)	0.459	0.692	0.692
GST	Correlation coefficient		0.557	0.557
	Sig. (2-tailed)		0.001**	0.001**
UDPGT	Correlation coefficient			1.000
	Sig. (2-tailed)			0.000
Kidney				
		-	UDPGT	MDA
GST	Correlation coefficient		0.101	0.196
	Sig. (2-tailed)		0.596	0.300
UDPGT	Correlation coefficient			0.342
	Sig. (2-tailed)			0.065
Lung				
		-	UDPGT	MDA
GST	Correlation coefficient		0.431	0.505
	Sig. (2-tailed)		0.017*	0.004**
UDPGT	Correlation coefficient			0.232
	Sig. (2-tailed)			0.217

Discussion

A general lack of information concerning wild mammalian sex-related differences in xenobiotic metabolism can be observed. Those studies, involving both wildlife and domestic animals, are of interest, as they can help to understand the different susceptibility of male and female to poisoning and general environmental contamination. In this sense, no differences associated to sex have been observed in the present study with respect to the normal levels of different biotransformation enzymes. Similar results have been reported in other domestic ruminant species, as for example sheep and goat, where mean values of biotransformation enzymes were very close in both sexes [29-30]. Nevertheless, when other mammal species are considered the results are contradictory. For example, some authors reported no differences because of sex in these biotransformation enzymes in rats [31]. In contrast, other studies reported that male rats metabolize foreign compounds more rapidly than females and they are therefore more susceptible to poisoning [11]. A more recent study related to GST activity confirm this phenomenon. An important difference ($P< 0.05$) between sexes in the levels of GST activity in liver of brown Norway rats has been found [32]. However, this difference was found in young animals, being less important or absent in adults. Hormonal factors have been proposed as underlying this phenomenon. When studying the effects of sex-hormones on GST activity, certain authors found that the total GST activity levels in castrated mice dropped to that of their female counterparts [33]. Such dimorphism could be explained in terms of sex-hormone regulation. The ageing may lead more analogous sex-hormonal composition of both sex and more similar levels in liver enzymatic activities [32]. To our knowledge this is the first study about the influence of sex in the levels of UDPGT activity in ruminants. These results seem to indicate that the tendency is the same as in the other biotransformation enzymes, but further studies are necessary to establish a conclusion.

When the effect of sex in the MDA levels was considered, an important difference ($P<0.05$) between males and females was found in liver. Similar results were found in previous studies with male and female Wistar rats. The MDA levels in the control male animals were higher than in females, and that difference was statistically significant at level $P<0.001$ [34]. Some studies have demonstrated the important antioxidant properties of some steroids such as estriol, 17α-estradiol and 17β-estradiol whereas testosterone and other tested steroids lack such properties. It has been observed that the significant differences in antioxidant potency of 17α-estradiol and 17β-estradiol are noteworthy [35-36]. This result suggests that minor differences in the molecule can affect considerably the antioxidant properties of the compound. It may explain the significant difference in MDA levels between males and females. The presence of these steroids in females can reduce the MDA levels, in contrast to male. Our results agreed with previous findings suggesting that ROS production is most likely related to sex hormone levels. Testosterone has been reported to promote H2O2 production, while lipid peroxidation and estrogens suppress it [37]. However, these hypothesis is in contrast with the one published by Grueso et al. [38] which support that the low concentration of estrogens in females makes it unlikely that they exhibit significant antioxidant capacity in the organism. The effects observed are related to the capacity of estrogens to up-regulate expression of antioxidant enzymes.

The effect of gender in other oxidative stress biomarker, including lipid peroxidation, has been studied before in deer species. For instance, Reglero et al. [39] also found in red deer higher levels of lipid peroxidation, as well as oxidized glutathione levels and glutathione peroxidase activity, in liver of males that in females. Rodriguez-Estival et al. [40, 41] did not detect differences by gender in lipid peroxidation in the liver or spleen of red deer, but there was again a higher level of oxidized glutathione in males, which also indicates a higher oxidative stress in males. Moreover, females had more total glutathione than males in spleen.

There was a positive correlation between GST activity and both UDPGT activity and MDA levels in liver and lung. In general, both phase II biotransformation enzymes catalyze the conjugation of an endogenous group -glucuronic acid, sulphate, methyl, acyl, or reduced gluthatione (GSH)- with xenobiotic which may or may not have previously undergone phase I reactions. The conjugates are usually of markedly increased water solubility and are readily excreted through the bile or the urine. Apart from a few exceptions of limited interest in veterinary toxicology, phase II reactions are generally responsible for the detoxification of xenobiotics, so the induction/inhibition of both enzymatic activities could indicate the presence of certain pollutants in the tissues, and in this sense an increase on the LP levels could help to confirm it [10].

The liver was found to be the organ with the highest level of xenobiotic metabolizing enzymes, which is consistent with the fact that the highest concentrations of the enzymes are found precisely in liver; hence, this organ is the most important site of biotransformation [42]. Previous studies made in ruminants (sheep and goat) had obtained the same results, with the highest values of biotransformation enzymes and MDA levels being associated to liver [5, 29]. A similar result was observed in other species like camels, rats, or also humans [43-45]. However, with respect to MDA levels, a study made in dogs under different conditions of ventilation discovered higher MDA levels in plasma than in liver [45]. In many diseases and toxic conditions, membrane damage often occurs in some organs or tissues, which triggers LP and accelerates the disorders in membrane structure and function [46]. When lipid peroxides are accumulated to a certain degree, they leak from the organ or tissue into the blood, consequently raising the lipid peroxide levels in plasma. The increased lipid peroxides in blood are not readily excreted in the urine and their blood level is regulated by the rates of their formation and enzymatic decomposition in the body. It has been suggested that the highest concentration of MDA in plasma, compared to liver, may indicate that MDA was originated in other tissues or organs besides the liver, thus indicating possible damage in other parts of the organism [47].

The relatively lower activities in extra-hepatic tissues have been attributed either to the production of inhibitors of enzyme activity by these tissues, or to the fact that they may produce different types and amounts of metabolites. In some instances, however, extra-hepatic tissues may have enzymes of the microsomal mixed function oxidase (MFO) system that are equal to or slightly higher than the corresponding activities in liver [29, 48-49]. However, in low-pollution exposed animals, it is normal that the enzimatic activity in kidney and lung was lower or also absent than that from liver in some samples.

Conclusions

The reference levels of EROD, GST, UDPGT activity and MDA levels and their differences among sex have been established for the first time in wild roe deer. The highest levels of these parameters have been quantified in liver. No differences associated to sex were found in the biotransformation enzymes. Nevertheless a significant difference between males and females was observed when MDA levels were determined. These results should be taken into account in future biomonitoring studies, in order to determine if changes of such biomarkers could be associated or not to an increase in the surrounding pollution levels.

Acknowledgements

Irene de la Casa-Resino was supported by fellowship PRE09001 from Consejería de Empleo, Empresa e Innovación (Gobierno de Extremadura, Spain).

References

Five "key references", selected by the authors, are marked below (Three recommended (●) and two highly recommended (●●) papers).

1. ●● Van der Oost, R., Beyer, J., Vermeulen, N.P. 2003. Fish bioaccumulation and biomarkers in environmental risk assessment: a review. Environ Toxicol Pharmacol. 13: 57-149. http://dx.doi.org/10.1016/S1382-6689(02)00126-6
2. Froslie, A., Norheim, G., Rambaek, J.P., Steinnes, E. 1984. Levels of trace elements in liver from Norwegian moose, reindeer and red deer in relation to atmospheric deposition. Acta Vet Scand 25: 333-345.
3. Linnell, J.D.C., Duncan, P., Andersen, R. 1998. The european roe deer: a portrait of a successful species. In: Andersen, R., Duncan, P., Linnell, J.D.C. (eds), The European Roe Deer: the Biology of Success. Scandinavian University Press, Oslo, pp. 11-22.
4. Kierdorf, U., Kierdorf, H. 2002. Assessing regional variation of environmental fluoride concentrations in western Germany by analysis of antler fluoride content in roe deer (*Capreolus capreolus*). Arch Environ Contam Toxicol 42: 99-104. http://dx.doi.org/10.1007/s002440010297
5. Lenartova, V., Holovska, K., Martinez-Lara, E., Lopez-Barea, J., Barcena, J., Rosival, I. 1996. Changes in GST-isoenzyme pattern of some organs of sheep exposed to different levels of pollution. Comp Biochem Physiol C Pharmacol Toxicol Endocrinol 114: 153-158. http://dx.doi.org/10.1016/0742-8413(96)00025-4
6. Kierdorf, H., Kierdorf, U., Sedlacek, F. 1999. Monitoring regional fluoride pollution in the Saxonian Ore mountains (Germany) using the biomarker dental fluorosis in roe deer (*Capreolus capreolus* L.). Sci Total Environ 232: 159-168. http://dx.doi.org/10.1016/S0048-9697(99)00102-3
7. Hewinson, A.J.M., Vicent, J.P., Reby, D. 1998. Social organisation of European roe deer. In: Andersen, R., Duncan, P, Linnell, J.D.C. (eds.). The European Roe Deer: the Biology of Success. Scandinavian University Press, Oslo, pp. 189-219.
8. ● Peakall, D.B. 1994. The role of biomarkers in environmental assessment (1). Introduction. Ecotoxicology 3: 157-160. http://dx.doi.org/10.1007/BF00117080
9. Word Health Organization (WHO). 1993. Biomarkers and Risk Assessment: Concept and Principles. International Program of Chemical Safety (IPCS): Geneva, pp. 1-57.
10. ●● Nebbia, C. 2001. Biotransformation enzymes as determinants of xenobiotic toxicity in domestic animals. Vet J 161: 238-252. http://dx.doi.org/10.1053/tvjl.2000.0561
11. Timbrel, J.A. 1991. Principles of Biochemical Toxicology, 4th ed. Taylor and Francis.
12. Goksøyr, A. 1995. Use of cytochrome P450 1A (CYP1A) in fish as a biomarker of aquatic

pollution. Arch Toxicol 17: 80-95.

13. Berdikova Bohne, V.J., Hamre, K., Arukwe, A. 2007. Hepatic metabolism, phase I and II biotransformation enzymes in Atlantic salmon (*Salmo Salar*, L) during a 12 week feeding period with graded levels of the synthetic antioxidant, ethoxyquin. Food Chem Toxicol 45: 733-746.
http://dx.doi.org/10.1016/j.fct.2006.10.020

14. Owens, I. 1977. Genetic regulation of UDP-glucuronosyltransferase induction by polycyclic aromatic compounds in mice. Co-segregation with aryl hydrocarbon (benzo (alpha) pyrene) hydroxylase induction. J Biol Chem 252: 2827-2833.

15. Pickett, C.B., Lu, A.Y. 1989. Glutathione S-transferases: gene structure, regulation, and biological function. Annu Rev Biochem 58: 743-764.
http://dx.doi.org/10.1146/annurev.bi.58.070189.003523

16. Livingstone, D. 2001. Contaminant-stimulated reactive oxygen species production and oxidative damage in aquatic organisms. Mar Pollut Bull 154: 427-430.

17. Valavanidis, A., Vlahogianni, T., Dassenakis, M., Scoullos, M. 2006. Molecular biomarkers of oxidative stress in aquatic organisms in relation to toxic environmental pollutants. Ecotoxicol Environ Saf 64: 178-189.
http://dx.doi.org/10.1016/j.ecoenv.2005.03.013

18. Porter, N.A., Caldwell, S.E., Mills, K.A. 1995. Mechanisms of free radical oxidation of unsaturated lipids. Lipids 30: 277-290.
http://dx.doi.org/10.1007/BF02536034

19. Pedrajas, J., Peinado, J., Lopez-Barea, J. 1995. Oxidative stress in fish exposed to model xenobiotics. Oxidatively modified forms of Cu, Zn-superoxide dismutase as potential biomarkers. Chem Biol Interact 98: 267-282.
http://dx.doi.org/10.1016/0009-2797(95)03651-2

20. Rodriguez-Ariza, A., Peinado, J., Pueyo, C., Lopez-Barea, J. 1993. Biochemical indicators of oxidative stress in fish from polluted littoral areas. Can J Fish Aquat Sci 50: 2568-2573.
http://dx.doi.org/10.1139/f93-280

21. Pellerin-Massicotte, J. 1997. Influence of elevated temperature and air-exposure on MDA levels and catalase activities in digestive glands of the blue mussel (*Mytilus galloprovincialis* L.). J Rech Oceanogr 22: 91-98.

22. Viarengo, A., Canesi, L., Pertica, M., Livingstone, D. 1991. Seasonal variations in the antioxidant defence systems and lipid peroxidation of the digestive gland of mussels. Comp Biochem Physiol C Pharmacol Toxicol 100: 187-190.

23. Burke, M.D., Mayer, R.T. 1974. Ethoxyresorufin: direct fluorimetric assay of a microsomal O-dealkylation which is preferentially inducible by 3-methylcholanthrene. Drug Metab Dispos 2: 583-588.

24. Sole, M., Porte, C., Barcelo, D. 2000. Vitellogenin induction and other biochemical responses in carp, Cyprinus carpio, after experimental injection with 17 alpha-ethynylestradiol. Arch Environ Contam Toxicol 38: 494-500.

25. Cohen, A., Smith, J., Turbert, H. 1964. Comparative detoxication. 10. The enzymic conjugation of chloro compounds with glutathione in locusts and other insects. Biochem J 90: 457.

26. Habig, W.H., Pabst, M.J., Jakoby, W.B. 1974. Glutathione S-transferases the first enzymatic step in mercapturic acid formation. J Biol Chem 249: 7130-7139.

27. Clarke, D.J., George, S.G., Burchell, B. 1992. Multiplicity of UDP-glucuronosyltransferases in fish. Purification and characterization of a phenol UDP-glucuronosyltransferase from the liver of a marine teleost, *Pleuronectes platessa*. Biochem J 284: 417-423.

28. Recknagel, R.O., Glende, E.A., Walker, R.L., Lowery, K. 1982. Lipid peroxidation biochemistry measurement and significance in liver cell injury. In: Plaa, G.L., Hewitt, W.R. (eds.). Toxicology of the Liver. Raven Press, New York, pp. 213-241.

29. El Sheikh, H., Ali, B., Homeida, A., Hapke, H. 1991. Activities of glutathione-S-transferase and ethoxycoumarin-O-deethylase in tissues of camels, sheep, goats and rats. Comp Biochem Physiol C Pharmacol Toxicol 98: 293-297.
http://dx.doi.org/10.1016/0742-8413(91)90208-B

30. Van't Klooster, G.A., Horbach, G., Natsuhori, M., Blaauboer, B.J., Noordhoek, J., Van Miert, A.S. 1993. Hepatic cytochrome P450 induction in goats: effects of model inducers on the metabolism of alkoxyresorufins, testosterone and ethylmorphine, and on apoprotein and mRNA levels. Biochem Pharmacol 45: 113-122.
http://dx.doi.org/10.1016/0006-2952(93)90383-8

31. Poon, R., Chu, I., LeBel, G., Yagminas, A., Valli, V. 2003. Effects of dibromoacetonitrile on rats following 13-week drinking water exposure. Food Chem Toxicol 41: 1051-1061.
http://dx.doi.org/10.1016/S0278-6915(03)00042-5

32. Coecke, S., Vandenberghe, Y., Callaerts, A., Sonck, W., Verleye, G., Van Bezooijen, C., Vercruysse, A., Rogiers, V. 1990. Hepatic cytosolic glutathione S-transferase activities in ageing Brown Norway rats—Importance of sex differences and phenobarbital treatment for studies of ageing. Mech Ageing Dev 55: 189-198.
http://dx.doi.org/10.1016/0047-6374(90)90026-C

33. Hatayama, I., Satoh, K., Sato, K. 1986. Developmental and hormonal regulation of the major form of hepatic glutathione S-transferase in male mice. Biochem Biophys Res Commun 140: 581-588.
http://dx.doi.org/10.1016/0006-291X(86)90771-0

34. Kamper, E., Chatzigeorgiou, A., Tsimpoukidi, O., Kamper, M., Dalla, C., Pitychoutis, P., Papadopoulou-Daifoti, Z. 2009. Sex differences in oxidant/antioxidant balance under a chronic mild stress regime. Physiol Behav 98: 215-222.
http://dx.doi.org/10.1016/j.physbeh.2009.05.011

35. Begoña Ruiz-Larrea, M., Ma Leal, A., Liza, M., Lacort, M., de Groot, H. 1994. Antioxidant effects of estradiol and 2-hydroxyestradiol on iron-induced lipid peroxidation of rat liver microsomes. Steroids 59: 383-388.
http://dx.doi.org/10.1016/0039-128X(94)90006-X

36. Mooradian, A.D. 1993. Antioxidant properties of steroids. J Steroid Biochem Mol Biol 45: 509-511.
http://dx.doi.org/10.1016/0960-0760(93)90166-T

37. Sullivan, J.C., Sasser, J.M., Pollock, J.S. 2007. Sexual dimorphism in oxidant status in spontaneously hypertensive rats. Am J Physiol Regul Integr Comp Physiol 292: 764-768.
http://dx.doi.org/10.1152/ajpregu.00322.2006

38. Vina, J., Gambini, J., Lopez-Grueso, R., Abdelaziz, K.M., Jove, M., Borras, C. 2011. Females live longer than males: role of oxidative stress. Curr Pharm Des 17: 3959-3965.
http://dx.doi.org/10.2174/138161211798764942

39. Reglero, M.M., Taggart, M.A., Monsalve, L., Mateo, R. 2009. Heavy metal exposure in large game from a lead mining area: effects on oxidative stress and fatty acid composition in liver. Environ Pollut 157: 1388-1395.
http://dx.doi.org/10.1016/j.envpol.2008.11.036

40. ● Rodríguez-Estival, J., Martínez-Haro, M., Monsalve-González, L., Mateo, R. 2011. Interactions between endogenous and exogenous antioxidants against Pb-induced oxidative stress in wild

ungulates from a Pb polluted mining area. Sci Total Environ 409: 2725-2733.
http://dx.doi.org/10.1016/j.scitotenv.2011.04.010

41. ● Rodríguez-Estival, J., Pérez de la Lastra, J.M, Ortiz-Santaliestra, M.E., Vidal, D., Mateo, R. 2013. Expression of immunoregulatory genes and its relationship with Pb exposure and Pb-mediated oxidative stress in wild ungulates from an abandoned mining area. Environ Toxicol Chem 32: 876–883.
http://dx.doi.org/10.1002/etc.2134

42. Buhler, D.R. Williams, D.E. 1988. The role of biotransformation in the toxicity of chemicals. Aquatic Toxicol 11: 19-28.
http://dx.doi.org/10.1016/0166-445X(88)90004-5

43. Krajka-Kuźniak, V., Szaefer, H., Baer-Dubowska, W. 2005. Modulation of cytochrome P450 and phase II enzymes by protocatechuic acid in mouse liver and kidney. Toxicology 216: 24-31.
http://dx.doi.org/10.1016/j.tox.2005.07.013

44. Raza, H., Lakhani, M., Ahmed, I., John, A., Morgenstern, R., Montague, W. 1997. Tissue specific expression of glutathione S-transferases, glutathione content and lipid peroxidation in camel tissues. Comp Biochem Physiol B Biochem Mol Biol 118: 829-835.
http://dx.doi.org/10.1016/S0305-0491(97)00234-4

45. El-Bassiouni, E., Abo-Ollo, M., Helmy, M., Ismail, S., Ramadan, M. 1998. Changes in the defense against free radicals in the liver and plasma of the dog during hypoxia and/or halothane anaesthesia. Toxicology 128: 25-34.
http://dx.doi.org/10.1016/S0300-483X(98)00045-6

46. Freeman, B.A., Crapo, J.D. 1982. Biology of disease: free radicals and tissue injury. Lab Invest 47: 412-426.

47. Cheeseman, K.M. 1982. Effect of scavengers and inhibitors on lipid peroxidation in rat liver microsomes. In: Mac Brien, D.C., Salter, T.F. (eds.). Free Radicals, Lipid Peroxidation and Cancer. Academic Press, New York, pp. 196-211.

48. Bend, J.R., Hook, G.E., Easterling, R.E., Gram, T.F., Fouts, J.R. 1972. A comparative study of the hepatic and pulmonary microsomal mixed-function oxidase systems in the rabbit. J Pharmacol Exp Ther 183: 206-217.

49. Helferich, W., Silva, M., Flueck, W., Hammock, B., Shull, L. 1987. Xenobiotic biotransformation in livers and lungs of adult black-tailed deer: Comparison with domestic goat and sheep. Comp Biochem Physiol C Comp Pharmacol Toxicol 88: 145-149.
http://dx.doi.org/10.1016/0742-8413(87)90058-2

50. Bradford, M.M. 1976. A rapid and sensitive method for the quantification of microgramquantities of protein utilizing the principle of proteindye binding. Anal Biochem 72: 248-254.
http://dx.doi.org/10.1016/0003-2697(76)90527-3

12

Age-Specific Changes in Body Mass and Delayed Physical Development of a Known-Aged Sample of White-Tailed Deer

K. L. Gee[1,3], S. L. Webb[1,*] & P. D. Jones[2]

[1] The Samuel Roberts Noble Foundation, 2510 Sam Noble Parkway, Ardmore, OK 73401, USA.

[2] Department of Forest and Wildlife Ecology, University of Wisconsin-Madison, 1630 Linden Drive, Room 226, Madison, WI 53706, USA.

[3] Present address: 6438 Gene Autry Road, Ardmore, OK 73401, USA.

* Corresponding author email: slwebb@noble.org.

Keywords

Body growth; Compensatory development; Environmental effects; Longitudinal study; Management; *Odocoileus virginianus;* Oklahoma; Ontogeny.

Abstract

Ontogeny is important to life history processes and evolution because it can influence fitness through survival and reproduction. However, most studies related to ontogeny of white-tailed deer (*Odocoileus virginianus*) have been carried out under controlled settings (in captive herds) where deer do not experience the same levels of environmental variability as wild deer. We used a 23-year data set of wild, known-aged deer to study changes in body mass across sexes and ages. We found that body mass increased with age in both females and males and was best described by 3rd and 2nd order polynomial functions, respectively. During the first year of life, females gained 51.3% mass while males gained 45.0% mass from fawns to yearlings. Body mass gain slowed significantly after 2 years of age (decreasing on average at 3 years of age) for females (average gain <5.2%), and by 2 years of age, females had reached 92.7% of their maximum lifetime body mass. For males, body mass continued to increase through 4 years of age. Body mass of males was largely influenced by their year of birth; however, this was not the case for females. We used best linear unbiased predictions (BLUP) of birth year for males, which served as a surrogate for a wide range of potential effects (e.g., variable forage, density, temperature and precipitation), to classify males to either optimal (i.e., positive BLUP) or suboptimal (i.e., negative BLUP) birth year. Males born during suboptimal years showed signs of delayed physical development; males had less body mass as fawns and yearlings than males born during good years, but at 2 years of age, deer born during suboptimal years exhibited compensatory growth and had body mass similar to deer born during good years. These data make a significant contribution to ontogeny of white-tailed deer because these longitudinal data were collected on a known-aged population under variable environmental conditions; a scenario for which a paucity of ontogenetic data exists.

Introduction

Ontogenetic development is influenced by a wide-range of factors (e.g., maternal, genetic and environmental), and these factors may influence sex- and age-specific development in different ways. Post-natal growth and development of white-tailed deer (*Odocoileus virginianus*) is influenced by milk production of the dam, which is influenced by a wide-range of factors such as population density, forage quality and quantity, and number of offspring [1]. Early developmental factors can have carryover effects at older ages, and may be important determinants of survival and lifetime reproductive success in polygynous cervids [2]. Among sexually dimorphic cervids, females generally reach sexual maturity and maximum body growth earlier than males. Males however continue to accumulate body mass because it may influence dominance [3-5], and perhaps because greater body mass or fat reserves allow more energy to be expended on mating activities; albeit greater body mass itself has greater metabolic requirements. Yearling white-tailed deer females typically produce only one offspring because they are still diverting resources to body development, but past this age, most body development has been achieved allowing them to invest more resources, given available nutrition, toward fetal and offspring development whereas most adults (≥2 years of age) carry 2 fetuses [6]. Therefore, knowing when maximum body development is reached may influence the potential for population growth, at least in females. On the other hand, male cervids do not contribute resources directly to their offspring. Rather, males divert more energy to breeding activities (e.g., searching, finding and tending mates), and greater body mass may help males sustain the necessary levels of energy to actively participate in rutting activities. However, a paucity of data exists on ontogenetic development of wild white-tailed deer, and all components of ontogenetic development (e.g., body mass, skeletal size, antler size, etc.) are important when designing and evaluating management programs.

Data for many studies of growth in deer exhibit shortcomings, primarily because most data on phenotypic and morphometric (e.g., antler size and body mass) characters come from captive or farmed deer populations [7-10]. In captive populations (i.e., primarily farmed deer), supplemental or *ad libitum* feeding of high-quality rations does not reflect the more dynamic environment faced by wild deer. Therefore, studies of captive deer may not accurately depict conditions in wild populations, which would be of interest to managers [11,12]. Another shortcoming of field studies using wild populations is that age determination is not accurate for older (≥2.5 years) age classes [13-15], which can result in less accurate and precise estimates of body development during ontogeny. Despite this fact, most field studies that estimate age of deer make no mention of how the inaccuracy of aging deer potentially biases interpretation of age-specific measures of health (e.g., body mass, antler size, conception rates, etc.). Some studies have used data from wild captured or harvested deer of known age (i.e., fawns or yearlings), but did not follow individuals to older age classes [e.g., 16-19]. To our knowledge, there are not any longitudinal studies of white-tailed deer growth using wild, known-age subjects, a situation that would best reflect natural processes and pressures.

Environmental conditions during early development are associated with life history effects in ungulates [20]. The year in which a mammal is born can have as much of an effect on similarity of traits as heritable genetic effects because of shared environmental conditions [21]. Variable resource availability caused by density-dependent or extrinsic factors, such as climate or land use, can cause subsequent physical and reproductive variation [22-27]. For example, rainfall influences vegetation quality and quantity, which affects nutrition of the dam as well as birth weight and mass gain of fawns [17,28,29]. Therefore, fawns born during optimal conditions may exhibit greater early development compared to fawns born during suboptimal conditions, as evidenced by low birth weights or delayed physical development. Despite being born during suboptimal years, it is unknown whether wild deer born during restrictive environmental conditions can compensate for poor early growth later in life. In this same population of deer, year of birth, as a form of permanent environmental effect, influenced body mass of males but not females, and showed patterns similar to maternal effects early in life [30]. In this study, we used the best linear unbiased predictions (BLUP) of body mass across birth years [30] to examine whether deer exhibit compensatory development and overcome delayed physical growth due to being disadvantaged when born during a suboptimal year.

We used a 23-year data set of wild, known-aged white-tailed deer that were captured and marked, and subsequently recaptured later in life to determine physical development (i.e., body mass). Our objectives were to: 1) estimate sex- and age-specific body mass, 2) characterize sex-specific growth curves in relation to ontogeny, and 3) examine how permanent environmental effects (i.e., year of birth) influence physical development and compensation for male deer born during suboptimal years. To accomplish objective 3, we use data from a variance component analysis [30] and the predictions for year of birth (BLUP) to classify the overall conditions that male deer experienced during their first year of life and how these conditions could influence ontogeny.

Methods

Study area

We conducted our study on the Samuel Roberts Noble Foundation Wildlife Unit (NFWU) located in south-central Oklahoma approximately 8.0 km south of Allen, Oklahoma, which was part of the Cross Timbers and Prairies ecoregion [31]. The NFWU was 1,214 ha, of which rangeland comprised 40% of the surface area and the remaining habitat was wooded. Prescribed fire was used annually to control overabundant brush, rejuvenate herbaceous plant communities, and reduce ground litter; an average of 316 ha were burned annually (range: 92–553 ha). Habitat management projects included livestock grazing (and rest from grazing) and selective control of noxious woody plants via single-stem herbicide treatments. Water was readily available in ponds (1 pond/33 ha) and streams, but artificial feeding did not occur on a year-round basis; shelled corn was only available during camera surveys and trapping efforts (i.e., from 28 December [earliest] to 10 April [latest]). Long-term, total annual precipitation averages 105.5 cm (Ada, Pontotoc County, Oklahoma; Oklahoma Mesonet, https://climate.mesonet.org), and ranged from 64.9 to 162.5 cm during the course of the study.

During the course of the study, the study area underwent changes in terms of harvest levels and spatial control of the deer population [32]. Deer were free-ranging from 1983 to 1987, mostly free-ranging from 1987 to 1992, and spatially confined beginning in 1993 by a 2.44-m high, 15-strand high-tensile electric fence (HTEF) [33]. From 1987 to 1992, the 20.1 km HTEF was constructed in 2.4–4.8 km segments per year. The fence was erected to facilitate white-tailed deer management and research, and discourage human trespass (i.e., illegal harvest). Harvest was permitted on both sexes from 1983 to 1992, but research during 1985–1987 required intense harvest of females, after which time harvest was minimal until completion of the HTEF. During 1993–1999, harvest was allowed on both sexes but with restrictions on numbers and ages. Beginning in 2000, harvest of males on the study area was eliminated through the end of the study (i.e., 2005); some harvest of females was allowed through 2001 (7 females were harvested each year in 2000 and 2001) and then discontinued through the end of the study. Trends in population size indicated an average estimated deer density of 1 deer/9.9 ha and ranged from 1 deer/19.0 ha to 1 deer/5.9 ha [30]. Potential natural predators to deer on the study area were bobcats (*Lynx rufus*) and coyotes (*Canis latrans*).

Capture and Handling

We conducted annual trapping primarily during February and March (1983–2005) using a drop-net [34, as modified by 35] baited with corn. We sedated deer using xylazine (3–6 mg/kg, Phoenix Scientific, St. Joseph, Missouri, USA) or a mixture of Telazol® and xylazine (4.4 mg/kg Telazol®, Fort Dodge Animal Health, Fort Dodge, Iowa, USA, plus 2.2 mg/kg xylazine). We used yohimbine (Abbott Laboratories, North Chicago, Illinois, USA) at 0.125 mg/kg or tolazine at 0.4 mg/kg as an antagonist to the xylazine. Beginning in 1986, deer were weighed to the nearest pound (0.45 kg) and then converted to kg. We aged deer according to tooth replacement and wear [36], and marked deer with 2 plastic livestock ear tags (i.e. 1 in each ear) with unique color and number combinations; deer could further be identified at time of capture or harvest by permanent ear tattoos, or metal ear clips that were placed on a sample of deer. All capture, handling, and marking procedures followed guidelines approved by the American Society of Mammalogists [37] and the Oklahoma Department of Wildlife Conservation (permit no. 2243, 2374, 2701, 046, 195, 493, 627, 796, 1000, 1224, 1432, 1654, 1912, 2141, 2424, 2517, 2723, 2941, 3065, 3174, 3239, 3364, 3591, and 3726).

Age-specific body mass

We only used data from known-aged deer that were first captured as a fawn (7 – 9 months) or yearling (19 – 21 months) because these age classes can be aged definitively based on tooth replacement patterns [13] and because of inaccuracy of aging deer ≥2.5 years of age [13,14]. We considered all subsequent recaptures of deer (i.e., ≥2 years) that were captured first as a fawn or yearling to be of known age. Because we did not start collecting body mass until 1986, some deer of known-aged status did not have body mass data recorded until ≥2 years of age.

We took a hierarchical approach to analyzing the data. First, we calculated descriptive statistics (mean, standard error, and confidence intervals) to assess age-specific body mass for each sex. Next, we modeled the age-specific growth curve for each sex using the most appropriate functional relationship. We used general linear mixed models (GLMM; MIXED procedure) in SAS® 9.3 (SAS Institute, Inc., Cary, NC) [38] and an information theoretic approach (i.e., AIC) [39] to choose the most appropriate functional relationship (i.e., linear, quadratic, or 3rd order polynomial) for each sex. We used a maximum age of 4 years for males because no known-aged males were captured after 4 years of age. We modeled 3 age structures for females: 1) age-specific from 0 to ≥8 years, with data ≥9 years of age being grouped within the ≥8 year class, 2) age-specific from 0 to ≥6 years, with data ≥7 years of age being grouped within the ≥6 year class, and 3) age-specific from 0 to ≥5 years, with data ≥6 years of age being grouped within the ≥5 year class. After the most appropriate functional relationship was identified for each sex, we tested whether the addition of covariates improved model fit. We developed covariates to adjust for date of capture on body mass measurements, where the first date of capture was set to day zero, and all subsequent captures were adjusted to day zero. We also created a covariate for days since peak conception (30 November), and an interaction between day zero adjusted mass and days since peak conception. After addition of covariates, we considered any model within 2 ΔAICc units of the best model (i.e., the most appropriate functional relationship model) as a plausible model [39]. Deer and year were considered random effects in all GLMMs, and the Kenward-Roger denominator degrees of freedom adjustment [38,40] was applied to all models. Last, we calculated all age-specific pairwise combinations (e.g., growth from 0 to 1, 1 to 2, 2 to 3, 3 to 4, etc.) to assess absolute (kg) and relative (%) changes in body mass for each sex.

Delayed physical development

We used data from [30] to test whether deer born during suboptimal years exhibited delayed physical development and compensatory growth later in life. We only examined male deer because year of birth had little influence on female body mass; thus females may have been less sensitive to environmental variation during year of birth relative to males [30]. We created a new variable (PE_{effect}, for permanent environmental effect) that classified male deer based on whether they were born during a 'bad' year or during a 'good' year. The prediction of good years versus bad years of birth was based on a variance component analysis and exportation of best linear unbiased predictions (BLUP) [30]. Good years were defined as BLUPs >0, and bad years when BLUPs <0. We used general linear models in SAS® 9.3 to test whether body mass (kg; dependent variable) of male deer at each age (i.e., 0, 1, 2, and ≥3) was different between years of birth (independent categorical variable = PE_{effect} with 2 levels [good and bad]).

Results

Body mass of females increased with age (Table 1a) and was characterized best by a 3rd order polynomial function of age (Fig. 1a). However, the addition of covariates (i.e., to adjust for capture date and days since peak conception) did not improve model fit, except for the addition of capture date, which was only 0.4 ΔAICc units from best model. Therefore, we chose to only interpret the simpler, 3rd order polynomial model. The growth curve for females was best fit by: $32.03 + 8.82(\text{age}) + 1.42(\text{age}^2) - 0.55(\text{age}^3)$, and explained 88.1% of the variation in body mass (Fig. 1a). Female fawn body mass was 30.16 kg and increased to 56.31 kg when ≥5 years (Table 1a).

Table 1: Descriptive statistics (mean, sample size, standard error [SE], and 95% confidence intervals [CI]) for age-specific body mass (kg) of wild, known-aged female (A) and male (B) white-tailed deer (*Odocoileus virginianus*) in southern Oklahoma from 1986 to 2005.

A

	Age (years)					
Parameter	**0-1**	**1-2**	**2-3**	**3-4**	**4-5**	**≥5**
Mean	30.16	45.64	52.46	51.48	55.18	56.31
n	123	74	32	18	17	14
SE	0.31	0.45	0.76	1.04	0.81	0.95
Lower 95% CI	29.54	44.74	50.92	49.29	53.47	54.25
Upper 95% CI	30.78	46.53	54.00	53.67	56.89	58.37

B

	Age (years)				
Parameter	**0-1**	**1-2**	**2-3**	**3-4**	**4-5**
Mean	33.85	49.07	59.47	64.02	68.95
n	139	94	19	14	3
SE	0.35	0.54	1.33	2.04	2.05
Lower 95% CI	33.16	47.99	56.68	59.61	60.15
Upper 95% CI	34.55	50.15	62.26	68.43	77.75

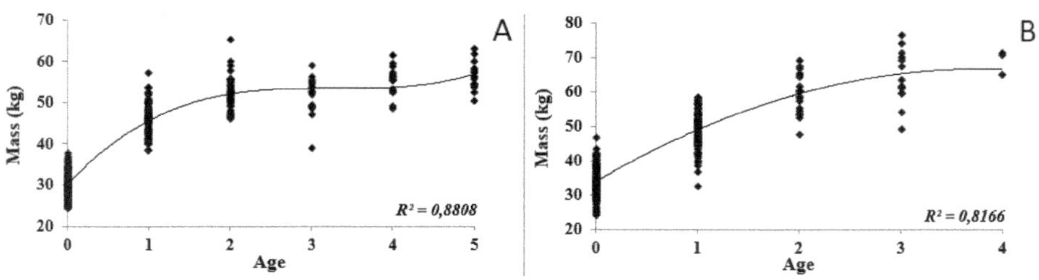

Figure 1: Age-specific (0 = fawns [0-1 year]; 1 = yearlings [1-2 years]; 2 = 2-3 years; 3 = 3-4 years; 4 = 4-5 years; ≥5 = ≥5 years) body mass measures (kg) of wild, known-aged female (a) and male (b) white-tailed deer (*Odocoileus virginianus*) in southern Oklahoma from 1986 to 2005. Changes in body mass were best described by a 3rd order polynomial function of age (black line) for females and a 2nd order polynomial function of age for males. Sample sizes for each sex and age are presented in Table 1.

Females gained the greatest percentage of body mass from fawns to yearlings (51.3%) followed by yearlings to 2-years (14.9%; Table 2a). By 2 years of age, females had reached 92.7% of their maximum body mass; body mass changed from -1.9 to 9.4% each year after 2 years of age (Table 2a).

Body mass of males also increased with age (Table 1b), but unlike females, the growth curve for males was best fit by a quadratic model for age (34.07 + 17.18(age) – 2.26(age^2)). Covariate models did not improve model fit of the quadratic model (ΔAICc $\geq$ 6.8). The quadratic equation for estimating body mass explained 81.7% of the variation observed in body mass (Fig. 1b). Males had greater body mass than females at all ages, but only slightly greater as fawns and yearlings (~3–4 kg difference; Table 1). Male fawns averaged 33.85 kg, whereas 4-year olds averaged 68.95 kg (Table 1b). We did not recapture any males ≥5-years-of-age, but one 5-year old male weighing 85.28 kg was harvested (Appendix 1). While females reached most of their lifetime body mass at 2 years of age, males continued to accumulate greater body mass annually (absolute and relative) through 4 years of age (Table 1b).

It appears that male deer in this population overcame delayed physical development and exhibited compensatory physical growth after being disadvantaged by being born during a suboptimal year. We found that male fawns born during good years (35.50

Table 2: Average, absolute (blue boxes above diagonal) and relative (%; gray boxes below diagonal) body mass (kg) changes for each pairwise combination of age for wild, known-aged female (A) and male (B) white-tailed deer (*Odocoileus virginianus*) in southern Oklahoma from 1986 to 2005.

A

Age	0-1	1-2	2-3	3-4	4-5	≥5
0-1		15.48	22.30	21.32	25.02	26.15
1-2	51.3		6.82	5.84	9.54	10.67
2-3	73.9	14.9		-0.98	2.72	3.85
3-4	70.7	12.8	-1.9		3.70	4.83
4-5	83.0	20.9	5.2	7.2		1.13
≥5	86.7	23.4	7.3	9.4	2.0	

B

Age	0-1	1-2	2-3	3-4	4-5
0-1		15.22	25.62	30.17	35.10
1-2	45.0		10.40	14.95	19.88
2-3	75.7	21.2		4.55	9.48
3-4	89.1	30.5	7.7		4.93
4-5	125.0	40.5	15.9	7.7	

kg ± 0.53 SE; n = 67) had 9.1% more body mass than male fawns born during bad years (32.54 kg ± 0.40; n = 75) ($F_{1,140}$ = 20.25, P < 0.001). One year later, as yearlings, males born during good years (50.15 kg ± 0.85; n = 55) still had 3.3% more body mass than yearlings born during bad years (48.53 kg ± 0.65; n = 51), but the difference was no longer significant ($F_{1,104}$ = 2.26, P = 0.136), indicating that male deer born during bad years may be exhibiting compensatory body growth. By 2 years of age, males that were born during bad years (59.22 kg ± 1.30; n = 9) had body mass measures similar ($F_{1,18}$ < 0.01, P = 0.987) to those deer born during good years (59.17 kg ± 2.15; n = 11). Similarly, at 3 years of age, body mass was similar ($F_{1,15}$ = 1.02, P = 0.329) between deer born during good (65.88 kg ± 1.80; n = 13) and bad years (61.69 kg ± 4.89; n = 4).

Discussion

Our data establish growth curves derived from a longitudinal study of known-age white-tailed deer under variable nutrition cycles, environmental conditions, and range improvement practices. Use of captive deer may allow greater survival and thus larger samples of older deer. However, results from captive situations may differ relative to deer experiencing variable environmental conditions because of differences in weather, population density, and food availability, particularly the use of high-quality pelleted feed in captive populations [41,42]. These data also have an advantage of clarity due to lack of error in aging, particularly in deer ≥2 years old, when accuracy of the tooth replacement and wear (TRW) method for aging begins to erode substantially [13].

Males in our study exhibited slower accretion of mass relative to samples of wild, harvested deer aged using TRW [43,44]. Error from using TRW raises the likelihood that misclassification of deer ≥2 years old in those studies could obscure differences among older age classes, explaining some portion of the difference in our findings. Aging deer by cementum annuli has been demonstrated to be more accurate than TRW in some instances [45], which may explain why body growth curves of wild deer from Minnesota [46] more closely resembled those from this study. The lack of data at older age classes for males underscores the difficulty of collecting large samples on known-aged, wild deer. Growth curves, especially of males, derived from captive animals or from harvested deer aged using TRW may not accurately reflect growth in large populations under variable conditions.

Females in our study reached 92.7% of their asymptotic mass around 2.5 years of age and maximum body mass around 4.5 to 5.5 years, which is similar to females from populations in Mississippi, South Carolina, and Minnesota [43,44,46]. Faster female development is common in sexually dimorphic cervids (e.g., mule deer (*Odocoileus hemionus*), [47]; moose (*Alce alces*), [48]; elk (*Cervus canadensis*), [49]; roe deer (*Capreolus capreolus*), [50]). Female deer apparently require attainment of a threshold body mass or condition to breed [51-53], and asymptotic body mass of females may be reached at a younger age than in males so that more energy can be devoted to the production of offspring [3,54]. However, once this threshold is crossed, mass appears to cause little difference in reproductive success among mature females compared with nutritional status [55,56] and maternal experience [57,58]. Primiparous females (typically yearlings) are still diverting substantial energy into body development, which may explain why smaller yearlings tend to produce singletons (P. Jones, University of Wisconsin-Madison, unpublished data). However, pregnancy and fetal rates are relatively stable among iteroparous females [6,51,59], and females may attempt to influence fitness by manipulating litter sex ratio [60,61]. Therefore, increasing body mass after attaining sexual maturity may have relatively little effect on female fitness [25].

During this study, body mass of females and males was measured after the breeding season. Based on the mean conception date from this study area (30 November), females were captured, on average, 91.4 days (± 1.2 SE) after peak conception. Adult females (from 2.5 to ≥5.5 years) that were recaptured were 1.93 to 4.7 kg heavier than females of the same age that were harvested during hunting season, which occurred prior to conception. This may partially be explained by fetal development (i.e., fetal age and number of fetuses). We were unable to document pregnancy rates, or the number of fetuses per female. However, these data could be adjusted based on days since conception and average number of fetuses per female by age. As discussed earlier, yearling females usually carry one fetus, whereas adults (≥2.5 years) usually carry two [6]. Using data from [62,63], individual fetal mass, based on an average of 92 days post-conception, was <0.5 kg; so for 2 fetuses, total fetal weight would be <1 kg, not considering weights of placenta and amniotic fluid. In comparison to adult females, captured yearlings were 1.65 kg lighter than yearling females that were harvested. This may be due to the stress associated with being pregnant at a time when body development is also requiring substantial nutritional demands, and one explanation for why most first time dams only have one fetus.

Males in our study were still accumulating body mass through 4 years of age. We only had the opportunity to weigh one male that was ≥5.5 years, although many males survived to, or even past, this age [32]. This one male was harvested and weighed 85.28 kg (Appendix 1), which was 16.33 kg more massive than the average 4-year old males that were recaptured. It is therefore unknown when males reach asymptotic body mass in southern Oklahoma, but it is reasonable to expect that body mass will continue to increase past 4.5 years of age, similar to other studies [1,46,54]. Body mass is an important factor influencing reproductive success in male deer [3-5], largely affecting dominance hierarchies [64]. Greater size may also increase energy available to rutting activities, thereby allowing a longer period of active rut for larger males [65]. Continuous increase in mass could therefore confer a fitness advantage to males.

Like females, male deer were undergoing physiological changes at the time of capture, which occurred after the breeding season (i.e., post-rut); as a result of breeding activities, males may lose up to 30% of their pre-breeding body mass [66]. The mean date of measurement for male body mass was 87.5 days (±1.16 SE) after peak conception (30 November). Therefore, deer had some time to start accumulating body mass after loss of mass due to breeding activities. However, we found in all instances that 1.5, 2.5, 3.5, and 4.5 year old males were 4.45, 3.94, 7.65, and 14.51 kg more massive, respectively during the hunting season (based on harvest data; Appendix 1) compared to captures of these same age classes that occurred post-rut. Although based on relatively small sample sizes at older ages, it appears that as male deer mature, their participation in the rut may increase as evidenced by the increasing percentage of mass loss based on comparisons of harvested (pre-breeding to breeding) and captured males (post-rut) (1 yr: 8.3%; 2 yr: 6.2%; 3 yr: 10.7%; 4 yr: 17.4%). Greater body mass going into the rut may influence dominance, which could further influence reproductive success [1].

Variation in adult body mass is now widely known to be influenced by environmental conditions, particularly those experienced early in life, among a wide range of cervids of both sexes (e.g., red deer (*Cervus elaphus*), [67,68]; moose, [69]; roe deer, [20]; white-tailed deer, [30]). Our use of birth year substituted for an array of possible positive or negative environmental effects. For example, both rainfall and estimated density varied by a factor of 2.5 over the course of this study and may have constituted important environmental effects in this population [30]. Nutritional restrictions faced by the mother during the third trimester and subsequent lactation may reduce birth weight, milk production, and offspring growth rate [70-72]. Weather and climate can affect forage production, quality, and temporal availability [73,74]. Local considerations such as acorn cycle or changes in agronomic crop availability may also influence growth [24,26]. Deer density during year of birth presents another potential permanent environmental effect. Using harvest and road-kill samples from a high density population, male deer grew slower and reached lighter asymptotic body mass [54]; given our small sample of males ≥4 years old, we were unable to determine with certainty if males born during bad years reached lighter asymptotic body mass. However, using information on birth year effects [30], we could determine whether males continued to accrue mass at a slower rate when born during suboptimal years, or whether these males could compensate for a suboptimal start and accrue body mass quicker at later ages.

Where it occurs, compensatory growth in males may be explained by diet selection. It is to be expected that improved nutrition will result in greater growth rates following periods of restricted nutrition [75,76]. Smaller-bodied individuals have been shown to be more selective during foraging bouts [77], and it is possible that this selectivity could aid in compensatory growth. Another potential strategy to aid in compensatory growth of smaller males is to feed for longer periods, which would yield greater forage intake, and potentially more nutrients [78]. The ability to compensate may be limited by forage availability and quality. Shorter growing seasons or poor forage quality may limit the ability to gain substantial mass [28,79], possibly yielding reduced mature body mass for some cohorts [80,81]. We may observe a much different pattern of permanent environmental effects (cohort effects) in populations that receive supplementation, meaning that bad years may be mitigated by the availability of high quality, supplemental feed [82]. Red deer in Spain partially compensated for negative density-dependent effects during early development when supplementation was available [65]. In southern Texas, supplementation again mitigated some of the natural environmental variation (e.g., variable precipitation patterns that influence forage availability and quality) on antler size in male white-tailed deer [82]. In this study population, supplemental feed enhancement (e.g., pelleted protein) was not provided; minimal agricultural plantings were available on surrounding properties prior to the study area being fenced because most rangeland was used for cattle grazing. However, in this population of deer, males overcame delayed physical development and were able to compensate for a body mass deficit at birth starting around 1–2 years of age. This may be partially due to intensive site-specific rangeland management practices (i.e., prescribed fire, rotational cattle grazing, selective herbicide treatment, and access to permanent sources of water) that were conducted each year to promote healthy ecosystems and were specifically designed to meet deer nutritional demands. Similar to feed supplementation dampening environmental variation [65,82], these management practices also could have dampened environmental variability. The last 9 of 10 years on this study area had the greatest predicted values (BLUP) for body mass in females, indicating the potential cumulative effectiveness of range management practices [30]. When female deer are in better condition (e.g., greater rump fat, body mass, etc.), offspring also are likely to be advantaged from an early age because maternal effects, from better body condition of the dam, will lead to positive pre- and post-natal effects on offspring [23,83].

Acknowledgements

We thank J. Holman for his numerous contributions to the project; all interns and volunteers that assisted during deer captures; and several anonymous reviewers for improving earlier drafts of this manuscript. Funding was provided by The Samuel Roberts Noble Foundation.

References

Five "key references", selected by the authors, are marked below (Three recommended (●) and two highly recommended (●●) papers).

1. Ditchkoff, S.S. 2011. Anatomy and physiology. In: D. G. Hewitt (ed), Biology and management of white-tailed deer. CRC Press, Boca Raton, Florida, pp. 43–73.
2. Albon, S.D., Clutton-Brock, T.H., & Langvatn, R. 1992. Cohort variation in reproduction and survival: implications for population demography. In: Brown, R.D. (ed), The biology of the deer. Springer-Verlag, New York, pp. 15–21. http://dx.doi.org/10.1007/978-1-4612-2782-3_2
3. Clutton-Brock, T.H., Guinness, F.E. & Albon, S.A. 1982. Red deer: behavior and ecology of two sexes. The University of Chicago Press, Chicago, Illinois.

4. McElligott, A.G., Gammell, M.P., Harty, H.C., Paini, D.R., Murphy, D.T., Walsh, J.T. & Hayden, T.J. 2001. Sexual size dimorphism in fallow deer (*Dama dama*): do larger, heavier males gain greater mating success? Behav. Ecol. and Sociobiol. 49: 266–272. http://dx.doi.org/10.1007/s002650000293

5. ● Jones, P.D., Demarais, S., Strickland, B.K. & DeYoung, R.W. 2011. Inconsistent relation of male body mass with breeding success in captive white-tailed deer. J. Mammal. 92: 527–533. http://dx.doi.org/10.1644/10-MAMM-A-294.1

6. Jones, P.D., Strickland, B.K., Demarais, S. & Blaylock, A.C. 2010. Reproductive characteristics of white-tailed deer in Mississippi. Southeast. Nat. 9: 803–812. http://dx.doi.org/10.1656/058.009.0414

7. Williams, J.D., Krueger, W.F. & Harmel, D.H. 1994. Heritabilities for antler characteristics and body weight in yearling white-tailed deer. Heredity 73: 78–83. http://dx.doi.org/10.1038/hdy.1994.101

8. Schultz, S.R. & Johnson, M.K. 1992. Antler development of captive Louisiana white-tailed bucks. P. Southeast. Assoc. Fish Wildlife Agencies 46: 67–74.

9. Schultz, S.R. & Johnson, M.K. 1995. Effects of birth date and body mass at birth on adult body mass of white-tailed deer. J. Mammal. 76: 575–579. http://dx.doi.org/10.2307/1382366

10. Lukefahr, S.D. & Jacobson, H.A. 1998. Variance component analysis and heritability of antler traits in white-tailed deer. J. Wildlife Manage. 62: 262–268. http://dx.doi.org/10.2307/3802287

11. McEwan, E.H. 1968. Growth and development of the barren-ground caribou. II. Postnatal growth rates. Can. J. Zool. 46: 1023–1029. http://dx.doi.org/10.1139/z68-143

12. Parker, K.L, Gillingham, M.P., Hanley, T.A. & Robbins, C.T. 1993. Seasonal patterns in body mass, body composition, and water transfer rates of free-ranging and captive black-tailed deer (*Odocoileus hemionus sitkensis*) in Alaska. Can. J. Zool. 71: 1397–1404. http://dx.doi.org/10.1139/z93-193

13. Gee, K.L., Holman, J.H., Causey, M.K., Rossi, A.N. & Armstrong, J.B. 2002. Aging white-tailed deer by tooth replacement and wear: a critical evaluation of a time-honored technique. Wildlife Soc. B. 30: 387–393.

14. Lewis, J.S. 2010. Factors influencing antler size in free-ranging white-tailed deer and mark/recapture estimates of demographic traits. Ph.D. dissertation, Texas A&M University-Kingsville.

15. Gee, K.L., Webb, S.L. & Holman, J.H. 2014. Accuracy and implications of visually estimating age of male white-tailed deer using physical characteristics from photographs. Wildlife Soc. B. 38: 96–102. http://dx.doi.org/10.1002/wsb.359

16. Causey, M.K. 1990. Fawning date and growth of male Alabama white-tailed deer. P. Southeast. Assoc. Fish Wildlife Agencies 44: 337–341.

17. Knox, W.M., Bara, M.O. & Miller, K.V. 1991. Effect of fawning date on physical development in yearling male white-tailed deer. P. Southeast. Assoc. Fish Wildlife Agencies 45: 30–36.

18. Shea, S.M., Breault, T.A. & Richardson, M.L. 1992a. Relationship of birth date and physical development of yearling white-tailed deer in Florida. P. Southeast. Assoc. Fish Wildlife Agencies 46: 159–166.

19. Gray II, W.N., Ditchkoff, S.S., Causey, M.K. & Cook, C.W. 2002. The yearling disadvantage in Alabama deer: effect of birth date on development. P. Southeast. Assoc. Fish Wildlife Agencies 56: 255–264.

20. Douhard, M., Gaillard, J.-M., DeLorme, D., Capron, G., Duncan, P., Klein, F. & Bonenfant, C. 2013. Variation in adult body mass of roe deer: early environmental conditions influence early and late body growth of females. Ecology 94: 1805–1814. http://dx.doi.org/10.1890/13-0034.1

21. Kruuk, L.E.B. & Hadfield, J.D. 2007. How to separate genetic and environmental causes of similarity between relatives. J. Evolution. Biol. 20: 1890–1903. http://dx.doi.org/10.1111/j.1420-9101.2007.01377.x

22. Kirkpatrick, M. & Lande, R. 1989. The evolution of maternal characters. Evolution 43: 485–503. http://dx.doi.org/10.2307/2409054

23. Mech, L.D., Nelson, M.E. & McRoberts, R.E. 1991. Effects of maternal and grandmaternal nutrition on deer mass and vulnerability to wolf predation. J. Mammal. 72: 146–151. http://dx.doi.org/10.2307/1381989

24. Wentworth, J.M., Johnson, A.S., Hale, P.E. & Kammermeyer, K.E. 1992. Relationships of acorn abundance and deer herd characteristics in the southern Appalachians. Southern J. Appl. Forestry 16: 5–8.

25. ● Strickland, B.K., Demarais, S. & Gerard, P.D. 2008. Variation in mass and lactation among cohorts of white-tailed deer *Odocoileus virginianus*. Wildlife Biol. 14: 263–271. http://dx.doi.org/10.2981/0909-6396(2008)14[263:VIMALA]2.0.CO;2

26. Strickland, B.K., Demarais, S., Jones, P.D. & Dacus, C.M. 2013. Phenotypic and reproductive variation in female white-tailed deer: the role of harvest and environment. J. Wildlife Manage. 77: 243–253. http://dx.doi.org/10.1002/jwmg.470

27. Campbell, S.A.B. & Wood, T.C. 2013. Influences of precipitation, temperature, and acorn mast on white-tailed deer body weight in the northern Piedmont of Virginia. Northeast. Nat. 20: 469–477. http://dx.doi.org/10.1656/045.020.0310

28. Shea, S.M., Breault, T.A. & Richardson, M.L. 1992b. Herd density and physical condition of white-tailed deer in Florida Flatwoods. J. Wildlife Manage. 56: 262–267. http://dx.doi.org/10.2307/3808821

29. Clutton-Brock, T.H., Stevenson, I.R., Marrow, P., MacColl, A.D., Houston, A.I. & McNamara, J.M. 1996. Population fluctuations, reproductive costs and life-history tactics in female soay sheep. J. Anim. Ecol. 65: 675–689. http://dx.doi.org/10.2307/5667

30. ●● Webb, S.L., Gee, K.L., DeYoung, R.W. & Harju, S.M. 2013. Variance component analysis of body mass in a wild population of deer (*Odocoileus virginianus*): results from two decades of research. Wildlife Res. 40: 588–598. http://dx.doi.org/10.1071/WR12224

31. Gee, K.L., Porter, M.D., Demarais, S., Bryant, F.C. & Vreede, G.V. 1994. White-tailed deer: their foods and management in the Cross Timbers. Second edition. The Samuel Roberts Noble Foundation Publication, Ardmore, Oklahoma.

32. Webb, S.L. & Gee, K.L. 2013. Annual survival and site fidelity of free-ranging white-tailed deer (*Odocoileus virginianus* Zimmermann): comparative demography before (1983–1992) and after (1993–2005) spatial confinement. Integrative Zool. 9: 24–33. http://dx.doi.org/10.1111/1749-4877.12032

33. Webb, S.L., Gee, K.L., Demarais, S., Strickland, B.K. & DeYoung, R.W. 2009. Efficacy of a 15-strand high-tensile electric fence to control white-tailed deer movements. Wildlife Biol. Practice 5: 45–57.

34. Ramsey, C.W. 1968. A drop-net deer trap. J. Wildlife Manage. 32:187–190. http://dx.doi.org/10.2307/3798257

35. Gee, K.L., Holman, J.H. & Crelia, D.. 2003. The Noble Foundation drop-net system for capturing white-tailed deer–the next generation. P. Southeast Deer Study Group 26: 27–28.

36. Severinghaus, C.W. 1949. Tooth development and wear as criteria of age in white-tailed deer. J. Wildlife Manage. 13: 195–216. http://dx.doi.org/10.2307/3796089

37. Gannon, W.L., Sikes, R.S. & the Animal Care and Use Committee of the American Society of Mammalogists. 2007. Guidelines of the American Society of Mammalogists for the use of wild mammals in research. J. Mammal. 88: 809–823. http://dx.doi.org/10.1644/06-MAMM-F-185R1.1

38. Littell, R.C., Milliken, G.A., Stroup, W.W., Wolfinger, R.D., & Schabenberger, O. 2006. SAS for Mixed Models. 2nd edn. SAS Institute, Cary, NC.

39. Burnham, K.P. & Anderson, D.R. 2002. Model selection and inference: a practical information-theoretic approach. 2nd edn. Springer, New York, New York.

40. Kenward, M.G. & Roger, J.H. 1997. Small sample inference for fixed effects from restricted maximum likelihood. Biometrics 53: 983–997. http://dx.doi.org/10.2307/2533558

41. Ozoga, J.J. & Verme, L.J. 1982. Physical and reproductive characteristics of a supplementally fed white-tailed deer herd. J. Wildlife Manage. 46: 281–301. http://dx.doi.org/10.2307/3808640

42. Bartoskewitz, M.L., Hewitt, D.G., Pitts, J.S. & Bryant, F.C. 2003. Supplemental feed use by free-ranging white-tailed deer in southern Texas. Wildlife Soc. B. 31: 1218–1228.

43. ●● Strickland, B.K. & Demarais, S. 2000. Age and regional differences in antlers and mass of white-tailed deer. J. Wildlife Manage. 64: 903–911. http://dx.doi.org/10.2307/3803198

44. Leberg, P.L, Smith, M.H. & Brisbin, Jr., I.L. 1992. Influence of sex, habitat and genotype on the growth patterns of white-tailed deer. In: Brown, R.D. (ed), The biology of the deer. Springer-Verlag, New York, pp. 343–350. http://dx.doi.org/10.1007/978-1-4612-2782-3_72

45. Asmus, J. & Weckerly, F.W. 2011. Evaluating precision of cementum annuli analysis for aging mule deer from southern California. J. Wildlife Manage. 75: 1194–1199. http://dx.doi.org/10.1002/jwmg.133

46. Fuller, T.K., Pace, III, R.M., Markl, J.A. & Coy, P.L. 1989. Morphometrics of white-tailed deer in north-central Minnesota. J. Mammal. 70: 184–188. http://dx.doi.org/10.2307/1381685

47. Anderson, A.E., Medin, D.E. & Bowden, D.C. 1974. Growth and morphometry of the carcass, selected bones, organs, and glands of mule deer. Wildlife Monogr. 39: 3–122.

48. Sand, H., Cederlund, G. & Danell, K. 1995. Geographical and latitudinal variation in growth patterns and adult body size of Swedish moose (*Alces alces*). Oecologia 102: 433–442. http://dx.doi.org/10.1007/BF00341355

49. Bender, L.C., Carlson, E., Schmitt, S.M. & Haufler, J.B. 2003. Body mass and antler development patterns of Rocky Mountain elk (*Cervus elaphus nelsoni*) in Michigan. Am. Midl. Nat. 150: 169–180. http://dx.doi.org/10.1674/0003-0031(2003)150[0169:BMAADP]2.0.CO;2

50. Hewison, A.J.M., Gaillard, J.-M., Delorme, D., van Laere, G., Amblard, T. & Klein, F. 2011. Reproductive constraints, not environmental conditions, shape the ontogeny of sex-specific

mass–size allometry in roe deer. Oikos 120: 1217–1226.
http://dx.doi.org/10.1111/j.1600-0706.2011.19316.x
51. Roseberry, J.L. & Klimstra, W.D. 1970. Productivity of white-tailed deer on Crab Orchard National Wildlife Refuge. J. Wildlife Manage. 34: 23–28.
http://dx.doi.org/10.2307/3799487
52. Hesselton, W.T. & Sauer, P.R. 1973. Comparative physical condition of four deer herds in New York according to several indices. New York Fish Game J. 20: 77–107.
53. Verme L.J. & Ozoga, J.J. 1987. Relationship of photoperiod to puberty in doe fawn white-tailed deer. J. Mammal. 68: 107–110.
http://dx.doi.org/10.2307/1381052
54. Leberg, P.L. & Smith, M.H. 1993. Influence of density on growth of white-tailed deer. J. Mammal. 74: 723–731.
http://dx.doi.org/10.2307/1382294
55. Verme, L.J. 1967. Influence of experimental diets on white-tailed deer production. N. Am. Wildlife Nat. Resour. Conference 32: 405–420.
56. Post, E. & Klein, D.R. 1999. Caribou calf production and seasonal range quality during a population decline. J. Wildlife Manage. 63: 335–345.
http://dx.doi.org/10.2307/3802517
57. Ozoga, J.J. & Verme, L.J. 1986. Relation of maternal age to fawn-rearing success in white-tailed deer. J. Wildlife Manage.50: 480–486.
http://dx.doi.org/10.2307/3801110
58. Mech, L.D. & McRoberts, R.E. 1990. Survival of white-tailed deer fawns in relation to maternal age. J. Mammal. 71: 465–467.
http://dx.doi.org/10.2307/1381963
59. Dusek, G.L., Mackie, R.J., Herriges, Jr., J.D. & Compton, B.B. 1989. Population ecology of White-tailed Deer along the Lower Yellowstone River. Wildlife Monogr. 104: 1–68.
60. Burke, R.L. & Birch, J.M. 1995. White-tailed deer vary offspring sex-ratio according to maternal condition and age. Ecol. Res. 10: 351–357.
http://dx.doi.org/10.1007/BF02347861
61. Carroway, C.J. & Broders, H.G. 2007. Adjustment of reproductive investment and offspring sex ratio in white-tailed deer (*Odocoileus virginianus*) in relation to winter severity. J. Mammal. 88: 1305–1311.
http://dx.doi.org/10.1644/06-MAMM-A-212R2.1
62. Armstrong, R.A. 1950. Fetal development of the northern white-tailed deer (*Odocoileus virginianus borealis*). Am. Midl. Nat. 43: 650–666.
http://dx.doi.org/10.2307/2421857
63. Hamilton, R.J., Tobin, M.L., & Moore, W.G. 1985. Aging fetal white-tailed deer. P. Southeast. Assoc. Fish Wildlife Agencies 39: 389–395.
64. Townsend, T.W. & Bailey, E.D. 1981. Effects of age, sex, and weight on social rank of penned white-tailed deer. Am. Midl. Nat. 106: 92–101.
http://dx.doi.org/10.2307/2425138
65. Mysterud, A., Langvatn, R. & Stenseth, N.C. 2004. Patterns of reproductive effort in male ungulates. J. Zool. 264: 209–215.
http://dx.doi.org/10.1017/S0952836904005618
66. Hewitt, D.G. 2011. Nutrition. In: D. G. Hewitt (ed), Biology and management of white-tailed deer. CRC Press, Boca Raton, Florida, pp.75–105.

67. Bonenfant, C., Gaillard, J.-M., Klein, F. & Loison, A. 2002. Sex- and age-dependent effects of population density on life history traits of red deer *Cervus elaphus* in a temperate forest. Ecography 25: 446–458.
http://dx.doi.org/10.1034/j.1600-0587.2002.250407.x

68. Rodriquez-Hidalgo, P., Gortazar, C., Tortosa, F.S., Rodriguez-Vigal, C., Fierro, Y. & Vicente, J. 2010. Effects of density, climate, and supplementary forage on body mass and pregnancy rates of female red deer in Spain. Oecologia 164: 389–398.
http://dx.doi.org/10.1007/s00442-010-1663-8

69. Solberg, E.J., Loison, A., Gaillard, J.-M. & Heim, M. 2004. Lasting effects of conditions at birth on moose body mass. Ecography 27: 677–687.
http://dx.doi.org/10.1111/j.0906-7590.2004.03864.x

70. Verme, L.J. 1963. Effect of nutrition on growth of white-tailed deer fawns. T. N. Am. Wildlife Confer. 28: 431–443.

71. Arman, P.R. Kay, N.B., Goodall, E.D. & Sharman, G.A.M. 1974. The composition and yield of milk from captive red deer (*Cervus elaphus* L.). J. Reprod. Fertil. 37: 67–84.
http://dx.doi.org/10.1530/jrf.0.0370067

72. Loudon, A.S.I., McNeilly, A.S. & Milne, J.A. 1983. Nutritional and lactational control of fertility in red deer. Nature 302: 145–147.
http://dx.doi.org/10.1038/302145a0

73. Lenart, E.A., Bowyer, R.T., Ver Hoef, J. & Ruess, R.W. 2002. Climate change and caribou: effects of summer weather on forage. Can. J. Zool. 80: 664–678.
http://dx.doi.org/10.1139/z02-034

74. Smart, A.J., Dunn, B.H., Johnson, P.S., Xu, L. & Gates, R.N. 2007. Using weather data to explain herbage yield on three Great Plains plant communities. Rangeland Ecol. Manage. 60: 146–153.
http://dx.doi.org/10.2111/05-099R4.1

75. McEwen, L.C., French, C.E., Magruder, N.D., Swift, R.W. & Ingram, R.H. 1957. Nutrient requirements of the white-tailed deer. N. Am. Wildlife Nat. Resour. Confer. 22: 119–132.

76. Ashley, E.P., McCullough, G.B. & Robinson, J.T. 1998. Morphological responses of white-tailed deer to a severe population reduction. Can. J. Zool. 76: 1–5.
http://dx.doi.org/10.1139/z97-159

77. Luna, R.S., Duarte, A. & Weckerly, F.W. 2013. Influence of body size on dietary nutrition of white-tailed deer *Odocoileus virginianus*. J. Fish Wildlife Manage. 4: 53–62.
http://dx.doi.org/10.3996/092012-JFWM-085

78. Watkins, W.G., Hudson, R.J. & Fargey, P.L.J. 1991. Compensatory growth of wapiti (*Cervus elaphus*) on aspen parkland ranges. C. J. Zool. 69: 1682–1688.
http://dx.doi.org/10.1139/z91-233

79. Keyser, P.D., Guynn, Jr., D.C. & Hill, Jr., H.S. 2005. Population density – physical condition relationships in white-tailed deer. J. Wildlife Manage. 69: 356–365.
http://dx.doi.org/10.2193/0022-541X(2005)069<0356:PDCRIW>2.0.CO;2

80. Suttie, J.M., Goodall, E.D., Pennie, K. & Kay, R.N.B. 1983. Winter food restriction and summer compensation in red deer stags (*Cervus elaphus*). Brit. J. Nutr. 50: 737–747.
http://dx.doi.org/10.1079/BJN19830145

81. Pettorelli, N., Gaillard, J.-M., Mysterud, A., Duncan, P., Stenseth, N.C., Delorme, D., Van Laere, G., Toïgo, C. & Klein, F. 2006. Using a proxy of plant productivity (NDVI) to find key periods for animal performance: the case of roe deer. Oikos 112: 565–572.
http://dx.doi.org/10.1111/j.0030-1299.2006.14447.x

82. ● Foley, A.M., DeYoung, R.W., Lukefahr, S.D., Lewis, J.S., Hewitt, D.G., Hellickson, M.W., Draeger, D.A. & DeYoung, C.A. 2012. Repeatability of antler characteristics in mature white-tailed deer in south Texas: consequences of environmental effects. J. Mammal. 93: 1149–1157.
http://dx.doi.org/10.1644/11-MAMM-A-183.2

83. Monteith, K.L., Schmitz, L.E., Jenks, J.A., Delger, J.A. & Bowyer, R.T. 2009. Growth of male white-tailed deer: consequences of maternal effects. J. Mammal. 90: 651–660.
http://dx.doi.org/10.1644/08-MAMM-A-191R1.1

Supporting Information

Appendix 1: Descriptive statistics (mean, sample size, standard error [SE], and 95% confidence intervals [CI]) for age-specific body mass (kg) of wild, known-aged female (A) and male (B) white-tailed deer (*Odocoileus virginianus*) in southern Oklahoma that were harvested during October and November, 1986–2005 (preceding recapture events as compared to Table 1).

A

	Age (years)					
Parameter	**1-2**	**2-3**	**3-4**	**4-5**	**5-6**	**6-7**
Mean	47.29	48.12	49.55	51.26	46.27	57.61
n	8	12	4	2	1	1
SE	1.26	1.82	2.16	3.18	--	--
Lower 95% CI	44.31	44.11	42.67	--	--	--
Upper 95% CI	50.26	52.13	56.44	--	--	--

B

	Age (years)				
Parameter	**1-2**	**2-3**	**3-4**	**4-5**	**5-6**
Mean	53.52	63.41	71.67	83.46	85.28
n	1	5	1	1	1
SE	--	3.07	--	--	--
Lower 95% CI	--	54.89	--	--	--
Upper 95% CI	--	71.94	--	--	--

13

Capture Efficiency and Size Selectivity of Sampling Gears Targeting Ladyfish (*Elops Saurus*) in Florida Waters (USA)

J. C. Levesque[1]

[1] Geo-Marine, Inc., Environmental Resources Division, Marine Science Department, 2201 Avenue K, Suite A2, Plano, Texas 75074 USA. Present address: Environmental Resources Management, 10210 Highland Manor Drive, Suite 140, Tampa, Florida (USA); email: Juan.Levesque@erm.com or shortfin_mako_shark@yahoo.com

Keywords

Fish Sampling;
Gear Selectivity;
Length Frequency;
Relative Abundance.

Abstract

Marine fish are among the most challenging type of wildlife to examine in the field given their expansive range, spatial and temporal distribution, and specific early life-history traits. Different researchers have investigated the catch efficiency of various sampling gears and techniques; however, field-sampling information is mostly unavailable for ladyfish (*Elops saurus*), an economically valuable species in the southeastern United States. The main purpose of this investigation was to evaluate, for the first time, various sampling gears potentially useful for targeting ladyfish. The specific objectives were to examine and compare differences in capture efficiency and size selectivity of different sampling gears useful for collecting ladyfish in nine Florida water bodies. This investigation found that ladyfish relative abundance, size, and length-frequency distribution varied significantly by sampling gear and geographical location. The findings revealed the most efficient gear for collecting ladyfish smaller than 100 mm SL were seines with an offshore deployment method and the least effective were otter trawls. Overall, seine catch efficiency decreased and gillnet efficiency increased with ladyfish size. These finding indicate it's essential that researchers use a variety of sampling gears to reduce any potential gear or sampling bias when designing field studies.

Introduction

To elucidate, monitor, and predict wildlife patterns in space and time requires the selection of specific sampling gears for the community or species of interest [1]. Marine fish are among the most challenging type of wildlife to examine in the field given their expansive range, spatial and temporal distribution, and specific early life-history characteristics (e.g., habitat, environmental conditions, and recruitment patterns). To examine the early life-history of fish, researchers use either active or passive field sampling methods, and a variety of field gears and procedures [1]. Researchers must choose not only the most appropriate sampling method, but they also need to consider the selectivity of the sampling gear when designing field studies [1, 2]. For instance, various researchers have reported statistical differences in length-frequency distribution and catch efficiency for several types of field sampling gear used to collect fish [3–6]. Capture efficiency is a complex dynamic that includes the sampling gear, technique, and the availability/vulnerability of the target species [e.g., 7–9]. Capture efficiency and size selectivity bias is problematic for researchers because it can lead to under or over estimating a population [10]. It can also affect estimates of various population factors, such as recruitment, size structure, and natural morality. Thus, field-sampling (gear, methods, and techniques) should be at the core of applied fisheries research in terms of estimating relative abundance and determining the status of a stock.

Ladyfish are an important commercial and recreational species in Florida and other worldwide regions [11, 12]. Despite their socio-economic importance, the basic concepts of collecting ladyfish have not been examined nor reported, such as the capture efficiency or size selectivity of field-sampling gears. Thus, the purpose of this investigation was to evaluate, for the first time, various sampling gears potentially useful for targeting ladyfish in Florida waters. The specific objectives were to examine and compare differences in capture efficiency and size selectivity of different sampling gears useful for collecting ladyfish in nine Florida water bodies.

Methods

Study area

Field collections were conducted in nine different water bodies, including several major rivers encompassing a variety of bottom, shoreline, and offshore habitats throughout the State of Florida, USA (Fig. 1). Sampling sites (Figs. 2–7) were located in Charlotte Harbor (CH), Choctawhatchee Bay and Santa Rosa Sound (CB-SRS),

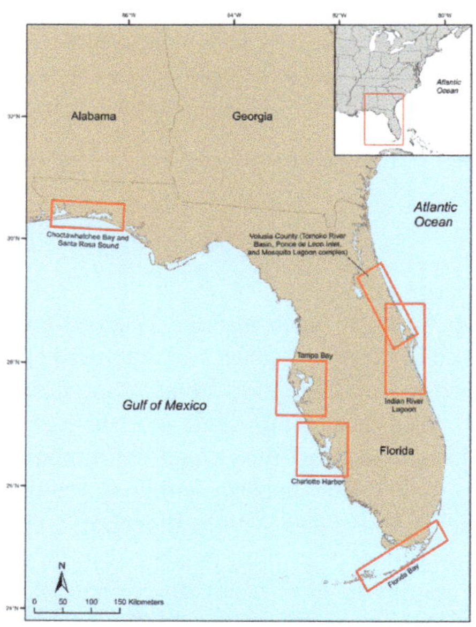

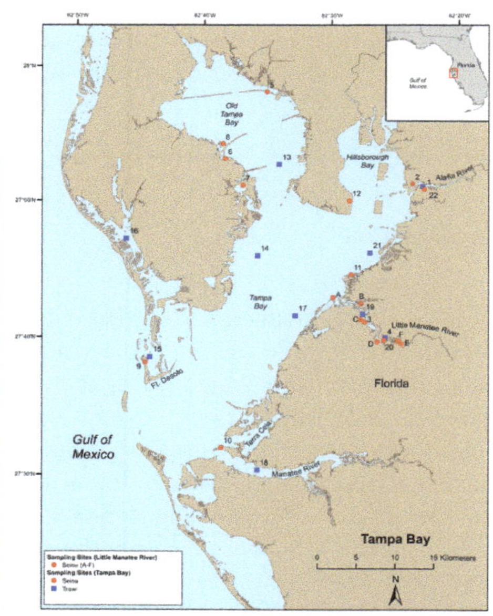

Fig. 1 (*left*): Map of field sampling locations in Florida (USA) during 1987 through 1995.

Fig. 2 (*right*): Map of fixed-station sampling sites in Tampa Bay during September 1987 through December 1995.

Florida Bay (FB)Indian River Lagoon (IRL), Little Manatee River (LMR), Tampa Bay (TB), and Volusia County (VC [Tomoko River Basin, Ponce de Leon Inlet, and Mosquito Lagoon complex]). Specific site descriptions are provided by [13–20].

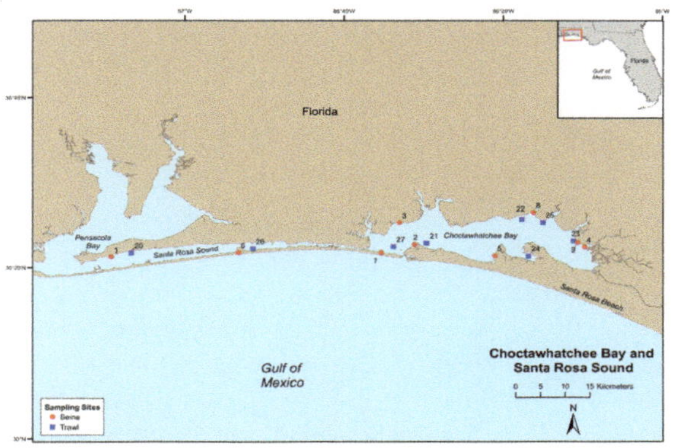

Fig. 3: Map of fixed-station sampling sites in Choctawhatchee Bay and Santa Rosa Sound during October 1992 through December 1995.

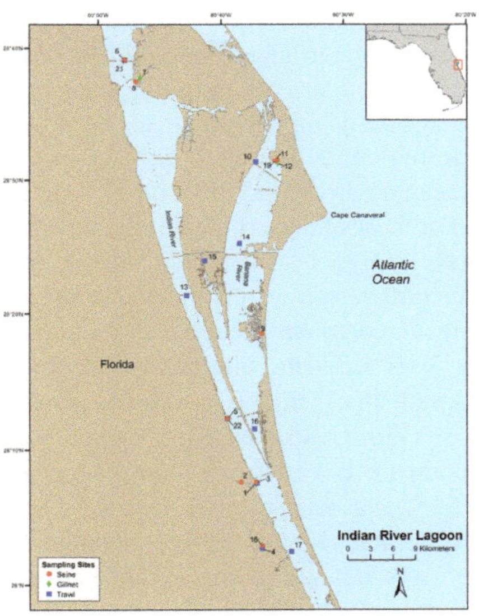

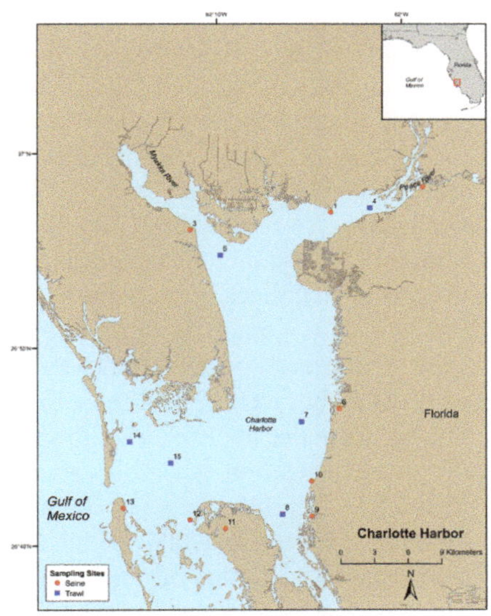

Fig. 4 (*left*): Map of fixed-station sampling sites in Charlotte Harbor during January 1991 through December 1995.

Fig. 5 (*right*): Map of fixed-station sampling sites in the Indian River Lagoon during January 1991 through December 1995.

Experimental Design, Sampling Methodology and Gear

Field sampling was conducted by Florida Fish and Wildlife Conservation Commission (FWC), Fisheries Independent Monitoring Program (FIM) personnel as part of a large-scale, statewide, multigear monitoring program designed to assess the population status and community structure of estuarine fishes throughout Florida [21]. The FIM used two different types of sampling approaches to provide comprehensive

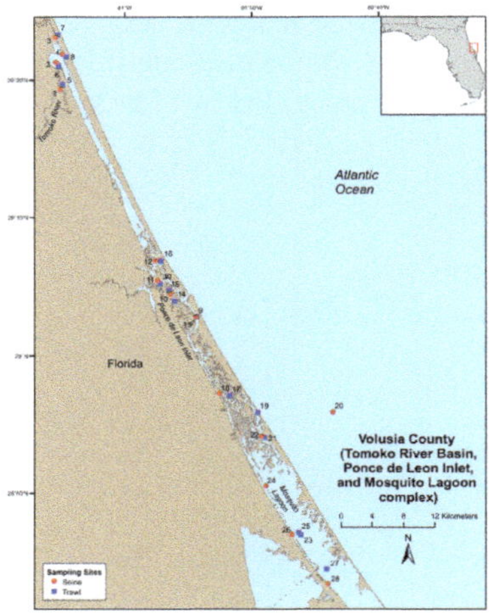

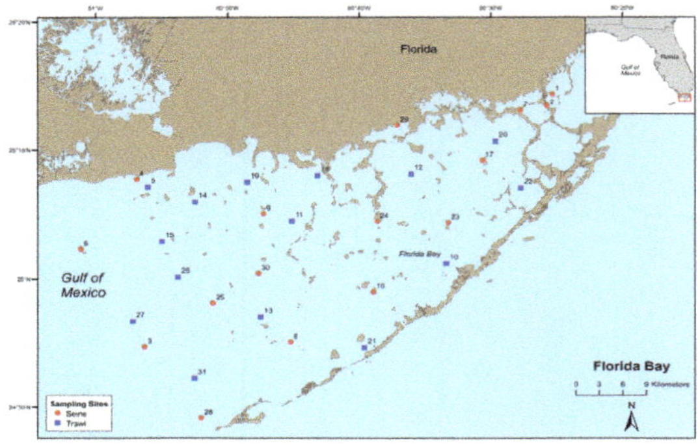

Fig. 6 (*left*): Map of fixed-station sampling sites in Volusia County during January 1993 through December 1995.

Fig. 7 (*right*): Map of fixed-station sampling sites in Florida Bay during January 1994 through December 1995.

distribution data on fishes found throughout Florida: Stratified-Random Sampling (SRS) and Fixed-Station Sampling (FSS). Under both sampling scenarios, each individual water body was divided into zones based on geographical, logistical, and hydrological characteristics [21]. Zones were defined as either bay or riverine, and stratified by depth and habitat (e.g., depth, seagrass beds, and shore type); riverine zones were also stratified by salinity gradient. For these analyses, data were restricted to monthly FSS collections because preliminary analyses of SRS data showed that most ladyfish collected under the SRS experimental design were collected primarily with only one gear type (gillnets), which prevented making comparisons among gears. Preliminary analyses also showed that most ladyfish collected with gillnets were not only larger (> 300 mm Standard Length [SL]) than those collected with other gears under the FSS, but the size-range was more restricted. Since the purpose of this investigation was to evaluate and compare a variety of sampling gears potentially useful for targeting ladyfish, the FSS data was selected because it was more appropriate and wide-ranging than the SRS data. Monthly FSS was conducted at pre-determined sites during the day, which was defined as the period between one hour after sunrise and one hour before sunset. Annual field collections varied slightly among geographical location: (TB [1987–1995], LMR [1988–1991], IRL [1991–1995], CH [1991–1995], CB-SRS [1992–1995], VC [1993–1995], and FB [1994–1995]).

The FIM monitored estuarine fishes throughout Florida using a suite of sampling gears (e.g., seines [small and large mesh], trawls, gillnets, and blocknets). The type of sampling gear the Agency selected was dependent upon the profile of the shoreline bank, water depth, and habitat (Table 1). Sampling gears were deployed using standardized procedures, deployment methods, and techniques. The number of repetitions (i.e., hauls) made at each site varied from one (blocknet) to four (gillnet), which was based on the gear type. To reduce gear bias (avoidance), sampling gears were used in specific habitats (e.g., shallow open-water and deep open-water environments). Depending on the gear, deployment method, and the number of hauls, the total sampling area varied (e.g., center-bag seine, beach set [340 m^2], otter trawl [1,130 m^2], and haul seine [4,120 m^2].

The primary gear the FIM used to monitor and assess estuarine fishes throughout Florida was a 21.3 m center-bag seine. Seines were deployment using one of three methods (beach, boat, and offshore), which depended on the shoreline bank profile, and water depth. A beach deployment was used when the water depth was shallow (< 1.5 m) and the bank had a gradual decline (< 45°). The beach deployment consisted of the seine being pulled parallel to shore by two biologists for a total distance of 9.1 m; a 15.5 m line stretched between each seine pole assured the net was being pulled the same inner-pole distance for every haul [21]. A boat deployment was used at sites with deep waters or when the bank profile was to steep to use a beach deployment. The boat deployment consisted of deploying the seine from the stern of the vessel in a semi-circular or rectangular pattern along the bank. Once the seine was fully deployed, two biologists would quickly work the seine bag on the shore. An offshore deployment was used when there was no apparent beach, or it was too shallow to reach the beach or bank by boat. The offshore deployment followed the same procedures as the beach deployment with one minor difference; at the end of the 9.1 m distance, two biologists worked the center-bag seine using a stationary pivot pole to ensure the catch did not escape [21].

The FIM also used blocknets, otter trawls, and gillnets to monitor and assess estuarine fishes [21]. A 61 m block net was used to monitor estuarine fishes along man-made seawalls or mangroves at high tide; fishes were collected at the ensuing low tide. A 6.1 m otter trawl was used to monitor demersal and cryptic estuarine fishes in deep waters. The otter trawl was towed at approximately 0.6 m/s for 5 and 10 min in river and bay zones, respectively [21]. To monitor open-water estuarine fishes, the FIM used two different size monofilament gillnets (184 and 198 m) constructed with varying stretch-mesh panels (50, 75, 100, 125, and 150 mm). At each site, the FIM simultaneously set four nets perpendicular to shore. To reduce any potential sampling bias, two gillnets with the larger stretch-mesh (150 mm) and two gillnets with the smaller stretch-mesh (50 mm) were set and oriented toward the shore.

Data Analyses

To test for normality and homoscedacity, each dataset was evaluated using Kolmogorov–Smirnov's [22] and Bartlett's tests [23, 24]. Normality was also checked by constructing a normal probability plot of the residuals to ensure robustness. As part of the data assessment process, all outlier observations were investigated using descriptive statistics and graphical exploratory techniques before being rejected or retained [25]. Outlier observations were discarded only on the basis that the data were in error or the data no longer represented valid observations of the original sample [25]. If the datasets passed the normality test, parametric procedures were employed; otherwise, data were transformed using an appropriate transformation process (e.g. log, square root, or arcsine square root) to meet the underlying assumptions of normality [22]. However, if the data failed to meet the assumptions of normality even after transformation, then non-parametric tests were applied. For all

Table 1: Total number of juvenile ladyfish collected and description of gear types used throughout Florida estuaries (Tampa Bay, Indian River Lagoon, Volusia County, Little Manatee River, Charlotte Harbor, Florida Bay) during 1987 through 1995.

Total Number of Juvenile Ladyfish Collected	Total Number of Hauls	Gear Number	Gear Description
941	2,728	23	21.3 m center-bag seine, 3.1 mm mesh, leads spaced every 150 mm, boat set
728	2,074	22	21.3 m center-bag seine, 3.1 mm mesh, leads spaced every 150 mm, onshore set without boat (beach set)
618	1,335	10	21.3 m center-bag seine, 3.1 mm mesh, 300 mm mesh lead spacing, onshore set, deployed from boat (boat set)
398	273	160	183 m center-bag seine, green dipped, 37.5 mm stretch-mesh (18.8 mm bar mesh), 150 mm lead spacing. Net set from skiff in a rectangular pattern on shore in water depth < 0.5 m
166	3,375	20	21.3 m center-bag seine, 3.1 mm mesh, leads spaced every 150 mm, offshore-circular set
157	6,297	300	6.1 m otter trawl w/ 3.1 mm liner & tickler chain. A straight tow
153	696	209	198 m large mesh gillnet with five panels: 15 m of 50 mm mesh (#139 twine), and 46 m each of 75 mm, 100 mm, 125 mm, and 150 mm mesh (#208 twine). NOT set in association with reverse set (large mesh to shore) gillnet. Normal set: Small mesh to shore
115	1,460	4	12.2 m center bag seine, 6.3 mm mesh, 1.2 m deep
97	51	159	183 m center-bag seine 25.4 mm stretch mesh (12.7 mm bar mesh), 150 mm lead spacing, set from skiff onshore (boat set)
95	861	204	184 m large mesh gillnet with four panels: 46 m each of 75 mm, 100 mm, 125 mm, and 150 mm mesh. Normal set: Small mesh to shore
77	137	153	61 m blocknet, 300 mm lead spacing. Used in association with mangroves
71	118	154	61 m blocknet, 300 mm lead spacing. Used in association with seawall
48	245	12	21.3 m center-bag seine, 3.1 mm mesh, 300 mm lead spacing, onshore without boat (beach set)
15	77	3	6.1 m center bag seine, 1.6 mm mesh, 1.2 m deep
9	8	157	183 m center-bag seine 25.4 mm stretch mesh (12.7 mm bar mesh), 150 mm lead spacing. Net set from skiff in a semicircular pattern onshore in water depth < 0.1 m
1	535	13	21.3 m terminal-bag seine, 3.1 mm mesh, leads spaced every 300 mm, offshore-circular set
1	420	301	6.1 m otter trawl w/ 3.1 mm liner & tickler chain. An arc tow
1	62	101	21.3 m terminal-bag seine, 300 mm lead spacing, offshore-circular set in association with seawall blocknet sets
0	79	1	6.1 m seine w/o bag, 3.1 mm mesh
0	81	2	6.1 m center bag seine, 3.1 mm mesh, leads spaced at 150 mm. Modified beach set
0	126	11	21.3 m center-bag seine, 3.1 mm mesh, 300 mm lead spacing, offshore set
0	72	20	21.3 m terminal-bag seine, 3.1 mm mesh, leads spaced every 150 mm, offshore-circular set
0	6	24	21.3 m center-bag seine, 3.1 mm mesh, 150 mm lead spacing, w/many ends line, onshore set without boat (beach set)
0	87	100	21.3 m terminal-bag seine, 300 mm lead spacing, offshore-circular set in association with mangrove blocknets
0	12	102	21.3 m terminal-bag seine, 150 mm lead spacing, offshore-circular set in association with seawall blocknets sets
0	12	103	21.3 m terminal-bag seine, 150 mm lead spacing, offshore-circular set in association with mangrove blocknet sets
0	18	104	21.3 m center-bag seine, 150 mm lead spacing, offshore-circular set in association with seawall blocknet sets
0	6	158	183 m center-bag seine 25.4 mm stretch-mesh (12.7 mm bar mesh), 150 mm lead spacing. Net set from skiff in a semicircular pattern onshore in water depth > 0.1 m
0	1	161	183 m center-bag seine, green dipped, 37.5 mm stretch-mesh (18.8 mm bar mesh), 150 mm lead spacing. Net set in a rectangular pattern onshore with water depth < 0.1m
0	3	306	6.1 m otter trawl w/ 3.1 mm liner & tickler chain. A combination tow (straight trawl/arc boat)

analyses, statistical significance was defined as $P < 0.05$. All analyses were conducted using Microsoft Excel® and SYSTAT® version 12.

To account for unequal sampling (area and effort), and permit comparison of nominal catch rates among gear types, catches of ladyfish were calculated and scaled (i.e., weighted proportionally) by the total area sampled for each gear type. To standardized sampling effort and area, catch was expressed as the number of ladyfish collected per unit area following the procedures by Wells *et al.* [9]. The catch-per-unit-area (CPUA) was calculated using the following formula:

$$\mathbf{CPUA = (catch_i / area\ sampled_i) / (\Sigma\ \ catch_i / \Sigma\ \ area\ sampled_i)}$$

where CPUA was expressed as a proportion of the catch per area sampled over all gears.

To determine whether capture efficiency differed among gears, nominal ladyfish catch rates (relative abundance) for each water body were standardized using the CPUA metric. It should be noted that the experimental design required several assumptions that could affect capture efficiency and size bias. For these analyses, it was assumed that all temporal/spatial biological (e.g., availability of prey) and environmental (e.g., water temperature, salinity, and dissolved oxygen) variables and interactions were equal within an area (water body). It was also assumed that capture probability on the basis of immigration and emigration was equal among water bodies. Using these general assumptions, separate Kruskal–Wallis tests were used to test the null hypothesis that mean ranks of ladyfish relative abundance, based on the metric CPUA, was equal among sampling gears by individual water body since the aim of the study was not to compare catches using habitat or any biological/environmental condition as a factor. Because each water body had unique habitats and environmental conditions, it was inappropriate to pool the data; thus, gear comparisons were made separately and reported by individual water body. Separate Kruskal–Wallis tests were also used to test the null hypothesis that mean ranks of ladyfish length was equal among gears by individual water body under the same assumptions as described above. A likelihood-ratio Kolmogorov-Smirnov two-sample test was used to test the null hypothesis that ladyfish length frequency-distribution, binned by 10 mm size classes, was equal between the two sampling gears with the greatest catches of ladyfish within each water body. A Kolmogorov-Smirnov two-sample test was also used to test the null hypothesis that ladyfish length-frequency distribution was equal between the two of the largest estuaries (Tampa Bay and Indian River Lagoon) in Florida.

Results

Data from 22,674 hauls and 3,692 ladyfish collections were used to evaluate and compare sampling gears in nine water bodies throughout Florida during 1987 through 1995 (Fig. 2). A total of 1,401 hauls were conducted at twelve stations over a 4-year period in the CB-SRS; however, since no ladyfish were collected these data were not considered further in this investigation.

Ladyfish were collected with a variety of field sampling gears, but some gear and methods were more successful than others (Table 1). Eighty-four percent ($n = 3{,}107$) of ladyfish were collected with seines, 8 percent ($n = 279$) with gillnets, 4 percent ($n = 148$) with blocknets, and another 4 percent ($n = 158$) with otter trawls. The mean relative abundance (CPUA) ranged from 4.58 x 10-14 ($n = 6720$) ladyfish per haul with otter trawl gear to 8.04 x 10-14 (n = 255) ladyfish per haul with blocknet gear. A significant difference in CPUA was detected by individual sampling gear and geographical location (CH [Kruskal–Wallis = 79.037, $P < 0.05$], IRL [Kruskal–Wallis = 184.379, $P < 0.05$], FB [Kruskal–Wallis = 7.3121, $P = 0.03$], LMR [Kruskal–Wallis = 7.2478, $P = 0.007$], TB [Kruskal–Wallis = 705.625, $P < 0.05$], and VC [Kruskal–Wallis = 13.5522, $P = 0.003$]), so findings were reported as bay-specific. Findings were also reported as bay-specific because sampling effort, habitat, and environmental conditions differed among location, which prevented pooling the data.

Ladyfish ranged from 1 to 300 mm SL ($n = 2{,}648$) with a mean of 93.3 mm SL

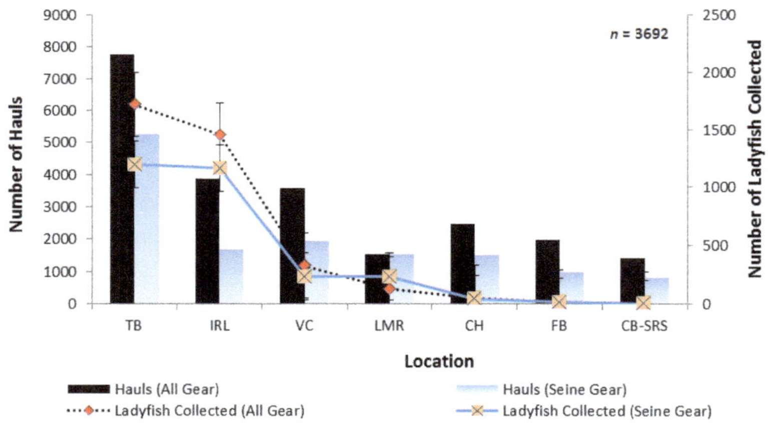

Fig. 8: The total number of hauls completed of ladyfish collected in Florida estuaries during 1987 through 1995. Tampa Bay (TB), Indian River Lagoon (IRL), Volusia County (VC), Charlotte Harbor (CH), Florida Bay (FB), and Choctawhatchee Bay and Santa Rosa Sound (CB-SRS). Mean ± Standard error are plotted.

(S.E. ± 1.7). The median and mode were 52.0 and 35.0 mm SL, respectively. The smallest specimens were collected with a 21.3 m center-bag seine and the largest with a 183m haul seine. Ladyfish collected with a 21.3 m center bag seine ranged from 1 to 291 mm SL with a mean of 60.0 mm SL (S.E. ± 1.1), while collections of ladyfish with a 183 m haul seine ranged from 131 to 300 mm SL with a mean of 264 mm SL (S.E. ± 4.7). Ladyfish collected with gillnets ranged from 3 to 298 mm SL with a mean of 114 mm SL (S.E. ± 2.1), and collections with otter trawls ranged from 2 to 276 mm SL with a mean of 15 mm SL (S.E. ± 0.27). A significant difference in size was detected for some sampling gears, methods, and geographical locations (CH [Kruskal–Wallis = 4.3374, $P = 0.22$], IRL [Kruskal–Wallis = 59.9191, $P < 0.05$], LMR [Kruskal–Wallis = 1.0044, $P = 0.32$], TB [Kruskal–Wallis = 799.289, $P < 0.05$], and VC [Kruskal–Wallis = 6.3773, $P = 0.04$]); thus, findings were reported as bay-specific.

Sampling Gear Efficiency

In TB, a total of 1,718 ladyfish were collected with various sampling gears during 1987 through 1995. Eighty-nine percent of ladyfish (n = 1,533) were collected with five types of gears and deployment methods (gear 10 [n = 599 or 35%], gear 160 [n = 398 or 23%], gear 23 [n = 363 or 21%], gear 155 [n = 97 or 7%], and gear 153 [n = 77 or 4%]). The average rank in relative abundance (CPUA) ranged from 3562.5 (n = 51) ladyfish per haul with gear 300 to 4828.2 (n = 51) ladyfish per haul with gear 155. A Kruskal–Wallis test detected there was a significant difference in mean ranks of CPUA among sampling gears and methods ([23, 7501] = 706.595, $P < 0.05$).

A total of 1,454 ladyfish were collected in the IRL during 1991 through 1995. Most ladyfish (98% or n = 1,430) were collected with five different sampling gears and deployment methods (gear 22 [n = 589 or 41%], gear 23 [n = 455 or 31%], gear 209 [n = 153 or 11%], gear 20 [n = 138 or 9%], and gear 204 [n = 95 or 7%]). The average rank in relative abundance (CPUA) ranged from 1750.66 (n = 589) ladyfish per haul with gear 300 to 2202.1 ($n = 253$) ladyfish per haul with gear 23. A Kruskal–Wallis test detected there was a significant difference in mean ranks of CPUA among sampling gears and methods ([9, 3857] = 184.379, $P < 0.05$).

In VC, a total of 326 ladyfish were collected during 1993 through 1995. Most juvenile ladyfish were collected with seines (gear 22 [n = 121 or 37%] and gear 23 [n = 89 or 27%]) and otter trawl (gear 300 [n = 116 or 36%]). The average rank in relative abundance (CPUA) ranged from 1861.5 ($n = 42$) ladyfish per haul with gear 301 to 1961.46 ($n = 880$) ladyfish per haul with gear 22 (Tables 1, 2). A Kruskal–Wallis test showed there was a significant difference in mean ranks of CPUA among sampling gears and methods ([4, 3857] = 13.5522, $P = 0.003$).

A total of 130 ladyfish were collected in the LMR during 1988 through 1995. Ladyfish were collected with two types of seine (3 [n = 78 or 60%]; 4 [n = 52 or 40%]). The average rank in relative abundance (CPUA) ranged from 767.023 (n = 1460) ladyfish per haul with gear 4 to 806.481 ($n = 77$) ladyfish per haul with gear 3 (Tables 1, 2). A Kruskal–Wallis test indicated there was a significant difference in mean ranks of CPUA among sampling gears and methods ([1, 1537] = 7.2478, P = 0.007).

In CH, a total of 50 ladyfish were collected during 1991 through 1995. Ladyfish were collected with three types of seines (gear 23 [n = 35 or 70%], gear10 [n = 9 or 18%], and gear 22 [n = 5 or 10%]) and otter trawl gear (gear 300 [n = 1 or 2%]). The average rank in relative abundance (CPUA) ranged from 1209 (n = 79, 81, 9, and 33) ladyfish per haul with gears 1, 2, 11, and 12 to 1315.38 ($n = 69$) ladyfish per haul with gear 10 (Tables 1, 2). A Kruskal–Wallis test demosntrated there was a significant difference in mean ranks of CPUA among sampling gears and methods ([9, 2446] = 79.0307, $P < 0.05$).

Only 14 ladyfish were collected in FB during 1994 through 1995; most were collected with a seine (gear 20). The average rank in relative abundance (CPUA) ranged from 981.5 (n = 641 and 298) ladyfish per haul with gears 300 and 301 to 989.143 ($n = 1031$) ladyfish per haul with gear 20. A Kruskal–Wallis test detected there was a weak significant difference in mean ranks of CPUA among sampling gears and methods ([2, 1970] = 7.3121, $P = 0.03$).

Sampling Gear Selectivity

Overall, average length and length-frequency distribution of ladyfish varied by sampling gear and water body. In TB, ladyfish ranged from 20 to 300 mm SL (Fig. 9) with a mean of 134.62 mm SL (S.E. ± 2.79). The smallest ladyfish (60.88 mm SL [S.D. ± 38.8, $n = 668$]) were captured with gear 23 and the largest (276.84 mm SL [S.D. ± 18.83, $n = 39$]) with gear 155. A Kruskal–Wallis test detected there was a significant difference in mean ranks of ladyfish length among sampling gears and methods ([12, 1139] = 799.289, $P < 0.05$). A Kolmogorov-Smirnov two-sample test showed there was a significant difference in ladyfish length-frequency distribution between the two gears that collected the most ladyfish (10 and 160). The smallest ladyfish (60.88 mm SL, S.D. ± 38.8) were collected with gear 10 and largest (264.1 mm SL, S.D. ± 24.7) with gear 160 (D [2, 590] = 0.9790, $P < 0.05$).

Ladyfish collected in the IRL ranged from 1 to 298 mm SL (Fig. 9) with a mean of 66.34 mm SL (S.E. ± 1.94). The smallest ladyfish (42.82 mm SL [S.D. ± 44.23, n

= 124]) were captured with gear 20 and the largest (122.28 mm SL [S.D. ± 111.53, n = 100]) with gear 209. A Kruskal–Wallis test indicated there was a significant difference in mean ranks of ladyfish length among sampling gears and methods ([8, 997] = 59.9191, $P < 0.05$). A Kolmogorov-Smirnov test showed there was a significant difference in ladyfish length-frequency distribution between the two gears that collected the most ladyfish (22 and 23). The smallest ladyfish (59.58 mm SL, S.D. 26.1) were collected with gear 23 and largest (61.0 mm SL, S.D. ± 49.4) with gear 22 (D [2, 702] = 0.2829, $P < 0.05$).

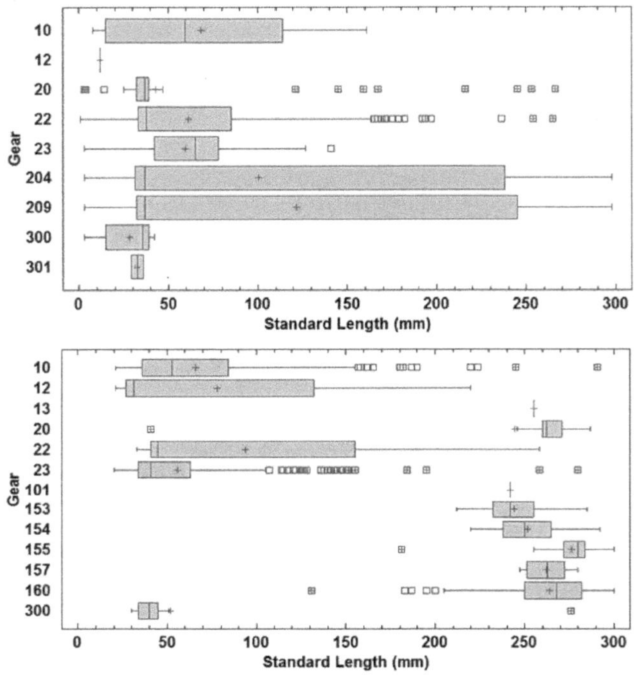

Fig. 9: Ladyfish length-frequency distribution in Tampa Bay (Top; 1987–1995) and Indian River Lagoon (Bottom; 1991–1995).

In VC, the ladyfish ranged from 2 to 258 SL mm SL with a mean of 39.87 mm SL (S.E. ± 2.04). The smallest ladyfish (36.58 mm SL [S.D. ± 35.8, n = 115]) were captured with gear 300 and the largest (43.78 mm SL [S.D. ± 36.6, n = 92]) with gear 22. A Kruskal–Wallis test detected there was a significant difference in mean ranks of ladyfish length among sampling gears and methods ([2, 296] = 6.3773, $P = 0.04$). A Kolmogorov-Smirnov test showed there was a significant difference in ladyfish length-frequency distribution between the two gears that collected the most ladyfish (22 and 23). The smallest ladyfish (40.69 mm SL, S.D. ± 34.7) were collected with gear 23 and largest (43.78 mm SL, S.D. ± 36.6) with gear 22 (D [2, 181] = 0.3042, $P < 0.05$).

Ladyfish collected in the LMR ranged from 1 to 295 mm SL with a mean of 68.05 mm SL (S.E. ± 4.59). A two-sample t-test indicated there was no significant difference in mean ladyfish length by gear (t [128] = -0.63, $P = 0.533$). The smallest ladyfish (60.07 mm SL, S.D. ± 45.5, n = 15) were collected with gear 22 and largest (69.09 mm

SL, S.D. ± 53.3, n = 115) with gear 23. A Kolmogorov-Smirnov test showed there was no significant difference in ladyfish length-frequency distribution between gears 22 and 23 [D (2, 130) = 0.2435, P = 0.361.

In CH, ladyfish ranged from 1 to 285 mm SL with a mean of 58.96 mm SL (S.E. ± 9.19). The smallest ladyfish (47.59 mm SL [S.D. ± 39.34, n = 44]) was captured with gear 23 and the largest (117.2 mm SL [S.D. ± 136.01, n = 5]) with gear 22. A Kruskal–Wallis test detected there was no significant difference in mean ranks of ladyfish length among sampling gears and methods ([3, 50] = 4.3374, P = 0.22).

Ladyfish length for individuals collected in FB ranged from 4 to 252 mm SL with a mean of 98.1 mm SL (S.E. ± 25.73). Ladyfish were collected primarily with only one gear (20), so no other tests were necessary.

Discussion

This study demonstrated that the total number and sizes of ladyfish collected varied by water body, gear type, and gear deployment method. Although each water body was stratified (e.g., bottom vegetation, water depth, and salinity) and sampled using standardize procedures [16], the findings demonstrated that researchers collected fewer ladyfish in some locations (Choctawhatchee Bay and Santa Rosa Sound, Charlotte Harbor, and Florida Bay) than in others (Tampa Bay and Indian River Lagoon). It is doubtful that the sampling gear or deployment method was the explanation since standardized gear and methods were used systematically in each water body. Thus, it is probable that the reason fewer ladyfish were collected in some locations than others was related to either restricted recruitment or inappropriate sampling site location. Based on other confronting factors and environmental conditions (e.g., salinity and micro-habitats), it is possible that site locations in Choctawhatchee Bay and Santa Rosa Sound, Charlotte Harbor, and Florida Bay were not located in preferred habitats conducive for collecting of ladyfish. The findings showed that preferred habitat for juvenile ladyfish (< 200 mm SL) were brackish riverine systems, while larger (200–300 mm SL) individuals preferred open-bay areas.

Overall, the findings showed that most ladyfish were collected with seines using a boat set seine deployment method, which agreed with previous studies for other estuarine fishes [13, 14, 26–31]. A boat set deployment method seemed reasonable since most ladyfish were found in habitats (e.g., bayous, creeks, and rivers with steep bank) that were too difficult to sample with traditional deployment methods (i.e., onshore beach seine sweep). Results also showed that more ladyfish were collected with seines (gear 22 and 23) constructed with leads spaced (i.e., leadline) every 150 mm than seines (gear 10) constructed with leads spaced every 300 mm. These finding suggested that gear constructed with leads spaced closer captured more ladyfish because the gear (i.e., leadline) sank quicker to the bottom, which prevented ladyfish from escaping. Since these seines (gear 22 or 23 *vs.* gear 10) were not deployed at comparable locations or simultaneously, they could not be statistically compared and tested. Nonetheless, these findings were practical since ladyfish were usually found in mid-water signifying the species could be easily startled (avoidance) during gear deployment (Levesque, pers. obs.).

This study also demonstrated that ladyfish relative abundance was highest with seines and lowest with otter trawls. Since seines and otter trawls were deployed on the same day in adjacent areas, it is possible that the lower catch rates for otter trawls was more related to depth (sampling location) or gear selectivity (i.e. seine gear capturing more animals) than lower relative abundance. Moreover, the results also showed that the total numbers of ladyfish collected varied by seine deployment method and water body, which pointed out how important it was to choose the proper deployment method for the specific sampling site. Ladyfish catch rates also varied by seine length and mesh size. The highest relative abundance was in the Indian River Lagoon with a 21.3 m seine using a boat set deployment method and second highest was with the 183 m seine; there are two possible explanations. The larger seine (183 m) not only encompassed a larger sampling area, it also created less site disturbance; only one haul was conducted with the larger seine instead of three with the smaller seine.

The present study also found that sampling efficiency was dependent upon fish length. The results showed that ladyfish length was significantly different between seines constructed with larger mesh sizes than those constructed with smaller mesh sizes, which indicated that larger individuals were likely capable of avoiding smaller length seines constructed with smaller meshes. While ladyfish length was significantly different among sampling gears, there was no clear or consist pattern evident among different water bodies. Overall, ladyfish length-frequency distribution was similar between Tampa Bay and the Indian River Lagoon; most ladyfish captured were smaller than 100 mm SL and larger than 220 mm SL. In both estuaries, ladyfish between 110 and 210 mm SL were the least sampled. Despite the variety of sampling gears, it was apparent that either the sampling gears were inappropriate (gear avoidance) or this size-class (110–210 mm SL) of ladyfish were not present at the sampling station or at the time of sampling. In Tampa Bay and the Indian River Lagoon, length-frequency distribution analyses indicated that a greater percentage of smaller ladyfish were collected with seines, whereas gillnets captured a greater percentage of larger individuals. Ladyfish length-frequency distribution was similar among some gears; however, seines captured a greater size range than otter trawls. Both gears captured similar minimal size individuals in the Indian River Lagoon; however, in Tampa Bay, ladyfish minimal sizes were larger with otter trawls than seines. It is difficult to explain why this was evident, but one possible explanation could be that it was due to the sampling site (e.g., gear inappropriate for the location) rather than any actual relative abundance difference. These findings corresponded with McBride *et al.* [14], which also reported that ladyfish length-frequency distribution was different among gears (i.e., seines, otter trawl, blocknet, and gillnets) in Tampa Bay.

Based on this study, sampling gear must be chosen with regard not only to fish life-stage, but according to sampling site characteristics (e.g., habitat, water depth, bank profile, and bottom substrate). Weinstein and Davis [4] concluded that catch efficiency differed between seines and blocknets because of the bottom type (mud vs. firmer mud/sand); they attributed their lower catch rates to fish escaping because of the mud bottom. It is difficult to recognize why otter trawl catch rates were inconsistent among estuaries, but seines captured more ladyfish over a greater period, indicating that capture rates and size selectivity were inversely correlated with time since fish were either able to avoid the gear or they had already migrated away from the sampling station. This observation was supported by gillnet data for the Indian River Lagoon, which pointed out that gillnets were more efficient at capturing a greater percentage of larger individuals, but this was dependent upon the sampling location and time of year.

Ladyfish recruitment is spring through summer in Tampa Bay and the Indian River Lagoon [14]; consequently, appropriate gear, deployment method, and mesh size are important factors that need to be considered when designing sampling protocols specifically for sampling ladyfish. More importantly, field sampling must coincide with recruitment periods or estimated abundance will be biased. Kingdom and Allison [32] found that gear efficiency for riverine fish (*Pellonula leonensis*) increased using a medium mesh size (12 mm), but primarily only in February, which the authors attributed to an increase in abundance.

Conclusion

As evident from this study, fishery managers must use caution prior to establishing standardized short or long-term fishery programs to avoid any potential bias associated with sampling gears, methods, or techniques, especially if the purpose and goal is to monitor specific species or life-stages. To avoid gear or habitat selectivity bias, it's recommended that fishery managers choose sampling gears, methods, and locations (e.g., estuaries or habitats) that capture the greatest proportion of a species' size range. Furthermore, sampling criteria should incorporate a variety of sampling gears and deployment methods according to a species' preferred habitat, recruitment period, and size-class. Despite evaluating various sampling gears potentially useful for collecting ladyfish in Florida waters, there were various field-sampling factors that were not evaluated in this study that could have influenced catch efficiency (i.e., avoidance and selectivity), such as operator's experience, speed of gear deployment, speed of haul, bottom type, water depth, atmospheric conditions (e.g., day, night, cloudy, sunny, and rainy) fish swimming speed, and fish avoidance (e.g., noise, vibration, pressure waves). As such, all of these factors should be considered when designing future sampling methodology for not only collecting ladyfish, but other species.

Acknowledgments

A great debt of gratitude is owed to the FIM staff from the FWC. I thank the FIM staff for their dedicated field sampling, sorting, identifying, and gear maintenance efforts. In addition, I thank P. Gehring and K. Knight from Geo-Marine Inc. for GIS and graphics support. More importantly, I thank B. McMichael and T. McDonald for kindly providing access to the FIM data. Lastly, I thank the three anonymous reviewers for their strict review and critique. This work was supported in part by funding from Florida saltwater fishing license sales and the Department of Interior, U.S. Fish and Wildlife Service, Federal Aid for Sportfish Restoration Project Number F-43 to the Florida Fish and Wildlife Conservation Commission. Sampling in the Little Manatee River was made available through grants CM-254 and CM-280 from the Department of Environmental Regulation, Office of Coastal Management, with funds made available through the National Oceanic and Atmospheric Administration under the Coastal Zone Management Act of 1972, as amended.

References

Five "key references", selected by the authors, are marked below (Three recommended (●) and two highly recommended (●●) papers).

1. ● Nielsen, L.A., D.L. Johnson. Editors. 1983. Fisheries Techniques. American Fisheries Society, Bethesda, Maryland. 468pp.
2. Neumann, R., C. Guy, D. Willis. 1995. Precision and size structure of juvenile percichthyid samples collected with various gears from Lake Texoma. North American Journal of Fisheries Management 15: 956–962. http://dx.doi.org/10.1577/1548-8675(1995)015<0956:PASSOJ>2.3.CO;2
3. Myers, R., J. Hoenig. 1997. Direct estimates of gear selectivity from multiple tagging experiments. Canadian Journal of Aquatic Fishery Science 54: 1–9.
4. Weinstein, M., R. Davis. 1980. Collection efficiency of seine and retenone samples from tidal creeks, Cape Fear River, North Carolina. Estuaries 3(2): 98–105. http://dx.doi.org/10.2307/1351553
5. Beckman, D., J. Dean. 1984. The age and growth of young-of-the-year spot, *Leiostomus xanthurus* Lacepede, in South Carolina. Estuaries 7(4): 487–496. http://dx.doi.org/10.2307/1352052
6. Layman, C., D. Smith. 2001. Sampling bias in minnow traps in shallow aquatic habitats on the eastern shore of Virginia. Wetlands 21(1): 145–154. http://dx.doi.org/10.1672/0277-5212(2001)021[0145:SBOMTI]2.0.CO;2
7. Walsh, S.J. 1996. Efficiency of bottom sampling trawls in deriving survey abundance indices. NAFO Sci. Coun. Studies 28: 9–24.
8. ●● Steele, M.A., S.C. Schroeter, H.M. Page. 2006. Experimental evaluation of biases associated with sampling estuarine fishes with seines. Estuarine and Coasts 29 (6B): 1172–1184.
9. Wells, D, K.M. Boswell, J.H. Cowan, W.F. Patterson III. 2008. Size selectivity of sampling gears targeting red snapper in the northern Gulf of Mexico. Fisheries Research 89: 294–299. http://dx.doi.org/10.1016/j.fishres.2007.10.010
10. ● Ricker, W.E. 1975. Computation and interpretation of biological statistics of fish populations. Bulletin 191 of the Fisheries Research Board of Canada.382pp.
11. Ugwumba, O.A. 1989. Distribution and growth pattern of the ten-pounder *Elops lacerta* (Val.) in the freshwater, estuarine and marine environments of Lagos, Nigeria. Arch. Hydrobiology 115(3): 451–462.
12. Brinda, S., S. Bragadeeswaran. 2005. Influence of physico-chemical properties on the abundance of a few economically important juvenile fin-fishes of Vellar estuary. Journal of Environmental Biology. 26(1): 109–112.
13. Tremain, D.M., D.H. Adams. 1995. Seasonal variations in species diversity, abundance, and composition of fish communities in the northern Indian River Lagoon, Florida. Bulletin of Marine Science 57: 171–192.
14. McBride, R., T. MacDonald, R. Matheson, D. Rydene, P. Hood. 2001. Nursery habitats for ladyfish, *Elops saurus*, along salinity gradients in two Florida estuaries. Fishery Bulletin 99: 443–458.
15. Levesque, J.C. 2010. Habitat use patterns of juvenile ladyfish (*Elops saurus*) in Florida (USA) Estuaries. Wildl. Biol. Pract. 6(2): 39–56.
16. Rider, S., W. Weber. 1998. Fisheries-independent monitoring program: summary of fixed-station sampling in Choctawhatchee Bay and Santa Rosa Sound 1992–1996. Florida Marine Research Institute Report Series IHR 1998-002. St. Petersburg, Florida. 82pp.

17. Poulakis, G.R., M.E. Mitchell. 1999. Fisheries-independent monitoring program: summary of fixed-station sampling in Charlotte Harbor, Florida 1990–1996. Florida Marine Research Institute Report Series IHR 1999-003. St. Petersburg, Florida. 107pp.

18. ●● Kupschus, S. 2003. Development and evaluation of statistical habitat suitability models: an example based on juvenile spotted seatrout *Cynoscion nebulosus*. Marine Ecology Progress Series. 265: 197–212.
http://dx.doi.org/10.3354/meps265197

19. Harnden, C.W. 1998. Fisheries-independent monitoring program: summary of fixed-station sampling in Indian River Lagoon, Florida 1991–1996. Florida Marine Research Institute. Report Series IHR 1998-001. St. Petersburg, Florida. 131pp.

20. Brodie, R.B. 1999. Fisheries-independent monitoring program: summary of fixed-station sampling in the Tomoka River/Basin, Ponce De Leon Inlet, and Mosquito Lagoon, Volusia County Florida, 1993 to 1997. Florida Marine Res. Inst. In-house Report Series IHR 1999-005. St. Petersburg, Florida. 183pp.

21. McMichael, R.H., R. Paperno, B.J. McLaughlin, and M.E. Mitchell. 1995. Florida's marine fisheries-independent monitoring program: a long-term ecological dataset. Bulletin Marine Science 57: 282–285.

22. Zar, J. 1994. Biostatistical Analysis, 2nd edition. Prentice-Hall, Englewood Cliffs. 718pp.

23. Bartlett, M.S. 1937. Properties of sufficiency and statistical tests. Proceedings of the Royal Society of London, Series A, 160: 268–282.
http://dx.doi.org/10.1098/rspa.1937.0109

24. Bartlett, M. S. 1937. Some examples of statistical methods of research in agriculture and applied biology. Journal of the Royal Statistical Society, 4: 137–170.

25. Gotelli, N.J, A. M. Ellison, 2004. A primer of ecological statistics. Sinauer Associates, Inc. Publishers. Sunderland, Massachusetts. 510pp.

26. Adams, A., D. Blewett. 2004. Spatial patterns of estuarine habitat type use and temporal patterns in abundance of juvenile permit, *Trachinotus falcatus*, in Charlotte Harbor, Florida. Gulf and Caribbean Research 16(2): 129–139.

27. Switzer, T., B. Winner, N. Dunham, J. Whittington, M. Thomas. 2006. Influence of sequential hurricanes on nekton communities in a southeast Florida estuary: short-term effects in the context of historical variations in freshwater inflow. Estuaries and Coasts 29(6): 1011–1018.

28. Stevens, P., D. Blewett, G. Poulakis. 2007. Variable habitat use by juvenile common snook, *Centropomus undecimalis* (Pisces: Centropomidae): Applying a life-history model in a southwest Florida estuary. Bulletin of Marine Science 80: 93–108.

29. Nelson, G. 1998. Abundance, growth, and mortality of young-of-the-year pinfish, *Lagodon rhomboides*, in three estuaries along the gulf coast of Florida. Fishery Bulletin 96: 315–328.

30. Peters, K.M., R.H. McMichael. 1987. Early life history of the Red Drum *Sciaenops Ocellatus* (Pisces: Sciaenidae), in Tampa Bay, Florida. Estuaries 10(2): 92–107.
http://dx.doi.org/10.2307/1352173

31. ● McMichael, R.H., K.M. Peters. 1989. Early life history of Spotted Seatrout *Cynoscion nebulosus* (Pisces: Sciaenidae), in Tampa Bay, Florida. Estuaries 12(2): 98–110.
http://dx.doi.org/10.2307/1351501

32. Kingdom, T., M. Allison. 2007. Abundance and gillnet selectivity of *Pellonula leonensis* in the lower Nun River, Niger Delta, Nigeria. African Journal of Biotechnology 6(20): 2419–2423.

14

New Live-Trapping Method Improves Capture Rates for Nine-Banded Armadillos

J. A. Martin[1,*], C. Marshall[2], J. L. Belant[3], S. Cagle[4] & B. C. West[5]

[1] Warnell School of Forestry and Natural Resources, Athens, GA 30602, USA.

[2] Agricultural Ecology Lab, Department of Wildlife, Fisheries, and Aquaculture, Mississippi State University, Mississippi State, MS 39761, USA.

[3] Carnivore Ecology Lab, Mississippi State University, Mississippi State, MS 39761, USA.

[4] Mississippi State Extension Service, Mississippi State University, Houston, MS 38851, USA.

[5] Extension Regional Director, University of Tennessee Instituted of Agriculture, Jackson, TN 38301, USA.

* Corresponding author email: jmart22@uga.edu.

Keywords

Dasypus novemcinctus; Nine-anded armadillo; Nuisance; Trapping.

Abstract

Nine-banded armadillos (*Dasypus novemcinctus*) are often the target of wildlife mitigation efforts to reduce their ill-effects on property, biodiversity, and other resources; however, efficient techniques for humane removal evade managers. Most trapping techniques tested to date have used a food-based lure with minimal success. We tested the effects of a wooden trap baited with a conspecific attractant—armadillo scent—in an effort to improve capture rates. Compared to the reference trap type, the conspecific attractant trap was 8.3 times more effective (Odds Ratio = 8.30, CL= 2.38, 25). Our findings suggest that this trap type baited with a conspecific attractant will improve trapping efficiency. We encourage researchers to test additional trap and scent combinations to further improve efficiency.

Introduction

The distributional limits of the nine-banded armadillo (*Dasypus novemcinctus*; hereafter armadillo) have continued to expand north and east within the continental United States over the past 150 years [1-4]. Armadillos were previously restricted to the southern portion of Texas [1, 5], but due to anthropogenic influences on the landscape, climate change, and natural dispersal, their range within the continental United States has expanded significantly [3]. The current estimated breeding range of the armadillo includes the states of Florida, Alabama, Mississippi, Louisiana, Arkansas, and portions of Georgia, South Carolina, Tennessee, Kentucky, Illinois, Missouri, Kansas, Oklahoma, and Texas [4, 6]. Armadillos have also been recorded in New Mexico, Wisconsin, Iowa, Nebraska, South Dakota, and Colorado, but these isolated observations do not represent a permanent breeding population of the species [4, 7, 8].

Further expansion of the armadillo's range is currently hindered by lack of adequate precipitation in the west and intolerable temperatures in the north and northeast [4], but with the current change in climate, the distributional limits of the armadillo may continue to expand farther north, northeast, and west. Though not considered a nuisance in all states, the range expansion of the armadillo is of great concern to landowners and management personnel due to the potential economic damage (rooting up lawns, golf courses, and gardens and causing structural instabilities) caused by its foraging and burrowing habits [9, 10]. There is also concern that armadillos may potentially pose a threat to forest-floor communities that were historically uninhabited by armadillos [2]. Other concerns include nest-predation of ground-nesting birds and sea turtles [11-13] and the transmission of the bacterium (*Mycobacterium leprae*) that causes leprosy in humans, as armadillos are a known host of the bacterium [14-17].

Current armadillo control techniques include exclusion fencing, habitat modification, lethal removal (shooting), or live trapping [2]. However, exclusion fencing is ineffective as armadillos can burrow under or climb over the fences, habitat manipulation may be impractical, and lethal removal may not be a practical or desirable alternative [9]. Live trapping has been used as a non-lethal, economical armadillo removal method, however the success rates of live-traps are quite low [9]. Studies to determine a suitable attractant coupled with live trapping have been tested with minimal effect [9, 10]. However, Loughry and McDonough (1994) determined that young-of-the-year armadillos were able to discriminate among scents of themselves, siblings, and unknown conspecifics. With that said, there is the potential for the identification and development of an attractant, based on the scent of conspecifics, that would significantly increase the capture rates of armadillos. Our study evaluated the efficacy of a wooden armadillo trap baited with conspecific scent compared to a conventional wire trap that was not baited.

Methods

Eight pairs of traps were deployed on three study sites within Chickasaw and Holmes counties of Mississippi, US. The sites were located near the towns of Okolona (34° 0' 21" N, 88° 45' 2" W) and Egypt (33° 20' 28" N, 90° 16' 38"). The study areas were privately owned farms with dominant land cover types that included upland pine (*Pinus* spp.) forests, bottomland hardwood forests (e.g., *Quercus* spp.), and agricultural fields. The climate of the region is considered subtropical with hot, humid summers and mild winters.

The baited, with conspecific scent, wooden armadillo trap (hereafter conspecific trap) and the unbaited wire cage trap were both classified as double-door live traps. The conspecific trap was specifically designed for the capture of armadillos with dimensions of 30cm (h) x 30cm (w) x 81cm (l). This trap was baited with an armadillo attractant (Patent Pending) that was obtained by placing a live armadillo in the trap overnight. The second trap was an unbaited conventional commercial wire cage trap (Havahart large 2-door trap; Woodstream Corporation, Lititz, Pennsylvania, USA) with the dimensions of the trap being 25cm (h) x 25cm (w) x 90cm (l). The unbaited conventional wire cage trap acted as the control in this experiment as it is currently considered the most effective trap type for the capture of armadillos [18].

At each study site, armadillo traps were set from 3 April-30 November during 2009. The traps were set randomly in pairs (one of each type) in areas where armadillo sign was present, such as soil disturbance, tracks, and visual sightings. Each trap in a pair was set no more than 20m apart, with pairs of traps being set greater than 225m apart so they could be treated as independent sampling units. All traps were checked daily for captures and all captured armadillos were euthanized. For each trap type we calculated the number of adjusted trap nights, where traps that were sprung with or without capture were adjusted as representing 0.5 trap nights [19]. We used logistic regression, a likelihood-ratio test, and odds ratios with 95% confidence limits (CL) to compare the capture efficacy of the two trap types to capture armadillos.

Results

The conspecific traps captured 23 armadillos in 806 adjusted trap nights with two non-target species captures: one cottontail rabbit (*Sylvilagus* sp.) and one opossum (*Didelphis virginiana*). The conventional wire traps captured three armadillos and captured eight non-target individuals in 822 adjusted trap nights: seven raccoons (*Procyon lotor*) and one striped skunk (*Mephitis mephitis*). According to the likelihood-ratio test, we rejected the null hypothesis that the null model and model including trap type were equal (χ_1^2 = <0.001). Conspecific traps were 8.30 times (Odds Ratio=8.30, CL=2.38, 25) more likely to catch an armadillo than unbaited conventional wire traps (test that coefficient does not deviate from zero: z-value=-3.38, d.f.=26,1; and P = <0.001).

Discussion

Our study determined that conspecific traps had higher capture rates of armadillos relative to unbaited conventional wire cage traps, while also reducing the capture of non-target species. The factor influencing increased armadillo capture rates, whether it is the addition of conspecific scent or the design of the conspecific trap, cannot be determined from this study. Further testing needs to be conducted in order to determine which of the two factors or combination of factors significantly influences the capture rates of armadillos. Future studies should be designed with an unbaited conspecific trap and a conventional wire cage trap acting as controls for the two trap types (conspecific and conventional wire cage) baited with the conspecific scent.

A test assessing the efficacy of unbaited live-traps compared to baited live-traps (with a variety of baits being used) no difference in capture success was found; overall the capture methods resulted in one capture every 132 trap nights [9]. Gammons et al. (2005) determined that the success rate of live-traps was not dependent on the attractant used, but on the placement of the traps (e.g. near areas that have armadillo sign). Our study shows that trap placement, though important, was not the most important factor influencing capture success rates as conventional wire cage traps had lower capture success rates than conspecific traps with the placement of both trap types being in areas with armadillo sign. With regards to conspecific scent as an attractant, Ober et al. (2011) determined that the use of conspecific scent as an attractant yielded no significant results. However, the collection of the conspecific scent was performed after initial capture, which, as stated by the authors, may have elicited the release of odors that convey fear, thus influencing the results of the study. Future studies exploring the efficacy of conspecific scent should obtain samples from non-stressed individuals, thus reducing the chances of the odor acting as a deterrent.

Previous capture methodologies for armadillos have included using unbaited live-traps, baited live-traps, hand-nets, drift fence/live-trap combinations, and dog/live-trap combinations [9]. All of these methods varied in success rates of armadillo capture; however, performance was ultimately poor. The success rate of hand-nets as a capture method for armadillos is dependent on the habitat type and effort [10, 20, 21, 22]. Dense vegetation in some habitats and private property issues associated with urban areas greatly reduces the chances of capture using hand-nets [23]. Drift fences combined with live-traps had resulted in no captures over 90 trap days whereas the live-capture using dogs resulted in six captures in 20 attempts [23]. Dog/live-trap combination may be a suitable choice for studies conducted in rural areas, but likely would not be suitable in most urban areas. Additionally, this technique requires substantially greater effort and consequently, greater overall cost compared to trapping.

In conclusion, continued research is needed to evaluate whether the conspecific trap design without the addition of an attractant, the addition of a conspecific attractant to the trap, or a combination of both factors is influencing armadillo capture rates. Our study has shed light on a new technique that has the potential to revolutionize armadillo trapping. This new trapping method could potentially provide researchers who require live specimens, management personnel, and animal control agencies with an effective method for the live capture of armadillos.

Acknowledgements

We thank Kris Godwin for loaning us the wire cage traps and the 3 landowners who provided access to their properties. Support for this study was provided by the Mississippi State University Extension Service and the Department of Wildlife, Fisheries and Aquaculture and Center for Resolving Human-Wildlife Conflicts at Mississippi State University.

References

Five "key references", selected by the authors, are marked below (Three recommended (●) and two highly recommended (●●) papers).

1. Galbreth, G.J. 1982. Armadillo. In: Chapman, J.A. & Feldhamer, G.A. (eds), Wild Mammals of North America. The Johns Hopkins University Press. Baltimore, Maryland, pp, 71-79.
2. Dolbeer, R.A., Holler, N.R., & Hawthorne, D.W. 1994. Identification and control of wildlife damage. In: Bookhout, T.A. (eds), Research and Management Techniques for Wildlife and Habitats. Allen Press. Lawrence, Kansas, pp, 474-506.
3. Taulman, J. F. & Robbins, L. W. 1996. Recent range expansion and distributional limits of the nine-banded armadillo (*Dasypus novemcinctus*) in the United States. Journal of Biogeography 23: 635-648.
http://dx.doi.org/10.1111/j.1365-2699.1996.tb00024.x
4. ●● Taulman, J.F. & Robbins, L.W. 2014. Range expansion and distributional limits of the nine-banded armadillo in the United States: an update of Taulman & Robbins (1996). Journal of

Biogeography 41: 1626-1630.
http://dx.doi.org/10.1111/jbi.12319

5. Taber, W.F. 1945. Contribution on the life history and ecology of the nine-banded armadillo. Journal of Mammalogy 26: 211-226.
http://dx.doi.org/10.2307/1374817

6. Hofmann, J.E. 2009. Records of nine-banded armadillos, *Dasypus novemcinctus*, in Illinois. Transactions of the Illinois State Academy of Science 102: 95-106.

7. Freeman, P.W. & Genoways, H.H. 1998. Recent northern records of the nine-banded armadillo (Dasypodidae) in Nebraska. The Southeastern Naturalist 43: 491-504.

8. Frey, J.K. & Stuart, J.N. 2009. Nine-banded armadillo (*Dasypus novemcinctus*) records in New Mexico, USA. Edentata 8-10: 54-55.
http://dx.doi.org/10.1896/020.010.0105

9. Gammons, D. J., Mengak, M. T., & Conner, M. L. 2005. Evaluation of attractants for live-trapping nine-banded armadillos. University of Georgia Warnell School of Forest Resources - WSFR-WMS 05-04. 4 pp.

10. ●● Ober, H. K., DeGroote, L. W., McDonough, C. M., Mizell III, R. F. and Mankin, R. W. 2011. Identification of an attractant for the nine-banded armadillo, *Dasypus novemcinctus*. Wildlife Society Bulletin 35: 421-429.
http://dx.doi.org/10.1002/wsb.79

11. ● Engeman, R.M., Martin, R.E., Constantin, B., Noel, R., & Woolard, J. 2003. Monitoring predators to optimize their management for marine turtle nest protection. Biological Conservation 113: 171-178.
http://dx.doi.org/10.1016/S0006-3207(02)00295-1

12. ● Staller, E. L., Palmer, W. E., Carroll, J. P., Thornton, R. P., & Sisson, D. C. 2005. Identifying predators at Northern Bobwhite nests. Journal of Wildlife Management 69: 124-132.
http://dx.doi.org/10.2193/0022-541X(2005)069<0124:IPANBN>2.0.CO;2

13. Dreibelbis, J. Z., Melton, K. B., Aguirre, R., Collier, B. A., Hardin, J., Silvey, N. J., & Peterson, M. J. 2008. Predation of Rio Grande Wild Turkey nests on the Edwards Plateau, Texas. The Wilson Journal of Ornithology 120: 906-910.
http://dx.doi.org/10.1676/07-183.1

14. Burchfield, H. 1999. Armadillo leprosy and a failed vaccine. World Journal of Microbiology and Biotechnology 15:653–667.
http://dx.doi.org/10.1023/A:1008983515373

15. Truman, R. 2005. Leprosy in wild armadillos. Leprosy Review 76: 195–208.

16. ● Loughry, W. J., Truman, R. W., McDonough, C. M., Tilak, M. K., Garnier, S., & Delsuc, F. 2009. Is leprosy spreading among nine-banded armadillos in the southeastern United States? Journal of Wildlife Diseases 45: 144–152.
http://dx.doi.org/10.7589/0090-3558-45.1.144

17. Truman, R. W., Singh, P., Sharma, R., Busso, P., Rougemont, J., Paniz-Mondolfi, A., Kapopoulou, A., Brisse, S., Scollard, D. M., Gillis, T. P., & Cole, S. T. 2011. Probable zoonotic leprosy in the southern United States. New England Journal of Medicine 364: 1626–1633.
http://dx.doi.org/10.1056/NEJMoa1010536

18. Hawthorne, D.W. 1994. Armadillos. In: Hygnstrom, S.E., Timm, R.M., & Larson, G.E. (eds), Prevention and Control of Wildlife Damage. Volume II. University of Nebraska. Lincoln, Nebraska, pp, D1-D3.

19. Belant, J. L. 1992. Efficacy of three types of live traps for capturing weasels, *Mustela* spp. Canadian Field-Naturalist 106: 394-397.

20. McDonough, C. M. & Loughry, W. J. 2005. Impacts of land management practices on a population of nine-banded armadillos in northern Florida. Wildlife Society Bulletin 33: 1198-1209.
http://dx.doi.org/10.2193/0091-7648(2005)33[1198:IOLMPO]2.0.CO;2

21. McDonough, C. M, Lockhart, J. M., Loughry, W. J. 2007. Population dynamics of nine-banded armadillos: Insights from a removal experiment. Southeastern Naturalist 6: 381-392.
http://dx.doi.org/10.1656/1528-7092(2007)6[381:PDONAI]2.0.CO;2

22. Gammons, D. J., Mengak, M. T., & Conner, M. L. 2009. Translocation of nine-banded armadillos. Human–Wildlife Conflicts 3:64–71.

23. Bergman, D. L., Bluett, R. D., & Tipton, A.R. 1995. An alternative method for capturing armadillos. The Southwestern Naturalist 40: 414-416.

15

FAECAL STABLE ISOTOPE ANALYSIS IN WHITE-TAILED DEER (*ODOCOILEUS VIRGINIANUS*), AN ALTERNATIVE METHOD FOR DIETARY ECOLOGY STUDIES

E. Nájera-Hillman[1,*] & S. Mandujano[2]

[1] COSTASALVAJE A.C. Blvd. Las Dunas 160-203. Fracc. Playa Ensenada, 22880 Ensenada, Baja California, México.

[2] Red de Biología y Conservación de Vertebradas, Instituto de Ecología A.C. (INECOL). Carretera antigua a Coatepec No. 351. El Haya, 91070 Xalapa, Veracruz, México.

E-mail: salvador.mandujano@inecol.edu.mx

* Corresponding author e-mail: eduardo.najera@wildcoast.net

Keywords

Carbon;
Diet;
Food sources;
Mammalian herbivores;
Mixing model;
Nitrogen;
Trophic shift.

Abstract

Stable isotope analyses of tissues and faeces offer an alternative to study the dietary ecology in wild animal populations. Faecal isotope analyses are advantageous because samples are easy to obtain in the field, their isotopic composition is related to the food ingested and allow for documenting short-term diet variability. Therefore, we estimated diet-faeces trophic shifts in white-tailed deer (*Odocoileus virginianus*) fed on a constant diet. The observed diet-faeces trophic shifts (0.81‰ for $\delta^{13}C$ and 2.13‰ for $\delta^{15}N$) were different to the trophic shifts reported for other herbivores. However, our estimates proved to be useful when applied to a mixing model to quantify the contributions of multiple food sources to the deer's diet. Our diet-faeces trophic shift estimates are reliable measurements that can be used to expand the potential of stable isotope analyses in white-tailed deer dietary ecology studies.

Introduction

Stable carbon and nitrogen isotopes analyses are a widely accepted tool for studying the dietary ecology of wildlife [1-3]. In particular the carbon isotopic composition ($\delta^{13}C$) of a primary consumer (herbivore) is a reflection of the photosynthetic mechanism of the primary producers on which it subsists [4]. Therefore, stable carbon isotope analysis is useful for investigating feeding strategies (such as browsing vs. grazing) of mammalian herbivores in environments where the available food sources such as C3 (e.g. shrubs, forbs; $\delta^{13}C$ = -35 to -21‰) or C_4 plants (e.g. warm season grasses; $\delta^{13}C$ = -14 to -10‰) have distinct carbon isotopic compositions [5,6]. In the other hand, although one of the primary uses of nitrogen isotope analysis is to provide trophic level information for particular species within an ecosystem [7], additional information about mammalian herbivore dietary ecology can be obtained from nitrogen isotope studies. For instance, it has been suggested that variation on nitrogen isotopic composition ($\delta^{15}N$) of animal tissues may be related to the contribution of nitrogen-fixing plants to an herbivore's diet [1,8,9]. Nevertheless, carbon and nitrogen isotope values of animal tissues do not exactly reflect diet isotope values but differ consistently due to a phenomenon known as trophic shift ($\Delta\delta^{13}C$, $\Delta\delta^{15}N$, where Δ represents the difference between diet and tissue isotope values). Trophic shifts are not constant since they may be influenced by some factors including the species, the tissue analysed [10,11], the food sources [12] and physio-anatomical differences [9]. Consequently it becomes essential to count for properly defined trophic shift values when using stable isotopes in research of animal dietary ecology. For example, when isotopic values are determined for the tissues of a consumer and its food sources a variety of mixing models can give an index of the relative importance of each food source in the consumer's diet [13-15]. However, the isotopic values of the food sources must be adjusted by adding the trophic shift value specific to the tissue and species under examination [e.g. 16].

In particular, trophic shifts for carbon and nitrogen isotopes have been experimentally determined between diet and some white-tailed deer (*Odocoileus virginianus*) tissues such as antler ($\Delta\delta^{13}C$ = 4.5‰; $\Delta\delta^{15}N$ = 4.29‰), hair ($\Delta\delta^{13}C$ = 0.27‰; $\Delta\delta^{15}N$ = 5.15‰) and red blood cells ($\Delta\delta^{13}C$ = 0.35‰; $\Delta\delta^{15}N$ = 4.41‰) [10]. Moreover, it has been suggested that these trophic shift values allow for white-tailed deer dietary reconstruction all over its distributional range, because latitude itself does not influence isotopic composition in white-tailed deer [1]. However, obtaining samples of blood, hair or antler can become problematic in areas of limited access or where white-tailed deer occur at low densities. Furthermore, sampling some deer tissues may involve unwanted invasive manipulation of wild animals. Alternatively isotope analyses of faecal samples have proven, despite representing undigested material with microbial products and endogenous materials, to be a reliable method for studying dietary trends in free ranging mammalian herbivores [17-20]. Faeces isotope analyses are advantageous because samples are fairly easy to obtain in the field and their isotopic composition is related to the feed ingested [16]. What is more, faeces have a short (2 to 14 days) isotopic turnover time (time required to replace molecular components, including the isotopes of these components) that allows for documenting short-term dietary variability [12,21].

Given the importance of having properly defined trophic shifts, some mammalian herbivore (e.g. *Capra hircus*, *Lama glama*) controlled feeding studies have shown diet-faeces trophic shifts to vary from -0.3 to -5.8‰ for carbon isotopes [12,16,22], with species accounting for most of the trophic shift variability rather than the food source or physio-anatomical differences (foregut vs. hindgut fermenters) [12]. Additionally, experimental determinations of nitrogen diet-faeces trophic shifts indicate that faeces of herbivore mammals of various sizes and physiologies (e.g. *Peromyscus maniculatus*, *Bos taurus* and *Myodes gapperi*) are consistently more enriched in ^{15}N (mean trophic shift = 2.34‰ ± SD 0.68) than their diets [18,22,23]. However, because diet-faeces trophic shifts can vary among taxonomic classes, it is preferred to obtain the trophic shift value for the species of interest [10]. Therefore, since white-tailed deer diet-faeces trophic shifts have not yet been reported, we estimated the $\delta^{13}C$ and $\delta^{15}N$ trophic shifts from a semi-captive white-tailed deer herd, which has been feeding on an isotopically homogeneous diet. Further, since failure of an isotopic mixing model to estimate the relative contribution of a food source in the consumers diet can be attributed to the use of incorrect trophic shift values [14,24];

we assessed the utility of our diet-faeces trophic shift estimates by implementing them in the concentration-weighted mixing model of Phillips and Koch [15] to determine the relative contribution of three known food sources to the diet of a captive white-tailed deer group.

Methods

Our study was conducted at the Rancho Ecológico el Mirador (REEM); a Wildlife Conservation, Management and Sustainable Utilization Unit (UMA, acronym in Spanish) located in the northeast hillside of the Cofre de Perote Mountain in the state of Veracruz, Mexico. This UMA holds all permits required by Mexican legislation (registry number: DGVS-CR-IN-0337-VER/98) to maintain white-tailed deer in captive and semi-captive conditions, to be used as a breeding stock and for potential exchange of deer individuals with other UMAs in the region. Within the REEM six white-tailed deer individuals (one male and five females) are held captive in an open-air enclosure of about 1500 m^2, while another six individuals (two males and four females) are kept in semi-captive conditions in a fenced area of about 300 ha of a temperate pine-oak forest.

During our study, from June 27th to July 17th 2011, captive deer were fed (*ad libitum*) with commercial pelleted feed for bovine cattle (18% protein; Nu-3 Balanced Foods®, Mexico), as they did before our investigation. Additionally, there were some common meadow-grass patches (*Poa* sp.) within the enclosure that were also consumed by deer. Moreover, captive deer were observed stripping the bark (deer shaving off tree-bark with their teeth for food) of some pine trees located within the enclosure. In contrast semi-captive animals were released from the enclosure 10 months prior to our investigation and since then have been feeding exclusively on understory wild vegetation. Water was always available for both captive and semi-captive deer, either from a water reservoir or streams in the pine-oak forest, or from a water trough within the enclosure.

Samples of all food sources and deer faeces were collected from both, the enclosure and the pine-oak surrounding forest, and were analysed for $\delta^{13}C$ and $\delta^{15}N$. In the pine-oak forest a total of 29 plant species were collected at the beginning of the study; there were 19 shrubs and forbs (shrubs, from now on), five trees, three grasses and sedges (grasses, from now on), and two ferns; one sample of each plant species was analysed. Plant flowers were not present during collection time and therefore were not sampled.

Ten white-tailed deer faecal samples were collected at the beginning of the study from the pine-oak forest; faecal samples consisted of independent faecal droppings separated by more than three metres and were not differentiated by sex or individual. Within the enclosure five samples of commercial feed, common meadow-grass and pine bark were collected at the beginning of the study, while 10 fresh faecal sample of captive white-tailed deer was collected on 10 days along the course of our investigation (one each day) in order to assess for faecal isotopic variability (Fig. 2).

After collection all samples were oven-dried for 72 h at 60°C. Once dried, samples were ground in an Agatha stone mortar to a fine, homogeneous powder and stored in a freezer in independent glass vials. Individual faecal samples resulted from the grinding and mixing all the faecal pellets from an independent faecal dropping; while plant samples resulted from grinding and mixing the twigs and leaves of each plant species. Isotopic analyses were carried out at the Texas A&M Stable Isotope Geosciences Facility using a dual-inlet system on a continuous-flow stable isotope-ratio mass spectrometer (Finnigan DELTA plus XP), to a precision better than 0.2‰, for both $\delta^{13}C$ and $\delta^{15}N$. Additionally the carbon and nitrogen concentrations (%C and %N) of all samples were obtained simultaneously during isotopic analyses.

Statistical tests to assess isotopic and elemental concentrations differences between sample groups were performed using R 2.13.1 (http://www.r-project.org). The isotope data was not normally distributed (Shapiro-Wilk's test, $p < 0.05$), and nonparametric statistical tests were therefore applied. A Mann-Whitney U-test was used for testing differences between two independent samples and a Kruskal-Wallis test was used for comparing more than two sample groups.

Diet-faeces trophic shift estimations can be miscalculated when herbivores eat mixed-diets containing two or more components with different isotopic compositions due to differential contribution of each food source to the diet [12]. Therefore, we chose to use faecal samples from semi-captive deer, which fed exclusive in C_3 forest plants with similar $\delta^{13}C$ and $\delta^{15}N$ signatures (see results), to calculate white-tailed deer diet-faeces trophic shifts. Trophic shifts are commonly expressed in Δ notation ($\delta^{13}C_{faeces}$ - $\delta^{13}C_{diet}$); nevertheless, ideally trophic shifts should be expressed as ε^* values $\{[(1000 + \delta^{13}C \text{ or } \delta^{15}N_{faeces})/(1000 + \delta^{13}C \text{ or } \delta^{15}N_{diet}) - 1] \times 1000\}$, which are sometimes but not always equivalent to Δ values; considerable differences between Δ and ε^* values have been observed and are theoretically possible given the non-linearity of the estimation [12,25]. Therefore, we used ε^* values in this investigation.

Additionally, during our study captive white-tailed deer fed on unknown quantities of commercial feed, meadow-grass and pine bark. Therefore, we were interested in using our trophic shift estimates to adjust the food sources $\delta^{13}C$ and $\delta^{15}N$ values in order to determine the relative consumption of each food source. Since these food sources can be different with respect to their C and N concentrations (e.g. commercial feed has high protein content), we used the concentration-weighted linear mixing model of Phillips and Koch [15] to ascertain the relative contribution of each food source to the captive white-tailed deer diet and to validate the use of the trophic shifts values calculated herein. This approach is outlined for two isotopic compositions (e.g. $\delta^{13}C$ and $\delta^{15}N$) and three food sources (e.g. commercials pellets, meadow-grass and pine bark) and employs the elementary difference of C and N as an indicator of the quality of the food source [26]. In other words, the model incorporates both isotopic composition ($\delta^{13}C$ and $\delta^{15}N$) and elemental concentration (%C and %N) of the samples to determine the proportional dietary contributions of C, N, and biomass for each food source [15]. When all available food sources are considered in the model and it fails to determine proportional dietary contributions, it may be an indication that the trophic shift values utilized were incorrect (14,15) and therefore can be used as a validation method for newly developed trophic shift estimates.

Results

Overall understory vegetation at the pine-oak forest presented $\delta^{13}C$ values for plants following the C_3-photosynthetic pathway with a mean $\delta^{13}C$ value of -30.77 ±

SD 1.51 and a mean $\delta^{15}N$ value of -2.87 ± SD 0.98‰. Statistical comparison among plant groups was possible only between shrubs and trees (n = 19 and 5 for shrubs and trees, respectively) due to low sample numbers in the other plant groups ($n \leqslant 3$; Table 1). Shrubs and trees were not significantly different for $\delta^{13}C$ and $\delta^{15}N$ (Mann-Whitney U-test, p = 0.36 and p = 0.78, respectively). Additionally, mean $\delta^{13}C$ and $\delta^{15}N$ values for grasses and ferns (Table 1) fell within the isotopic range values of shrubs and trees ($\delta^{13}C$ = -32.93 to -27.07‰ and $\delta^{15}N$ = -4.29 to -0.35‰). Likewise, carbon and nitrogen concentrations of shrubs and trees were also similar (Mann-Whitney U-test, p = 0.10 and 0.13, respectively) and the mean carbon and nitrogen concentrations of grasses and ferns (Table 1) fell within the range of values for carbon and nitrogen concentrations of shrubs and trees (%C = 36.53 to 54.27 and %N = 0.91 to 3.8); indicating that there were no discernible plant groups, in terms of isotopic composition or elemental concentration, within the understory vegetation at the pine-oak forest (Table 1).

Table 1: $\delta^{13}C$ and $\delta^{15}N$, and %C and %N values of understory plant species collected from the pine-oak forest in the Rancho Ecológico el Mirador, Veracruz, Mexico. SD = standard deviation.

Growth form	**Family; *spp***	$\delta^{13}C$‰	$\delta^{15}N$‰	%C	%N
Shrubs	Aricaceae; *Pernettya ciliata*	-31.52	-1.94	36.54	0.92
	Asteraceae; *Baccharis* sp.	-32.49	-3.42	53.03	2.30
	Asteraceae; *Cirsium* sp.	-32.38	-2.65	44.57	1.93
	Asteraceae; *Eupatorium* sp.	-32.93	-3.39	44.40	3.71
	Asteraceae; *Roldana barba-johannis*	-31.77	-2.22	45.02	1.87
	Asteraceae; sp.	-31.80	-3.44	47.42	3.65
	Caprifoliaceae; *Lonicera japonica*	-29.96	-2.40	49.99	1.64
	Fabacea; sp.	-32.09	-1.56	47.88	3.79
	Lamiaceae; *Lepechinia schiedeana*	-32.09	-0.99	46.78	2.41
	Lamiaceae; *Prunella vulgaris*	-31.85	-3.59	39.56	1.04
	Oxalidaceae; *Oxalis latifolia*	-32.16	-4.30	45.11	3.20
	Rosaceae; *Alchemilla pectinata*	-29.76	-3.19	45.80	1.61
	Rosaceae; *Alchemilla procumbens*	-31.07	-2.85	47.26	1.42
	Rosaceae; *Alchemilla procumbens*	-30.90	-2.72	46.35	1.81
	Rosaceae; *Rubus trilobus*	-29.48	-2.61	48.24	1.87
	Salicaceae; *Salix paradoxa*	-28.59	-1.89	52.51	1.71
	Scrophulariaceae; *Buddleia parviflora*	-29.54	-3.23	50.20	2.11
	Smilaceae; *Smilax moranensis*	-27.08	-3.68	49.71	2.69
	Verbenacea; sp.	-31.53	-2.66	51.26	1.83
	Mean ± SD	-30.99±1.5	-2.77±0.8	46.92±4.1	2.18±0.8
Trees	Fagaceae; *Quercus crassifolia*	-28.70	-2.65	41.14	2.19
	Pinaceae; *Abies religiosa*	-30.63	-2.58	54.28	1.67
	Pinaceae; *Pinus ayacahuite*	-31.84	-3.63	52.20	1.54
	Pinaceae; *Pinus patula*	-30.83	-3.25	50.99	1.60
	Pinaceae; *Pinus pseudoestrobus*	-31.09	-0.35	52.41	1.58
	Mean ± SD	-30.62±1.2	-2.49±1.3	50.20±5.2	1.72±0.3
Grasses	Poaceae; sp. 1	-31.66	-4.63	54.94	1.21
	Poaceae; sp. 2	-30.88	-4.44	48.22	1.65
	Cyperaceae; sp.	-30.08	-2.32	47.46	1.57
	Mean ± SD	-30.88±0.8	-3.80±1.3	50.20±4.1	1.48±0.2
Ferns	Pteridaceae: sp.	-27.74	-3.73	47.46	2.84
	Pteridaceae; *Pteris* sp.	-28.15	-2.48	47.30	2.35
	Mean ± SD	-27.95±0.3	-3.10±0.9	47.38±0.1	2.59±0.3

The stable isotope composition of captive deer food sources (meadow-grass, pine bark and commercial pelleted feed) was distinct for both $\delta^{13}C$ and $\delta^{15}N$ (Kruskal-Wallis test, p = 0.006 and p = 0.001, respectively). Nevertheless, we found that the commercial feed was ^{13}C-enriched (-19.92 ± SD 2.12‰) by about 11.2‰ on average, compared to meadow-grass and pine bark, which had similar carbon isotopic compositions (-31.44 ± SD 0.37‰ and -31.09 ± 0.53‰, respectively). In terms of nitrogen isotopic compositions meadow-grass had the higher $\delta^{15}N$ values (6.03 ± SD 0.24‰) followed by the commercial feed (3.53 ± SD 0.81‰) while the pine bark was the most ^{15}N-depleted food source (-2.45 ± SD 1.47‰). On the other hand, although meadow-grass and commercial feed had similar carbon (%C = 41.84 ± SD 0.82 and 41.47 ± SD 0.96, respectively) and nitrogen concentrations (%N = 4.66 ± SD 0.44 and 4.33 ± SD 0.65, respectively), they were significantly different to pine bark (Kruskal-Wallis test, $p \leqslant 0.01$ for both %C and %N), which had a higher carbon concentration (%C =52.47 ± SD 1.18) but a lower nitrogen concentration (%N = 1.60 ± SD 0.05%). Thus, captive deer food sources were different in terms of isotopic composition and elemental concentrations (Fig. 1).

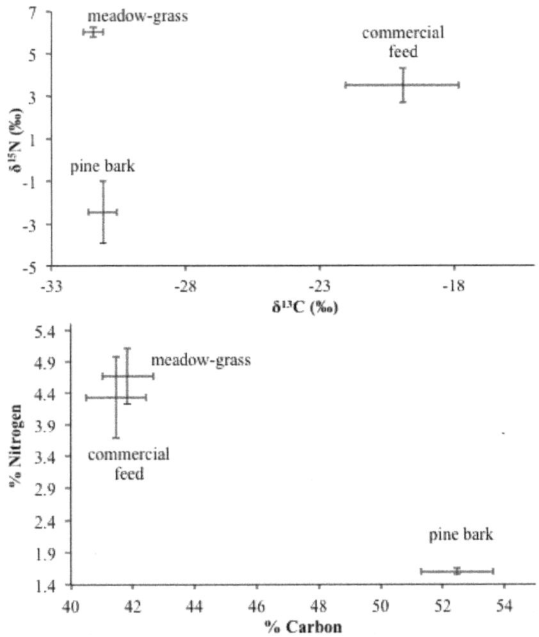

Fig. 1: Mean $\delta^{13}C$, $\delta^{15}N$, %C and %N for food sources of captive white-tailed deer (*Odocoileus virginianus*) in the Rancho Ecológico el Mirador, Veracruz, Mexico. Error bars = standard deviation.

Captive deer faeces showed strong isotopic variability during the 21 days of our study (Fig. 2) and were significantly different, in terms of isotopic compositions and elemental concentrations, to semi-captive white-tailed deer faeces (Mann-Whitney U-test, $p < 0.001$, for both isotopes and both elements). Semi-captive white-tailed deer faecal were collected in the same day but consisted on independent samples, which were ^{13}C and ^{15}N-depleted and less variable ($\delta^{13}C$ = -29.98 ± SD 0.65‰; $\delta^{15}N$ = -0.74 ± SD 0.55‰) in comparison to faeces of captive white-tailed deer ($\delta^{13}C$ = -24.30 ± SD 1.29‰; $\delta^{15}N$ = 5.20 ± SD 0.60‰), indicating isotopic homogeneity among semi-captive deer faeces. Semi-captive deer faeces had higher carbon and nitrogen concentrations (%C = 49.6 ± SD 2.7; %N = 2.8 ± SD 0.4) than captive deer faeces (%C = 41.5 ± SD 2.1; %N = 1.9% ± SD 0.2).

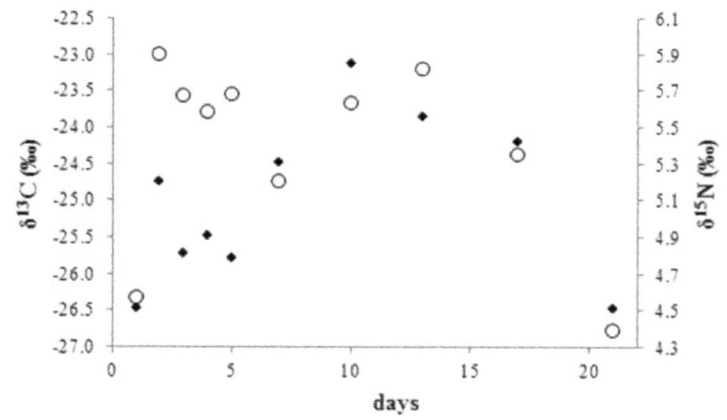

Fig. 2: Faecal isotopic values for captive white-tailed deer (*Odocoileus virginianus*) during the 21 days of our study. White circles = $\delta^{13}C$; black diamonds = $\delta^{15}N$.

Diet-faeces trophic shifts, determined with the mean isotopic values of semi-captive deer faeces and pine-oak forest understory vegetation, resulted in ε* values of 0.81‰ and 2.13‰ for $\delta^{13}C$ and $\delta^{15}N$, respectively. The isotopic signatures of all captive deer faecal samples fell within the triangular space (mixing triangle) formed by the lines connecting the three food sources in a bivariate plot of the two isotopes (Fig. 3), validating the trophic shift values calculated in this investigation.

In the mixing triangle, the line between meadow-grass and commercial feed was nearly straight because the C and N concentrations were similar in those two foods (Fig. 2). In contrast since the commercial feed and the meadow-grass had higher N concentrations than the pine bark, the lines connecting pine bark with commercial feed and meadow-grass tended to be curved or have asymmetrical distribution of iso-diet lines (Fig. 3). Thus, even a small fraction of commercial feed or meadow-grass in the deer's diet causes the $\delta^{15}N$ of the faeces to shift towards the $\delta^{15}N$ of those foods.

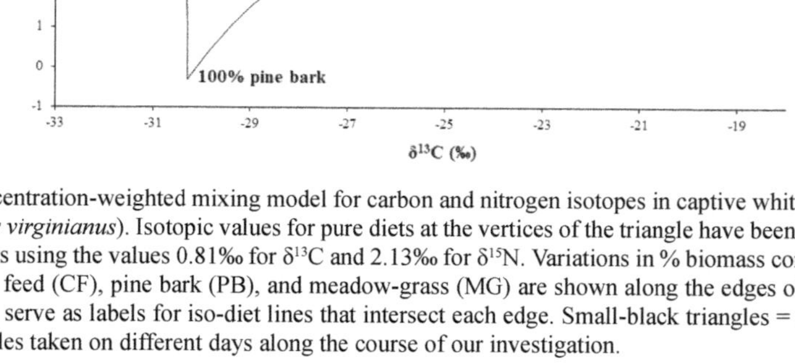

Fig. 3: Concentration-weighted mixing model for carbon and nitrogen isotopes in captive white-tailed deer (*Odocoileus virginianus*). Isotopic values for pure diets at the vertices of the triangle have been adjusted for trophic shifts using the values 0.81‰ for $\delta^{13}C$ and 2.13‰ for $\delta^{15}N$. Variations in % biomass contribution of commercial feed (CF), pine bark (PB), and meadow-grass (MG) are shown along the edges of the mixing triangle and serve as labels for iso-diet lines that intersect each edge. Small-black triangles = captive deer faecal samples taken on different days along the course of our investigation.

The output of the concentration-weighted mixing model suggested that commercial feed was the main contributor to the deer diet biomass (57.8 ± 11.2%) while the contribution of meadow-grass was low (12.3 ± 8.4%) and the contribution of pine bark was intermediate (29.8 ± 14.1%) (Figs. 3 and 4). The relative contribution in terms of carbon from each food source was similar to the relative contribution of biomass (Fig. 4). On the other hand the nitrogen relative contribution, in relation to biomass, was different for the commercial feed and the pine bark and similar for the meadow-grass. The nitrogen relative contribution of the commercial feed (70.1 ± 10%) was higher compared with its biomass contribution, while the relative nitrogen contribution of pine bark (14.2 ± 7.9%) was lower compared with its biomass contribution (Fig. 4).

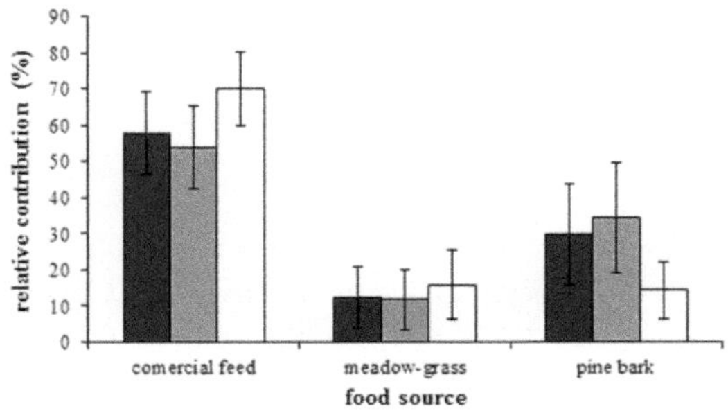

Fig. 4: Relative biomass (dark grey), carbon (light grey) and nitrogen (white) contributions of three different food sources to the diet of captive white-tailed deer during the 21 days of our investigation. Error bars = standard deviation.

Discussion

In order to determine isotopic trophic shifts for a given tissue or material, the animals under examination must feed on a constant diet for a period of time longer than the isotopic turnover time for the tissue or material in question [e.g. 9,10,12,16]. Semi-captive deer in our investigation fed exclusively on understory vegetation with similar $\delta^{13}C$ and $\delta^{15}N$ signatures (Table 1) and for a much longer period of time (~10 months) than the isotopic turnover time for herbivore faeces (< 14 days) [12]. In contrast, captive deer fed on different food sources and do so in different proportions during the study period showing strong isotopic variability confirming that when sources differ in digestibility, tissues may over represent intake of more digestible items and faeces may over- represent less digestible items [27]. Thus, some samples may have indicated high consumption of pine bark in a given day while low consumption in another day (Fig. 2).

Our trophic shift estimates were successfully applied in the concentration-weighted mixing model because all the isotopic values of the captive deer faeces fell within the mixing triangle shaped by the model [15]. If a sample does not fall within the mixing triangle it also may be assumed that the trophic shift values that were utilized were incorrect because all food sources were known for the captive deer [14,15]. Thus, it can be deliberated that since we considered all food sources our estimates are reliable values of diet-faeces trophic shifts for white-tailed deer.

Nevertheless, the $\delta^{13}C$ diet-faeces trophic shift value observed herein was different in relation to the values reported for other large (-0.84 ± SD 0.27‰) [12,16] and some small (-3.76 ± SD 1.27‰) [22] mammalian herbivores. Faeces were previously believed to become ^{13}C-depleted in relation to the diet, although it is still unknown why this negative trophic shift occurs [12,22]. A possible explanation for the positive trophic shift observed herein is that the fraction of indigestible plant material (IPM) in forage is ^{13}C-enriched in comparison to the whole plant; consequently, since the IPM is found in a higher proportion in faeces than in diets, it may be expected that faeces result in elevated $\delta^{13}C$ values [12]. Available evidence indicates very few carbon isotopic shifts along the alimentary tract [28], re-enforcing the plausability of the interpretation that the observed trophic shift values in faeces are influenced by differences in digestibility of food sources. However, an accurate explanation requires comprehensive understanding of the processes causing isotopic trophic shifts of carbon in the digestive system of mammalian herbivores.

Our diet-faeces $\delta^{15}N$ trophic shift estimate is similar to other mammalian herbivore diet-faeces trophic shift values (2.34 ± SD 0.7‰) and consistent with the notion that the faeces of herbivores are enriched in ^{15}N relative to diet [18,22,23,29]. The mechanisms responsible for ^{15}N enrichment of faeces during their passage through the gastrointestinal tract are poorly understood. However, it has been suggested that the activity of digestive enzymes in the stomach, preferential absorption of ^{14}N in the small intestine and colon, and presence of ^{15}N-enriched microorganisms in the caecum, may all contribute to ^{15}N enrichment in faeces of mammalian herbivores [22,30]. Nevertheless, our results and those of other investigations [10,11,12] indicate that the diet-faeces trophic shifts are not always the same for all large mammalian herbivores, especially for carbon. Therefore, the ability to reconstruct diet from the $\delta^{13}C$ and $\delta^{15}N$ of faeces depends on the correct determination of species-specific isotopic trophic shifts.

On the other hand, captive deer fed on three isotopically distinct different food sources (Fig. 1) while semi-captive deer fed only on C_3 plants with similar carbon and nitrogen isotopic compositions (Table 1). Although, in the enclosure, meadow-grass and pine bark presented $\delta^{13}C$ values similar to the understory vegetation (Table 1; Fig. 1), the commercial pelleted feed was significantly ^{13}C-enriched and variable on $\delta^{13}C$ values (standard deviation = 2.12‰). Therefore, since the isotopic compositions of animal tissues or excreta are a reflection of the isotopic composition of the diet [4,18], semi-captive and captive white-tailed deer faeces were also isotopically different. Moreover, it has been suggested that populations that consume a wider variety of food sources (in terms of isotopic composition) will exhibit wider isotopic variation in their tissue or excreta than those consuming a narrower variety of food sources [2]. In this study captive deer faeces exhibited a wider standard deviation in $\delta^{13}C$ values (1.29) than the semi-captive deer faeces (0.65), suggesting that variability in stable isotope compositions of faeces may be an alternative to measure the dietary niche width of some mammalian herbivores.

Additionally, the output of the mixing model indicates that the most important food source to the captive deer's diet was the commercial feed followed by the pine bark (Fig. 4). However, the nitrogen contribution of pine bark was low, indicating the poor nutritional value of this food source [18]. Furthermore, the differential contribution of each food source to the captive deer's diet on a given day resulted in the lack of isotopic equilibrium of their faeces (Fig. 2) and substantial variation on the relative contribution of each food source along the course of our investigation (Fig. 3 and 4). Thus, illustrating the fidelity with which faecal $\delta^{13}C$ and $\delta^{15}N$ can reflect the composition of food consumed by white-tailed deer over a short period of time, enabling investigations of short-term dietary fluctuations where the available food sources have distinct isotopic compositions as previously suggested in other investigations [e.g. 12,21]. For example, some white tailed deer populations inhabit some dry tropical forests where they may feed on plants with distinct photosynthetic pathways (C_3, C_4 and CAM) or nitrogen fixation capabilities, and where abundance and nutritional contributions to the deer's diet may vary seasonally [31,32].

In summary, the diet-faeces isotopic trophic shift values we produced expand the potential uses of stable isotopes in white-tailed deer because they can be used in stable isotope investigations in areas where white-tailed deer occur at low densities and invasive manipulation of wild animals is not desired. Carbon and nitrogen stable isotopes of faeces may be used to measure the relative consumption of isotopically distinct food sources like C_3 and C_4 forages, or nitrogen isotopes in faeces can be used to help partition contributions of food sources if they too differ in $\delta^{15}N$ as well as to measure the consumption of supplemental feeds in areas where the habitat is managed for proper nutrition. Moreover, since white-tailed deer faecal isotopic analyses record dietary trends is short periods of time, they can be used to measure seasonal diet variability (e.g. rainy vs. dry seasons in the tropics).

Acknowledgments

We thank Carolina Hernández and Teresa de Jesús Pérez for field and laboratory assistance; the 'The Rancho Ecologico el Mirador' (REEM) people provided support to undertake research at their facilities, especially Arturo Orozco and Luis Bello. We are also thankful to Carlos M. Durán for all plant identifications. This research was conducted with financial support from CONACYT (project No.130702) and the INECOL postdoctoral scholarship granted to the primary author. Logistic support was provided by the 'Red de Biología y Conservación de Vertebrados' and valuable scientific input was provided during the 'Academic Retreat' organized by INECOL.

References

Five "key references", selected by the authors, are marked below (Three recommended (●) and two highly recommended (●●) papers).

1. ●● Kelly, J.F. 2000. Stable isotopes of carbon and nitrogen in the study of avian and mammalian trophic ecology. Can. J. Zool. 78: 1–27. http://dx.doi.org/10.1139/z99-165
2. Bearhop, S., Adams, C.E., Waldron, S., Fuller, R.A. & Macleod, H. 2004. Determining trophic niche width: a novel approach using stable isotope analysis. J. Anim. Ecol. 73: 1007–1012. http://dx.doi.org/10.1111/j.0021-8790.2004.00861.x
3. Inger, R. & Bearhop, S. 2008. Applications of stable isotope analyses to avian ecology. Ibis 150: 447–61. http://dx.doi.org/10.1111/j.1474-919X.2008.00839.x

4. ● Wittmer, M.H.O.M., Auerswald, K., Schönbach, P., Shäufele, R., Müller, K., Yang, H., Bai, Y.F., Susenbeth, A., Taube, F. & Schnyder, H. 2010. Do grazer hair and faeces reflect the carbon isotope composition of semi-arid C_3/C_4 grassland? Basic Appl. Ecol. 11: 83–92. http://dx.doi.org/10.1016/j.baae.2009.10.007

5. O'Leary, M.H. 1981. Carbon isotope fractionation in plants. Phytochemistry 20: 553–567. http://dx.doi.org/10.1016/0031-9422(81)85134-5

6. Fleming, T.H., Nuñez, R.A. & da Silveira, L.S.L. 1993. Seasonal changes in the diets of migrant and non-migrant nectarivorous bats as revealed by carbon stable isotope analysis. Oecologia. 94: 72–75. http://dx.doi.org/10.1007/BF00317304

7. DeNiro, M. & Epstein, S. 1981. Influence of diet on the distribution of nitrogen isotopes in animals. Geochim. Cosmochim. Ac. 45: 341–351. http://dx.doi.org/10.1016/0016-7037(81)90244-1

8. Heaton, J.H.E., Vogel, J.C., Von la Chevallerie, G. & Collet, G. 1986. Climatic influence on the isotopic omposition of bone nitrogen. Nature 322: 822–23. http://dx.doi.org/10.1038/322822a0

9. Sponheimer, M., Robinson, T., Ayliffe, L., Roeder, B., Hammer, J., Passey, B., West, A., Cerling, T., Dearing, D. & Ehleringer, J. 2003b. Nitrogen isotopes in mammalian herbivores: hair $\delta^{15}N$ values form a controlled feeding study. Int. J. Osteoarchaeol. 13: 80–87. http://dx.doi.org/10.1002/oa.655

10. ●● Darr, R. & Hewitt, D.G. 2008. Stable isotope trophic shifts in white-tailed deer. J. Wildlife Manage. 72(7): 1525–1531.

11. McCutchan, J.H.J., Lewis, W.M., Kendall, C. & McGrath, C.C. 2003. Variation in trophic shift for stable isotope ratios of carbon, nitrogen, and sulfur. Oikos 102: 378–90. http://dx.doi.org/10.1034/j.1600-0706.2003.12098.x

12. ● Sponheimer, M., Robinson, T., Ayliffe, L., Passey, B., Roeder, B., Shipley, L., Lopez, E., Cerling, T., Dearing, D. & Ehleringer, J. 2003. An experimental study of carbon-isotope fractionation between dier, hair, and feces of mammalian herbivores. Can. J. Zool. 81: 871–76. http://dx.doi.org/10.1139/z03-066

13. Schwarcz, H.P. 1991. Some theoretical aspects of isotope paleodiet studies. J. Archaeol. Sci. 18: 261–76. http://dx.doi.org/10.1016/0305-4403(91)90065-W

14. Phillips, D.L. 2001. Mixing models in analyses of diet using multiple stable isotopes: a critique. Oecologia 127: 166–70.

15. Phillips, D.L. & Koch, P.L. 2002. Incorporating concentration dependence in stable isotope mixing models. Oecologia 130: 114–25.

16. ● Codron, D., Codron, J., Sponheimer, M., Lee-Thorp, J.A., Robinson, T., Grant, C.C. & de Rutter, D. 2005b. Assessing diet in savanna herbivores using stable carbon isotope ratios of faeces. Koedoe. 48(1): 115–24. http://dx.doi.org/10.4102/koedoe.v48i1.170

17. Botha, M.S. & Stock, W.D. 2005. Stable isotope composition of faeces as an indicator of seasonal diet selection in wild herbivores in southern Africa. S. Afr. J. Sci. 101: 371–74.

18. Codron, D., Codron, J., Sponheimer, M. & de Ruiter, D. 2005a. Animal diets in the Waterberg based on stable isotopic composition of faeces. S. Afr. J. Wildl. Res. 35(1): 43–52.

19. Codron, D., Codron, J., Lee-Thorp, J.A., Sponheimer, M., de Ruiter, D., Sealy, J., Grant, R. & Fourie, N. 2007. Diets of savanna ungulates from stable carbon isotope composition of faeces. J. Zool. 273: 21–29.
http://dx.doi.org/10.1111/j.1469-7998.2007.00292.x

20. Codron, D. & Codron, J. 2009. Reliability of $\delta^{13}C$ and $\delta^{15}N$ in faeces for reconstructing savanna herbivore diet. Mammalian Biology-Zeitschrift fur Saugetierkunde. 74: 36–48.
http://dx.doi.org/10.1016/j.mambio.2007.12.005

21. Gannes, L.Z.C., Martinez del Rio, C. & Koch, P. 1998. Natural abundance variations in stable isotopes and their potential uses in animal physiological ecology. Comp. Biochem. Phys. A. 119: 725–737.
http://dx.doi.org/10.1016/S1095-6433(98)01016-2

22. Hwang, Y.T., Millar, J.S. & Longstaffe, F.J. 2007. Do $\delta^{15}N$ and $\delta^{13}C$ values of feces reflect the isotopic composition of diets in small mammals? Can. J. Zool. 85: 388–96.
http://dx.doi.org/10.1139/Z07-019

23. Steele, K.W. & Daniel, R.M. 1978. Fractionation of nitrogen isotopes by animals: a further complication to the use of variations in the natural abundance of ^{15}N for tracer studies. J. Agr .Sci. 90: 7–9.
http://dx.doi.org/10.1017/S002185960004853X

24. Ben-David, M. & Schell, D.M. 2001. Mixing models in analyses of diet using multiple stable isotopes: a response. Oecología 127: 180–184.

25. Cerling, T.E. & Harris, J.M. 1999. Carbon isotope fractionation between diet and bioapatite in ungulate mammals and implications for ecological and paleoecological studies. Oecologia. 120: 347–63.
http://dx.doi.org/10.1007/s004420050868

26. Oliveira, A.C.B., Marinelli, L.A., Moreira, M.Z., Soares, M.G.M., Cyrino, J.E.P. 2006. Seasonality of energy sources of *Colossoma macropomum* in a flood plain lake in the Amazon-lake Camaleaon, Amazonas, Brazil. Fisheries Manag. Ecol. 13: 135–42.
http://dx.doi.org/10.1111/j.1365-2400.2006.00481.x

27. Codron, D., Codron, J., Sponheimer, M., Bernasconi, S.M.. Clauss, M. 2011. When animals are not quite what they eat: diet digestibility influences ^{13}C-incorporation rates and apparent discrimination in a mixed-feeding herbivore. Can. J. Zool. 89: 453–465.
http://dx.doi.org/10.1139/z11-010

28. Codron, D., Sponheimer, M., Codron, J., Hammer, S., Tschuor, A., Braun, U., Bernasconi, S.M., Clauss, M. 2012. Tracking the fate of digesta ^{13}C and ^{15}N compositions along the ruminant gastrointestinal tract: does digestion influence the relationship between diet and faeces? Eur. J. Wildl. Res. 58: 303–313.
http://dx.doi.org/10.1007/s10344-011-0581-3

29. Sare, D.T.J., Millar, J.S. & Longstaffe, F.J. 2005. Tracing dietary protein in red-backed voles (*Clethrionomys gapperi*) using stable isotopes of nitrogen and carbon. Can. J. Zool. 83: 717–25.
http://dx.doi.org/10.1139/z05-064

30. Macko, S.A., Fogel, M.L., Hare, P.E. & Hoering, T.C. 1987. Isotopic fractionation of nitrogen and carbon in the synthesis of amino acids by microorganisms. Chem. Geol. 65: 79–92.
http://dx.doi.org/10.1016/0168-9622(87)90064-9

31. Ramírez, R.G., Quintanilla, J.B. & Aranda, J. 1997. White-tailed deer food habits in northeastern Mexico. Small Ruminant Res. 25: 141–46. http://dx.doi.org/10.1016/S0921-4488(96)00960-1

32. Arceo, G., Mandujano, S. & Gallina, S. 2005. Diet diversity of white-tailed deer (*Odocoileus virginianus*) in a tropical dry forest in Mexico. Mammalia 69(2): 159-168. http://dx.doi.org/10.1515/mamm.2005.014

16

On the Reliability of Expert Identification of Small-Medium Sized Mammals from Camera Trap Photos

P.D. Meek[1,2,*], K. Vernes[2] & G. Falzon[3]

[1] NSW Department of Primary Industries and Invasive Animal CRC, PO Box 530, Coffs Harbour, NSW 2450, Australia.

[2] Ecosystem Management, University of New England, Armidale, NSW 2351, Australia.

[3] C4D: Centre 4 Dimensions, University of New England, Armidale, NSW 2351, Australia

* Corresponding author e-mail: paul.meek@invasiveanimals.com.

Keywords

Automated identification;
Camera trapping;
Identification questionnaire;
Wildlife surveys.

Abstract

Camera trapping in scientific research has captivated practitioners globally and is now widely used as a primary survey method. Using photos to identify some species, especially coexisting small species of similar appearance and niche is questionable and if incorrectly identified can lead to poor conservation and management outcomes.

The aim of our investigation was to test how accurately mammalogists with expertise in mammal identification could identify a range of small-medium sized species recorded during camera trap surveys. This study presents the results of an Australian case study but the findings apply broadly and are relevant throughout the world's ecosystems.

Using an internet survey, 58 questions were posed to assess practitioner expertise in mammal trapping, and their accuracy in identifying 21 photos of 10 small-medium sized Australian mammal species. Particular focus was placed on the identification of the rare and threatened Hastings River Mouse (*Pseudomys oralis*) but other rodent species such as the Black Rat (*Rattus rattus*), the Bush Rat (*Rattus fuscipes*), and the Swamp Rat (*Rattus lutreolus*) were included. The survey indicated that the correct identification of some small mammals, in particular rodents, is highly variable between images of the same species, and that as a whole the professional wildlife community performs poorly at the identification of such species. Identification was more accurate where species were less likely to be confused with similar looking species, or where their identification was simple and/or obvious. Accurate identification of small-medium sized mammals, in particular rodents using camera trap imagery is difficult and this study highlights the potential discrepancies that may arise in wildlife research. Robust methods to assist in minimising errors and bias of animal identification need to be explored. In circumstances where similar looking species coexist, or in species rich ecosystems, a reliance on camera trapping alone may lead to mistaken identity.

Introduction

The use of camera traps to investigate wildlife ecology is being used widely as a cheaper, convenient, versatile and multi-faceted alternative to classic census methods such as cage traps, hair-tubes and pitfalls [1-5], and also allows for continuous monitoring across a greater spatial and temporal scale [3, 6-8]. Camera traps were predominantly designed for hunters to detect large northern hemisphere mammals such as deer, and the technology has been developed accordingly [9, 10], although some manufacturers have begun specifically servicing wildlife research needs [10, 11]. The scientific interest in using camera traps for smaller animals has posed detection probability issues. To complicate matters more, where small similar looking mammals are being studied, using photos to identify to species can be difficult, subjective, and biased by observer expertise regardless of their level of experience. There are no dichotomous keys for distinguishing species from images accurately and this is one of the "Achilles-heals" of camera trapping.

In recognition of the potential benefits of camera trapping and the possible consequences of a misidentification in mind, we ask the question "*With what degree of accuracy can wildlife professionals identify camera trap images of rodents and other small-to-medium sized mammals?*" In Northern New South Wales and Southern Queensland, Australia where there are several species of small-medium sized mammals, it can be challenging to use photos to identify rodents such as the Hastings River Mouse (*Pseudomys oralis*), Bush rat (*Rattus fuscipes*), Fawn-footed Melomys (*Melomys cervinipes*) and medium sized animals like Long-nosed Potoroo (*Potorous tridactylus*), Long-nosed Bandicoot (*Peremeles nasuta*) and Northern Brown Bandicoot (*Isoodon macrourus*). However, the issue is not specific to this study area or these species alone, the same issues can occur throughout the world with species such as civets, martens, shrews, rodents and a plethora of small-medium sized mammals.

Methods

For the purposes of this study and analysis of the anonymous group of professional mammal participants, we hereafter refer to them as the 'wildlife community'.

Camera Trap Image Collection

A sample of images of ten verified species of small-medium sized mammals, with a focus on *P. oralis* were collected by one of us during a number of camera trap investigations. Small-medium sized mammals are defined as those falling within the range of 35-5500 gm [12]. Species identification was confirmed for each image based on the known trapped records and/or repeated images of the animals of each species from NSW study sites in Carrai Plateau National Park, Mutton Bird Island, Fox and Berry Farm, and Mt Hyland Nature Reserve [7, 13, 14]. In the case of *P. oralis* and *R. fuscipes*, animals were trapped prior to camera trapping and did not coexist, so there was no potential for confusion. Colour images were taken using either Pixcontroller

DigitalEye (Pixcontroller, Pennsylvania) or Cuddeback Capture (Cuddeback, Wisconsin) camera traps whilst infrared images were taken using Reconyx HC600 (Reconyx, Wisconsin) camera traps. The Cuddeback Capture was set to take an image each trigger using the factory setting delay of 30 seconds, Pixcontroller camera traps were set to continuously trigger when the animal was in front of the device and the Reconyx HC600 was set to take 5 photos per trigger with no delay.

Survey Methodology Design

We designed an internet based survey targeting fauna surveyors who have worked in Northern NSW and Southern Queensland with specific experience trapping and handling small-medium sized mammals. The survey was deliberately biased in that our target cohorts were members of the Australian wildlife community, with expertise in mammal trapping and handling. We were seeking a generalist rather than a specialist audience in regards to familiarity with both camera traps and rodent species. Although we acknowledge that the cohort of experts that can identify images of particular species in our study (such as *P. oralis*) is restricted to a small number, the survey was intended to mimic the practical reality of a mammal ecologist attempting to reliably identify small mammal/rodent species captured by camera traps.

Using the internet tool SurveyMonkey™ (www.surveymonkey.com) respondents were canvassed using several mammal groups (Australasian Wildlife Management Society, Australian Mammal Society, Facebook Camera Trap Page, and the Yahoo Camera Trap LISTSERVER). Participants were invited to answer 16 questions which helped us establish the background profiles of the respondents in the wildlife community. A sequence of 21 camera trap photos of small-medium sized mammals was then displayed and the respondent asked to identify the species present within each image. All images contained only a single animal to ensure that only a unique species label could be assigned. Multiple examples of species such as *P. oralis* were presented for variability. There were 58 questions in total (Appendix 1) and a supplementary website of the same camera trap images was provided as an image library resource. A total of 10 unique species were presented in the image sequence: *Antechinus stuartii* (n = 2), *I. macrourus* (n = 1), *M. cervinipes* (n = 2), *P. nasuta* (n = 1), *P. tridactylus* (n = 1), *Pseudocheirus peregrinus* (n = 1), *P. oralis* (n = 6), *R. fuscipes* (n = 2), *R. lutreolus* (n = 2), *R. rattus* (n = 2). The individual level responses were censored to ensure that we could not associate incorrect answers with an individual respondent. This feature was incorporated by study design as we wanted to determine the overall species identification skills of the wildlife community as a whole rather than an individual. Because there were multiple respondents for each image, we had sufficient information to obtain overall species identification accuracy estimates for the wildlife community, particularly for our target species *P. oralis*.

Statistical Analysis

The response (accuracy estimate) was a percentage and hence defined on the interval [0,100] therefore we used compositional data analysis methodology. Mean classification accuracy estimates and their uncertainties for each species were calculated using the Aitchison simplex [15]. An isometric log transformation was applied to the wildlife community level species accuracy estimates and generalized linear models used for inference for rodent-non-rodent, *P. oralis*-other rodents, small mammal-medium mammal contrasts. Fligner-Killeen tests were also used to assess the null hypothesis of equal variances of the accuracies within the groups in these contrasts [16]. Multiple comparison corrections using the false discovery rate method were performed when statistical significance was detected [17]. All analysis was conducted using R v2.15.0 software [18] and the package compositions [19].

Results

Respondent Profile

A total of 158 respondents completed the community profile questions. Based on these responses a profile of the respondent wildlife community and a 'typical' member within it could be created. All respondents had participated in catch and release trapping of small-medium sized mammals for at least 1 year but the largest proportion of the wildlife community had been doing so for 10-20 years (28.5%). The biggest proportion of respondents had captured and handled rodents or small dasyurids (*Antechinus*, dunnarts) within the last 6 months (38.6 %) or one month (18.4%). A small percentage of the wildlife community claimed that they had never handled such species (4%). Most respondents had handled rodents (98.1%) and small dasyurids (97.5%) on multiple occasions.

Some respondents stated they had caught and handled medium-sized mammals (large dasyurids, bandicoots, possums, pademelons, and potoroos) in the last 5 years (42.4%) while 32.3% said they had within the last 6 months, with a small percentage (5.1%) having never handled such mammals. The majority (86.1%) of respondents had caught and handled multiple numbers of bandicoots, but only 58.2% had caught and handled multiple numbers of potoroos.

Test Rodent Experience

The respondents experience in handling the rodent species in the test image sequence was also recorded. Of the target species, only 12% had caught and handled *P. oralis*, and 29.7% *M. cervinipes* but 90.5% had caught and handled *R. fuscipes*, 69.0% *R. lutreolus*, and 78.5% *R. rattus*. Of those who have caught and handled *P. oralis* (n=22), 54.6% had only handled 1-4 animals with 9.1% stating they had caught and handled hundreds of this species.

Camera Trap Experience

Of all respondents 53.2% of respondents used camera traps for small mammal surveys (Fig. 1), while 74.5% used camera traps for medium sized mammal surveys and 60.0% for large mammals. In total 3.2% of respondents skipped the previous three questions adding further uncertainty to these estimates. Among those participants who use camera traps for mammal surveys, most had being doing so for 1-4 years (72.6%) but 14.4% had been doing so for less than 1 year, with a small minority (2.6%) having greater than 10 year experience (Fig. 2).

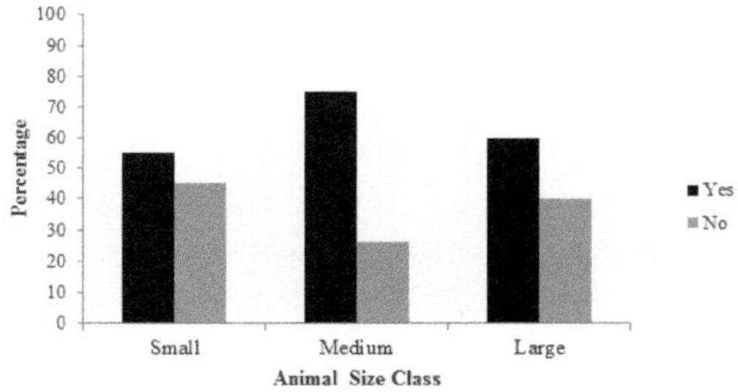

Fig. 1: The use of camera traps by respondents (n=158) for small, medium and large sized mammals.

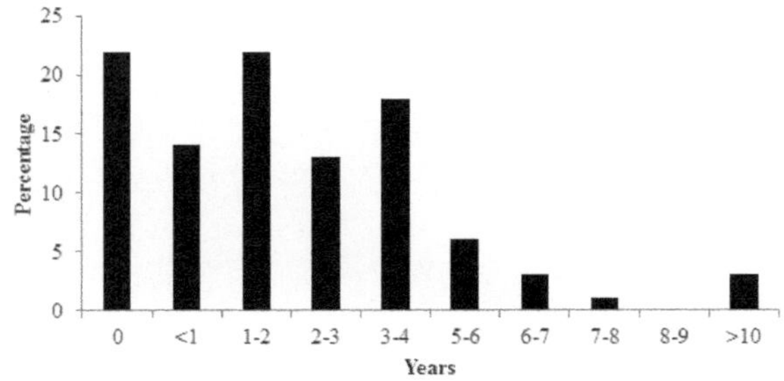

Fig. 2: The number of years respondents (n=158) have been using camera traps for fauna surveys.

Species Identification Methodology and Accuracy

Participants stated that they generally did not use an image reference library/database (57.1%) to assist in their identification of species caught on camera traps. The most common references used were: (a) reference books and guides (73.2%), (b) personal photo library (51.0%) and (c) internet photo library (30.1%). Note that multiple responses were possible to this question hence the percentage values sum to greater than 100%. A small proportion of users reported they use new technologies such as Smartphone applications (2% - 3 respondents), but some still use traditional methods like museum collections (11.8% - 18 respondents). Interestingly, in this survey only 10.7% of participants used the website of images provided to aid identification (Fig. 3) in contrast to 42.9% stating in previous questions that they do use resources.

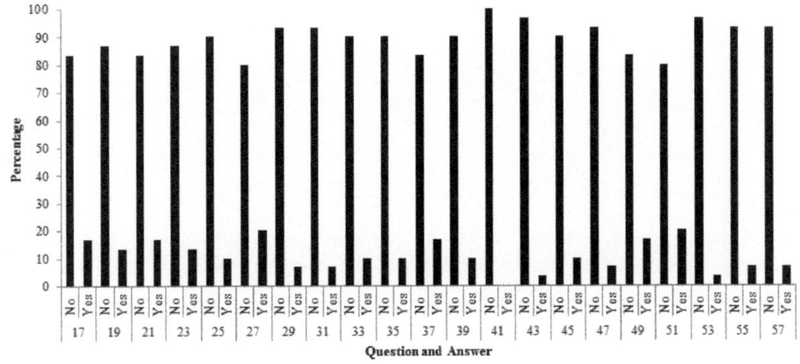

Fig. 3: The proportion of respondents (n=30) who used the supplementary blog site of images as an additional identification resource.

Thirty members (19%) of the wildlife community (n=158) completed the animal identification section of the survey. Average species classification percentages and their uncertainties are displayed in Table 1 and community level species accuracy percentages for specific contrasts are displayed in Tables 2, 3 and 4.

The accuracy of genus identification for the wildlife community on a whole was reasonable (71.1%) but the accuracy of species identification quite low (44.5%) (Table 1). The percentage of accurate genus identification was always greater than or equal to accurate species identification. This result was expected and comparisons between the genus and species accuracy percentages were not made further.

Table 1: Aitchison simplex species mean accuracies and uncertainties (square root of mean square distance). The symbol 'NA' means unable to compute as there were no replicates. N.B. Results on the simplex will be different from the arithmetic mean and the uncertainties are akin to standard deviations constrained to the interval [0,100].

Species	Genus Accuracy (%)	Genus Uncertainty (%)	Species Accuracy (%)	Species Uncertainty (%)
A. stuartii	72.11	1.54	38.88	1.54
I. macrourus	91.00	NA	84.00	NA
M. cervinipes	70.81	34.96	59.00	48.54
P. oralis	56.53	32.16	40.15	30.79
P. nasuta	14.00	NA	13.00	NA
P. peregrinus	100.00	NA	100.00	NA
P. tridactylus	83.00	NA	83.00	NA
R. fuscipes	77.28	27.00	34.00	30.07
R. lutreolus	84.00	51.48	34.57	33.78
R. rattus	85.04	7.89	33.98	31.66
Overall	71.05	17.07	44.46	27.21

Table 2: Aitchison simplex mean accuracies and uncertainties (square root of mean square distance) for selected groupings.

Groups	Accuracy (%)	Uncertainty (%)
P. oralis	40.15	30.79
Rodents excluding *P. oralis*	42.51	34.84
Rodents	41.56	33.65
Non-rodents	51.83	48.17
Small mammals	41.25	31.74
Medium size mammals	58.32	15.76

Table 3: Results of statistical testing for difference in species accuracy percentages selected contrasts. Notes [a]: a binomial family generalised linear model (BGLM) with a log link function fit, [b] BGLM with a cauchit link function, [c] BGLM with a logit link function.

Contrast	df	Slope	Std. Error	z-statistic	p-value
P. oralis – Other rodents[a]	13	0.2198	0.6601	0.333	0.7391
Rodent-Non-rodent[b]	19	1.8967	1.4686	1.292	0.1970
Small mammal – Medium mammal[c]	19	1.1289	0.9138	1.235	0.2167

Table 4: Results of Fligner-Killeen median tests for homogeneity of variances for selected contrasts. The symbol * indicates the p-value magnitude post multiple comparison correction.

Contrast	df	med χ^2	p-value
P. oralis – Other rodents	1	0.0807	0.7764
Rodent-non-rodent	1	5.9862	0.0144 (0.0433*)
Small mammal – Medium mammal	1	2.0445	0.1528

The identification accuracy of *P. oralis*, was estimated at 40.2% with a large uncertainty of 30.8% (refer Tables 2 and 3). This large uncertainty value was more due to the wide range of accuracies (17-70%)(Fig. 4) across the 6 different images of *P. oralis* rather than the limited number of image replicates. These results suggest that in some cases the wildlife community as a whole can reliably identify *P. oralis* in camera trap imagery and in other cases they cannot. This is likely to have been influenced by the small number of practitioners who have handled enough *P. oralis* to be confident with their distinguishing features. Statistical comparison between the mean species accuracies for *P. oralis* as compared to other rodents via a binomial generalized linear model (BGLM) did not reveal a statistically significant difference ($z = 0.333$, $p = 0.7391$), indicating that overall wildlife community accuracies in the identification of *P. oralis* from camera trap imagery compared to other rodents are similar, so poor identification accuracies irrespective of rodent species can be expected. The variability in the wildlife community accuracies for *P. oralis* when compared to other rodents did not vary significantly ($p = 0.7764$) (Table 4).

a) 70% b) 30% c) 17%

Fig. 4: Camera trap images of *P. oralis* and the estimated accuracy rates.

The accuracies of rodent species identification compared to non-rodent species identification showed an average expected accuracy of 41.6% when the species in the image was a rodent, with a wide range of values (33.6%). We found a non-significant result when we compared the species accuracies of the rodent-non-rodent contrasts (see Table 3) ($z = 1.292$, $p = 0.197$). There was no evidence in the survey results for a difference in the accuracies of species identifications irrespective of whether the true species was a rodent or not. Tables 2 and 4 indicate (even after a multiple comparison correction) that the variation in the accuracies for rodents is statistically significantly lower (uncertainty decrease =14.52%, $p = 0.04326$) than that of non-rodents. The general spread of values of accuracies for the rodent species was lower than that observed for non-rodents.

In comparing small mammal identification to medium sized mammal identification the mean accuracies appeared different (41.25% as compared to 58.32%) with considerable variation within these two groups (31.74% compared to 15.76%) resulting in overlap of the two classes (Table 2). Statistical inference via a BGLM (Table 3) revealed that there was no statistically significant difference in the mean values of these two groups ($z = 1.235$, $p = 0.2167$). On average the accuracies of small mammal identification compared to medium sized mammal identification was similar, although there was no statistically significant difference in the spread of accuracies between these two groups ($p = 0.1528$) (Table 3).

Participants commonly misidentified the genus of the species; *Rattus* was often confused with *Pseudomys* and *Melomys*, and for medium-sized mammals, *Peremeles* and *Isoodon* were commonly confused. Table 5 provides a list of most commonly confused genera for the suite of species used in the study.

Table 5: Genera that were misidentified using a series of camera trap images of small-medium sized mammals in North-eastern NSW and South-eastern Qld.

Species	Confused Genus
Peremeles nasuta	*Isoodon*
Antechinus stuartii	*Planigale*
Antechnius stuartii	*Rattus*
Isoodon macrourus	*Peremeles*
Melomys cervinipes	*Rattus*
Melomys cervinipes	*Antechinus*
Potorous tridactylus	*Isoodon*
Pseudocheirus peregrinus	*Rattus fuscipes*
Pseudomys oralis	*Rattus*
Pseudomys oralis	*Rattus*
Pseudomys oralis	*Rattus*
Pseudomys oralis	*Rattus*
Pseudomys oralis	*Rattus*
Pseudomys oralis	*Rattus*
Rattus fuscipes	*Melomys*
Rattus fuscipes	*Melomys*
Rattus fuscipes	*Pseudocheirus & Trichosurus*
Rattus lutreolus	*Hydromys*
Rattus lutreolus	*Antechinus & Isoodon*
Rattus rattus	*Mus*
Peremeles nasuta	*Isoodon*

Discussion

Camera trapping of wildlife is very popular worldwide and is appealing to the growing number of groups including citizen science programs [5, 20-23]. This study has shown the complexities of accurately identifying small-medium sized mammals even when the practitioners are trained and experienced; highlighting the risks involved where less experienced operators may be charged with identification responsibilities. These risks are clearly less concerning where the species identification is straight forward. Although where small-medium sized mammals of similar pelage colour and morphology coexist, the reliance on people to accurately identify these species can lead to some spurious identification.

We acknowledge that there is a difference between trying to identify species as a participant of an internet survey, and having to identify animals as a part of research project. In an internet survey, when a question is too taxing, the participant can choose to miss the question or answer with a non-serious attempt. However, when a practitioner needs to identify an animal as a part of an investigation, a range of options to assist in identification will likely be pursued over a greater time period. Similarly, when attempting to identify an animal, the practitioner will have full access to a range of tools including the original sequence of images upon which they can manipulate to aid identification. Despite providing access to the images on a website, participants chose not to use this additional resource in their answers.

For expediency, we tried to make the survey as quick to complete as possible which precluded the chance to extract individual responses. Limiting the length of the study compromised the extent to which detailed contrasts could be made for individual species. As such, we grouped the species into categories such as rodent and non-rodent, and small mammal and medium sized mammal. Nonetheless such categories were sufficient for our primary investigation which sought to gain a general insight into the overall accuracy of the wildlife community in identifying *P. oralis* specifically, and rodents generally. This analysis was sufficient to determine that the overall species identification of *P. oralis* and rodents in camera trap imagery was poor and *P. oralis* did not have a wildlife community level accuracy any different from any of the other rodents in the study. The apparent lack of replication in images for some species in the study (such as *I. macrourus*) does not underlie its true design. Given that for each image there were 30 responses each providing a likely classification of species, the study design was, in essence, a nested one in which respondent's species assignment was nested within image.

This design allowed us to estimate average classification accuracies for the surveyed population as a whole rather than focus on those of any particular individual. By design this was an essential aspect of the study allowing us to obtain the parameters of most critical interest without impinging on the privacy of the respondents. Further inference within an image level or for respondents was neither desirable nor possible and we do not draw any conclusions about the individual responses because to do so would be a logical flaw.

It is our observation that overall identification accuracy is image dependent and probably influenced by factors such as camera-to-target distance, animal size and shape, illumination intensity and type, image quality (infrared monochrome or colour), individual colourations, and animal pose. Some of these issues can be improved by careful choice of camera type, distance between camera trap and detection points, the use of baits to hold animals in front of the camera and, importantly, accurate placement of the detection zone in relation to the size and passage of the animal [6, 9, 24-28].

It is unfortunate that only 19% of participants attempted to identify the animals provided in the survey; perhaps a reflection of their available time to commit to the survey, or a manifestation of uncertainty in identifying animals from an image. Irrespective of the reasoning, our data confirm that mammal experts experience great difficulty in accurately identifying some smaller size class mammals from camera trap images.

That is not to say that species specialists (who will be probably targeting particular species with camera traps anyway) or very determined individuals with access to supporting resources do not identify the target rodents in their images accurately. Rather this study emphasizes that when mammalogist are presented with a range of species outside their specialist range, they are likely to make mistakes in identification. Such a situation reflects real-life camera trapping deployments (particularly surveys for species difficult to identify) in which a number of different species are often detected and there is a large number of images to catalogue thereby constraining the amount of time that can be spent on attempting accurate identification of species in a single image. In diverse and complex mammal communities where large-scale camera trap surveys are being carried out [2, 20-23], accurate identification of sympatric species poses similar issues to those discussed above; so the problem is not exclusive to Australia.

At the very least, this study has highlighted the need to carefully question the reliability of some mammal identification in camera trap imagery and has demonstrated the pressing need for the development and incorporation of alternative methodologies which increase species classification accuracy. Moreover, if fauna surveyors cannot reliably identify rodent species (and potentially other small to medium sized mammal species) in their camera trap images it casts doubt over the value of this survey method for some species surveys. In such circumstances we recommend that expert identification advice is sought. In some cases it may be necessary to run concurrent surveys to derive statistical correction factors for camera trap image identification accuracy, albeit costly. Alternatively, perhaps practitioners could acknowledge the constraints of identification of small mammals using camera traps and adopt more robust survey methods such as traditional trapping. Furthermore automated identification models that remove the subjectivity and need for human derived identification are being developed [29] and if successful will have enormous value in all aspects of management, especially pest animal bio-security and endangered species conservation. These models use key properties of the animal from photos and compare them to a reference collection, thereby providing the most likely species assignment [29]. The creation and implementation of such identification tools will increase the overall utility and reliability of species identification (particularly for rodents). These technological advances will revolutionize camera trapping research.

Acknowledgments

We thank Guy Ballard for advice on the design of the questions in this survey, and the Australasian Wildlife Management Society and Australian Mammal Society for disseminating the invitations to participate. We are grateful to the 158 mammalogists who started the survey and the dedicated 30 who completed the critical part of the survey. Surveys were carried out under Human Ethics Approval Number HE12-091 issued by the University of New England Human Ethics Committee. Two anonymous referees made helpful comments that improved this manuscript.

References

Five "key references", selected by the authors, are marked below (Three recommended (●) and two highly recommended (●●) papers).

1. Swann, D. E., Hass, C. C., Dalton, D. C., and Wolf, S. A. (2004). Infrared-triggered cameras for detecting wildlife: an evaluation and review, Wildlife Society Bulletin 32, 357-365. http://dx.doi.org/10.2193/0091-7648(2004)32[357:ICFDWA]2.0.CO;2

2. Kays, R., Kranstauber, B., Jansen, P. A., Carbone, C., Rowcliffe, M., Fountain, T., and Tilak, S. (2009). Camera traps as sensor networks for monitoring animal communities. The 34th IEEE Conference on Local Computer Networks, 811-818.

3. Kays, R. W., and Slauson, K. M. (2008). Remote Cameras, *In* Noninvasive Survey Methods for carnivores: Methods and Analyses (Long, R. A., MacKay, P., Zielinski, W. J., and Ray, J. C., Eds.), Island Press, Washington.

4. Kucera, T. E., and Barrett, R. H. (2011). A History of Camera Trapping, *In* Camera Traps in Animal Ecology (O'Connell, A. F., Nichols, J. D., and Karanth, K. U., Eds.), pp 9-26, Springer, New York.
http://dx.doi.org/10.1007/978-4-431-99495-4_2

5. O'Connell, A. F., Nichols, J. D., and Karanth, K. U. (2011). Camera Traps in Animal Ecology Methods and Analyses, Springer, New York.
http://dx.doi.org/10.1007/978-4-431-99495-4

6. De Bondi, N., White, J. G., Stevens, M., and Cooke, R. (2010). A comparison of the effectiveness of camera trapping and live trapping for sampling terrestrial small-mammal communities, Wildlife Research 37, 456-465.
http://dx.doi.org/10.1071/WR10046

7. Meek, P. D. (2010). Remote Camera Monitoring of The Hastings River Mouse (*Pseudomys oralis*): Trial of a Novel Technique for Monitoring Populations. Unpublished Report for Gondwana Rainforests of Australia.

8. Tobler, M. W., Carrillo-Percastegui, S. E., Leite Pitman, R., Mares, R., and Powell, G. (2008). An evaluation of camera traps for inventorying large- and medium-sized terrestrial rainforest mammals. Animal Conservation 11, 169-178.
http://dx.doi.org/10.1111/j.1469-1795.2008.00169.x

9. Meek, P. D., Ballard, A. G., and Fleming, P. J. S. (2012). An Introduction to Camera Trapping for Wildlife Surveys in Australia., In Pestsmart Toolkit, Invasive Animals CRC, Canberra, ACT

10. Meek, P. D. (2012). Refining and Improving the Use of Camera Trap Technology for Wildlife Management and Research in Australia and New Zealand, The Winston Churchill Memorial Trust of Australia.

11. Meek, P. D., and Pittet, A. (2012). User-based design specifications for the ultimate camera trap for wildlife research, Wildlife Research 39, 649-660.
http://dx.doi.org/10.1071/WR12138

12. Burbidge, A. A., and McKenzie, N. L. (1989). Patterns in the modern decline of western Australia's vertebrate fauna: Causes and conservation implications, Biol. Conserv. 50, 143-198.
http://dx.doi.org/10.1016/0006-3207(89)90009-8

13. Meek, P. D., Zewe, F., and Falzon, G. (2012). Temporal activity patterns of the swamp rat (*Rattus lutreolus*) and other rodents in north-eastern New South Wales, Australia, Australian Mammalogy 34, 223-233.
http://dx.doi.org/10.1071/AM11032

14. Zewe, F. (2010). Like a Rat up a Drainpipe: A Black Rat Station to Protect Native Rodents., *In* Dept Environmental and Rural Science, University of New England, Armidale.

15. Aitchison, J. (1986). The Statistical Analysis of Compositional Data Monographs on Statistics and Applied Probability, Chapman & Hall Ltd., London (UK).
http://dx.doi.org/10.1007/978-94-009-4109-0

16. Conover, W. J., Johnson, M. E., and Johnson, M. M. (1981). A comparative study of tests for homogeneity of variances, with applications to the outer continental shelf bidding data, Technometrics 351-361.
http://dx.doi.org/10.1080/00401706.1981.10487680

17. Benjamini, Y., and Hochberg, Y. (1995). Controlling the false discovery rate: a practical and powerful approach to multiple testing., Journal of the Royal Statistical Society Series B. 57, 289-300.

18. R Development Core Team. (2012). R: A language and environment for statistical computing.,

R Foundation for Statistical Computing, Vienna, Austria. ISBN 3-900051-07-0, URL http://www.R-project.org/.

19. van de Boogart, K. G., Tolosana, R., and Bren, M. (2011). Compositions: Compositional data analysis, R package version 1.10-2, http://CRAN.R-project.org/package=compositions.

20. Kays, R., Tilak, S., Kranstauber, B., Jansen, P. A., Carbone, C., Rowcliffe, J. M., Fountain, T., Eggert, J., and He, Z. (2010). Monitoring wild animal communities with arrays of motion sensitive camera traps, International Journal of Research and Reviews in Wireless Sensor Networks 1, 19-29.

21. ● O'Brien, T. G., Baillie, J. E. M., Krueger, L., and Cuke, M. (2010). The Wildlife Picture Index: monitoring top trophic levels, Animal Conservation 13, 335-343.
http://dx.doi.org/10.1111/j.1469-1795.2010.00357.x

22. ● Ahumada, J. A., Silva, C. E. F., Gajapersad, K., Hallam, C., Hurtado, J., Martin, E., McWilliam, A., Mugerwa, B., O'Brien, T., Rovero, F., Sheil, D., Spironello, W. R., Winarni, N., and Andelman, S. J. (2011). Community structure and diversity of tropical forest mammals: data from a global camera trap network, Philosophical Transactions of the Royal Society B: Biological Sciences 366, 2703-2711.
http://dx.doi.org/10.1098/rstb.2011.0115

23. Fegraus, E. H., Lin, K., Ahumada, J. A., Baru, C., Chandra, S., and Youn, C. (2011). Data acquisition and management software for camera trap data: A case study from the TEAM Network, Ecological Informatics 6, 345-353.
http://dx.doi.org/10.1016/j.ecoinf.2011.06.003

24. ●● Claridge, A. W., Paull, D. J., and Barry, S. C. (2010). Detection of medium-sized ground-dwelling mammals using infrared digital cameras: an alternative way forward? Australian Mammalogy 32, 165-171.
http://dx.doi.org/10.1071/AM09039

25. Paull, D. J., Claridge, A. W., and Barry, S. C. (2011). There's no accounting for taste: bait attractants and infrared digital cameras for detecting small to medium ground-dwelling mammals, Wildlife Research 38, 188-195.
http://dx.doi.org/10.1071/WR10203

26. Claridge, A. W., Misfud, G., Dawson, J., and Saxon, M. J. (2004). Use of infrared digital cameras to investigate the behaviour of cryptic species, Wildlife Research 31, 645-650.
http://dx.doi.org/10.1071/WR03072

27. Nelson, J. E., Menkhorst, P., Howard, K, Chick, R. and Lumsden, L. (2009). The status of smoky mouse populations at some historical sites in Victoria, and survey methods for their detection. Arthur Rylah Institute for Environmental Research Unpublished report number 2009/17. Department of Sustainability and Environment, Heidelberg, Victoria.

28. Smith, J. K., and Coulson, G. (2010). A comparison of vertical and horizontal camera trap orientations for detection of potoroos and bandicoots, Australian Mammalogy, 34, 196-201.

29. ●● Falzon, G., Meek, P. D., and Vernes, K. (In Press). Computer-Assisted Identification of Australian Rodents in Camera Trap Imagery, *In* Camera Trapping in Wildlife Management and Research (Meek, P. D., Ballard, A. G., Banks, P. B., Claridge, A. W., Fleming, P. J. S., Sanderson, J. G., and Swann, D. E., Eds.), CSIRO Publishing, Melbourne, Australia.

Appendix: A list of 58 questions provided on an internet survey of animal surveyors to assess their ability to identify mammals using camera trap images taken in north-eastern NSW and south-eastern Qld. *# an alternative website was provided with the same images as shown in this survey so that participants could access the images for closer inspection, no information was provided with these images.*

How long have you been doing catch and release trapping of small-medium sized animals in Australia?
How recently have you captured and handled small mammals in Australia? (small mammals include rodents, small dasyurids, or similar-sized mammals)
Please indicate how many small dasyurids (*Antechinus*, Dunnarts, and similar) you estimate you have caught and handled in Australia?
Please indicate how many rodents (native and introduced rats and mice) you estimate you have caught and handled in Australia?
Which of the following rodents have you captured and handled?
Have you ever captured a Hastings River Mouse (*Pseudomys oralis*)?
Please estimate how many Hastings River Mouse (*Pseudomys oralis*) you have handled?
How recently have you captured and handled medium sized mammals in Australia? (Examples of medium sized mammals include large dasyurid, bandicoot, possum, pademelon, potoroo)
Please estimate how many bandicoots you estimate you have caught and handled, in Australia?
Please estimate how many potoroo and pademelon etc (Macropods) you estimate you have caught and handled, in Australia?
Do you use camera traps for small sized mammal surveys in Australia?
Do you use camera traps for medium sized mammal surveys in Australia?
Do you use camera traps for surveying large mammals in Australia?
If you answered yes to questions 11, 12 or 13, please estimate how many years you have been using camera traps
Do you use an image reference library / database to aid you in identifying animals photographed using camera traps?
What image reference information do you use?
Questions 17-58 comprise 21 animal photos each followed by the question below
Did you use the http://camera-trap.blogspot.com.au site to decide on the species identification of the image above?

17

Calling Activity of Indian Giant Flying Squirrel (*Petaurista philippensis* Elliot, 1839) in the Tropical Deciduous Forests, India

V. K. Koli[1,*] & C. Bhatnagar[1]

[1] Wildlife Research Laboratory, Department of Zoology, University College of Science, Mohanlal Sukhadia University, Udaipur (Rajasthan), 313 001, India.

* Corresponding author email: vijaykoli87@yahoo.in.

Keywords
Calling bout;
GBH;
Petaurista philippensis;
Perch height.

Abstract

The calling behaviour of *Petaurista philippensis* was studied from March 2009 to February 2010, in the tropical deciduous forests of the Sitamata Wildlife Sanctuary, Rajasthan, India. Surveys were conducted from the dusk or 1900 hr to 0600 hr. A total of 268 call bouts were recorded. Call frequency was highest in March 2009. The most vocal periods were midnight and early night, just before entering into the nest cavities. Annual, seasonal and daily call frequencies were insignificant ($P > 0.05$) between male and female. 5-10 min call bout category contained highest account (40%). Calling was mostly performed in the mid canopy area of the trees. Perch height was found positively related to tree height.

Introduction

Detailed information on the ecology of the flying squirrels is scarce, due perhaps to their low population density, arboreal, nocturnal and cryptic or elusive habits. According to Corbet and Hill [1] a total of 44 species of flying squirrels belong to 14 genera present worldwide. Except three species (*Glaucomys volans, G. sabrinus* and *Pteromys volans*) all other (Pet*aurista, Eupetaurus cinereus, Hylopetes, Eoglaucomys, Petinomys, Trogopterus, Belomys, Pteromyscus, Petaurillus, Lomys, Biswamoyopterus* and *Aeretes*) species are found in Asia [2]. *Petaurista philippensis* is dominant among all species and widely distributed in the tropical and sub-tropical zones of south-east Asia [3-5]. It is a typical inhabitant of the forests in lowlands and mountains up to 4000 m [6].

A total of 13 flying squirrel species found in India, wherein two are recently identified in last decade namely *Petaurista nigra* [7] and *P. mishmiensis* [8]. *P. philippensis* has a broader distribution among them and restricted to the northeast region of Himalaya (Tropical moist deciduous forest) and the Western Ghats (Tropical wet evergreen forest) [9-13]. A few patches of occurrence have been reported from the southern peninsula, which fall under moist temperate forest [12], whereas southern Rajasthan (Tropical deciduous forest) is an isolated distributional patch in tropical deciduous forests of western India [5, 14, 15]. *P. philippensis* roosts in cavity nests [5] and feeds on fruits, leaves, buds, flowers and pith [16-18], but subsists on pith in the monsoon season when preferred food unavailable [18]. It is mainly a nocturnal, solitary animal, but sometimes in tropical forest, where the temperature highly fluctuates over the year, the species was seen sleeping out from its hiding sites in the day time to rid itself of high temperature during the summer season [12, 19]. While, in winter season, the species basks in the sunlight for thermoregulation [20]. During locomotion, the species can glide more than 50 m, but prefers short ones (between 11-20m) [21]. Early summer season identified as its breeding period in India [21, 22].

According to IUCN, *P. philippensis* categorized least concern (LC) globally [23], but recent survey by Molur et al. [24] in South Asia considered it near threatened (NT). Although various aspects of its ecology have been widely studied in various parts of the world i.e., distribution [5, 9, 10, 15], nesting [5, 25], food habits [16-18, 26], gliding [21], home range [27], activity pattern [27], predation [28], phylogenetic relationships [6, 29] and conservation problems [14, 5, 10, 13] of *P. philippensis*, but calling activity remains unstudied due perhaps to its shy and cryptic habit. Considering the lacuna, we started studying *P. philippensis* to examine calling behaviour of *P. philippensis* as a part of an extensive study [15] in the tropical deciduous forest of southern Rajasthan, India and hope that it will add an additional valuable information to the natural history of this species.

Study area

The study was conducted in the Sitamata Wildlife Sanctuary (74°25'- 40' E and 24°04'- 23' N), located in the world's oldest Aravalli mountains of southern Rajasthan, India. The sanctuary covers an area of 423 km^2, and configuration of the land is hilly and rugged with considerable altitudinal variation from 280 to 600 m. The general slope of the land is from north-west to south-east. According to Champion and Seth [30], the forest of the sanctuary is tropical deciduous, dominated by *Tectona grandis, Terminalia tomentosa, Terminalia arjuna, Boswellia serrata, Madhuca indica, Erahatia levis, Diospyros melanoxylon, Soymida febrifuga, Ficus religiosa, Ficus amplissima, Cordia dichotoma, Aegle marmelos, Mangifera indica* and *Emblica officinalis*. The average annual rainfall is 756 mm with an average 30 rainy days. The temperature varies from 5°C to 45°C. The area is characterized by distinct winter (November-February), summer (March-June) and monsoon (June-October) seasons.

Methods

The study was conducted from March 2009 to February 2010. The study site was visited twice each month, for 4 nights each. Trails were walked at dusk (1900 hr) randomly and continuous till 0600 hr. During the survey, first we search for the flying squirrel by using spotlights. Distinct red eye shine [31], calls and movement of the animals were used to locate the animals during the night. When a flying squirrel was spotted, the spotlight was switched off for two minutes and then we followed the animal. When the animal escaped from our view, we moved on to find another squirrel and observation was continued. The same procedure was conducted till 0500 hr, when flying squirrels retreat to nest cavities and stop their activities [27]. Since white light bothers the nocturnal animals and hinder their activities [32], red light was used when observing flying squirrels' activities. More than 5 m distance maintained from the animal during observations because at this distance flying squirrels can tolerate human presence and appeared undisturbed [33]. Just after the encounter, flying squirrels stop their activities and become motionless in most cases, or hide into the foliage, but after a few minutes, they resume and start activities normally. Data collection was started with their normal activities. Disturbed activities were excluded from the study data. It was very difficult to identified same animal regularly in the night without labeled them. The similar colour pattern of the squirrels also restricted us from following the same individuals consistently and maintaining separate records for animals. So, every encountered squirrel considered as a new squirrel during the study. Sex was identified by the presence of scrotal sacs or external genitalia. During the study, a cavity was declared the nest when a flying squirrel used it more than five time and also confirm with consulting to local people or tribes of that area. Because *P. philippensis* calls upon a significant height above the ground, therefore we were not able to record the sound of the animal by any instrument on the ground, but we were able to listen.

When the animal started to call, we observed it quietly without making any noise or disturbance and recorded its activity and response, to the presence or calls of other animals or conspecifics. A call bout, consisted of calls, was defined when it was separated by intervals of ≥ 2 min. For all calls, we recorded the date, time, call bout duration, perch height, tree height and GBH (Girth at breast height, 1.37 m above the ground) of the tree. Bout duration was recorded by stopwatch. Perch height of the animal and tree height were measured by clinometer. During the survey, encounter rate was calculated as number of animals observed/travelled km.

The active period of flying squirrels was divided into three sessions for precise description of calling activity: first 3 hrs as early-night (1900 to 2200 hr), subsequent 4 hrs as mid-night (2200 to 0200 hr) and remaining 3 hrs as late-night (0200 to 0500 hr).

Single call bout used as a sampling unit. We used the Chi square test (χ^2) to evaluate the hypothesis that there was no difference between male and female in calling frequency. Student *t* test was used to evaluate call bout duration, GBH of used trees and perch height difference between sexes. Correlation between GBH and perch height was estimated as Pearson's product moment correlation (*r*). All statistical analysis were performed on GraphPad Prism statistical software (version 3.02).

Results

A total of 268 call bouts were recorded during the study period. Flying squirrels were heard in all months but the frequency varied (Fig. 1). Calling activity increased from April, 2009 to July 2009 and declined after September, 2009 till January, 2010. Flying squirrels were most vocal during March, 2009 and lowest in January, 2010. Difference in call frequency of males and females was insignificant during the whole year ($\chi^2 = 13.71$, $df = 11$, $P > 0.05$) as well as in the all seasons (Summer, $\chi^2 = 7.41$, $df = 3$, $P > 0.05$; Monsoon, $\chi^2 = 2.74$, $df = 3$, $P > 0.05$; Winter, $\chi^2 = 3.55$, $df = 3$, $P > 0.05$).

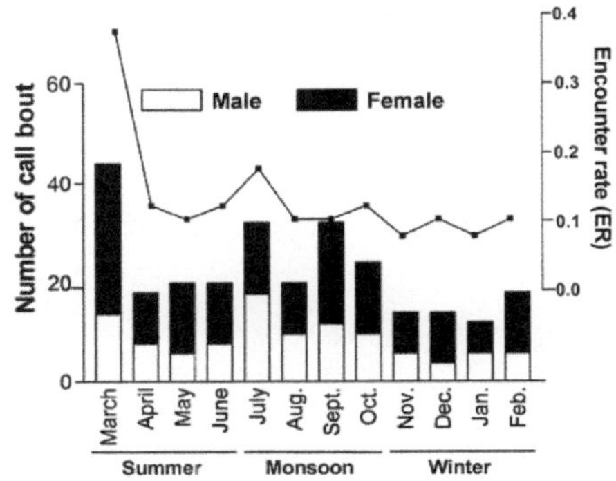

Figure 1: Monthly and seasonal variation in call frequency (in bars) and encounter rate (in lines) of *Petaurista philippensis* in the tropical deciduous forest, Rajasthan, India.

The highest encounter rate was in March 2009 while the lowest in November 2008 and January 2010 (Fig. 1).

During the night time, *P. philippensis* was found most vocal between 2200 to 0200 hr (Fig. 2). An initial small peak of calling activity was found between 1900-2000 hr and the second between 2300-0000 hr. The highest and third peak was found between 0400 to 0500 hr, just before animals entered in their nest cavaties. Difference in daily call frequency was insignificant between male and female (χ^2 = 12.029, *df* = 9, $P >$ 0.05).

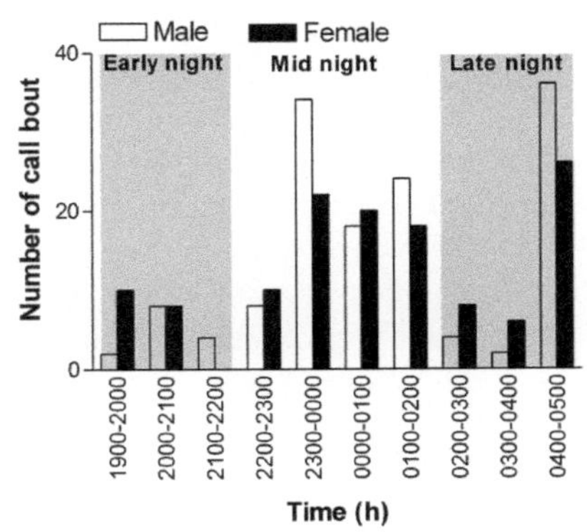

Figure 2: Call frequency of *Petaurista philippensis* during its active period (1900hr to 0500hr) in the tropical deciduous forest, Rajasthan, India.

Mean duration of calling bouts was 9.11±7.24 SD min (n = 268). 0-5 min category contained 32.08% (n = 86), 5-10 min category 40.29% (n = 108), 10-15 min category 12.68% (n = 34), 15-20 min category 4.47% (n = 12), 20-25 minute category 2.98% (n = 8) and 25-30 minute contained 7.46% (n = 20) calls respectively. Call bout duration was not significantly different between male and female (t = 0.007, *df* = 10, $P >$ 0.05).

Mean GBH of the used tree was 206.5 ± 58.80 *SD* cm. The most calls were recorded on the trees with 20-25m GBH trees. Highest was 320 cm and lowest was 80 cm. GBH of trees used by male and female was not significantly different (Table 1; t = 1.34, *df* =

5, $P > 0.05$). *P. philippensis* used mostly mid canopy area of trees during calling with mean used perch height was 19.40 ± 5.18 *SD* m and it was found positively related to tree height ($r = 0.87$, $n = 134$, $P < 0.05$). Highest perch height was 30 m and lowest was 7 m. A difference was insignificant in used perch height between male and female (Table 1; $t = 2.010$, $df = 5$, $P > 0.05$) and no significant correlation ($r = -0.045$, $n = 134$, $P > 0.05$) was found between perch height and GBH of trees that squirrels called from.

Table 1: Number of calls bout (n), GBH (cm) and perch height (m) (mean±SD) of used tree in different calling bout categories (minutes) of *Petaurista philippensis* in the tropical deciduous forest, Rajasthan, India.

Time	n		GBH		Perch height	
(min)	Male	Female	Male	Female	Male	Female
0-5	34	52	137.7±37.74	194.3±50.72	19.94±5.58	19.46±5.27
5-10	42	66	212.1±56.54	235.2±60.07	18.44±5.11	18.73±5.67
10-15	12	22	220.8±62.32	218.0±50	18.17±4.79	20.82±4.81
15-20	6	6	172.0±31.10	216±32.60	20.67±5.13	22.33±4.16
20-25	2	6	211.00±00.0	216.7±40.41	12.00±00.0	18.67±6.80
25-30	12	8	241.7±49.56	216±28.61	20.17±4.07	21.75±5.56

Without spectrographic analysis of the vocalization, it is difficult to separate different call types, but from the direct observation, we recorded some behavioural aspects of the animal and try to describe what we heard. In 48 cases, we observed that the calling animal got a response from a conspecific in the form of calls and reacted to these by changing its call intensity or by changing its position on the tree or by stopping its calling. Many times (n=26), the focal animal called repeatedly for over >20 minutes without any change in call intensity or its position, even if they got a response from other flying squirrels. This protracted unchanging calling was observed mostly (n=12) during mid-night (2200-0200 hr) and just before sun rise or entering into the nest cavities (n=14). During calling in the morning session, the all calling animals were always found on its nesting tree and approximately 5-10 m from the nest cavities.

Discussion

Petaurista species use distinct and unique vocalizations [34]. *P. petaurista* used low pitch sound while *P. alborufus* was found to be more vocal with high pitch whistle calls [33, 34]. Calls of flying squirrels have varied functions such as contact, territory defense, making pairs and parental care [33]. Ando and Kuramochi [35] stated that flying squirrels vocalize mainly for appealing about their existence to each other and it also seems adaptive to their nocturnal and arboreal life. Many species are used ultrasonic sound to increased predators avoidance [36]. Most frequent vocal species were found to defend its resources against other inter- and intraspecies [34], but to prove this hypothesis in the flying squirrels, more studies are require.

P. philippensis is mostly arborial solitary animal and only observed in parties during its breeding season. Nandini [9, 10] also frequently sighted *P. philippensis* alone during surveys in the different forest types *i.e.*, evergreen forests, moist deciduous forest, dry forests and different plantations (Coffee, Cardamon and Coconut plantation) of Western Ghats. In context to cryptic habits, Eisenberg [37] hypothesizes that arboreal folivores do not live in large social groups. Arboreal folivores are believed to have well developed communication for localizing their position in relation to each other [9]. Similarly, Lee et al. [35] found *P. petaurista* and *P. alborufus* in the solitary condition during the most study periods. So, because of their solitary and nocturnal habit, calls play a key role for localizing positions of their genetic relatives and making pairs during their breeding periods.

In the study period, calling in morning session at nesting tree seem related to some kind of social or territorial behaviour of *P. philippensis*. Some bird species defend their territories by territorial calls at morning before leaving their nests [38] and at evening after reaching at nest [39, 40]. Thus, just before entering into the nest cavity, this kind of high calling around the nest cavities seems likely related to territorial appeal by flying squirrels.

Since, early summer is identified as breeding season for *P. philippensis* in India [22], so high calling frequency in March, 2009 (Fig. 1) during the study period found likely related to its breeding activities. Frequently using calls during this period may be helping them to make contact between male and female and also to make mating pairs. Lower number of calls were recorded in winter season, possible due to temperature fall. Low temperature makes animals less active.

P. philippensis called frequently during the whole night, but found most vocal at mid-night and just before sunrise. The early-night period was found to be a less active period for calling. Bimodal activity pattern performed by *P. philippensis* [27]. Most activities are carried out at early-night and these activities are mainly devoted to foraging, feeding [17, 18] and gliding [21] but not to call. During the survey in the Western Ghats, Nandini [9] seen high calling individuals during the early night, which is inconsistent to the present study and suggests that the calling behaviour of flying squirrels is possibly affects by regional and topographic changes.

Arborality allows animals to escape predation, for gliding [41] and increase longevity in mammals [42]. Flying squirrels were found mostly in the middle canopy of the trees during making calls in the study period and perch heights were found to be increased with tree heights. Kumara and Singh [13] encounter *P. philippensis* in the mid canopy of 16 to 20 m tall trees of evergreen and deciduous forests of Western Ghats. In the mid canopy area, dense foliage and branches may act as obstacles to sound transmission, but also provide better cover for calling animals, whose sounds might otherwise expose them to an increased possibility of predation. Thus, preference to mid canopy during the calling seem likely related to minimize predation. Garstang et al. [43] stated that changes in ambient temperature with height above the ground also affects sound transmission and seems a possible reason that why *P. philippensis* use high perch heights with increasing tree height. Habitat selection of arboreal mammals is mainly influenced by the structure and composition of forest vegetation, feeding, nesting and mode of locomotion [9, 10].

Thus, from the above study it can be concluded that *P. philippensis* shows monthly and seasonal variations in the calling activity. Breeding season was found the most vocal period. Reproductive status, canopy cover and tree height play a key role in their calling. Preference to mid canopy during calling seems related to minimize predation.

Mid-night period and early hours before sunrise were the most vocal period of the animal.

Lee et al. [34] has suggested a spotlight count method, while Babu and Jayson [44] preferred call play back method for flying squirrels' census and from this study we recommend that mid-night and late-night or early morning are the optimum time for flying squirrels' census in tropical deciduous forest as they are most vocal at this time.

Acknowledgements

The study was funded by the University Grant Commission (UGC), Delhi, in the form of JRF. We thank Department of Forests, Rajasthan for permission to carry out this work in a protected area. We are also grateful to Dilip Mali for his help in field work.

References

Five "key references", selected by the authors, are marked below (Three recommended (●) and two highly recommended (●●) papers).

1. Corbet, G.B. & Hill, J.E. 1991. A World List of Mammalian Species, Third edition. Natural History Museum Publications & Oxford University Press, London and Oxford.p. 1-243.
2. Lee, P.F. & Liao, C.Y. 1998. Species richness and research trend of flying squirrel. J Taiwan Museum 51:1-20.
3. Wilson, D.E. & Reeder, A.E. 2005. Mammal species of the world: a taxonomic and geographic reference. 3rd ed. Johns Hopkins University Press, Baltimore, Maryland. p. 2142.
4. Corbet, G.B. & Hill, J.E. 1992. Mammals of the Indomalayan region. A systemic review. Oxford University Press, Oxford. p. 488.
5. ●● Koli, V.K., Bhatnagar, C. & Sharma, S.K. 2013. Distribution and status of Indian giant flying squirrel *Petaurista philippensis* Elliot) in Rajasthan, India. Natl Acad Sci Lett 36:27-33. http://dx.doi.org/10.1007/s40009-012-0105-z
6. Yu, F., Yu, F., Pang, J., Kilpatrick, C.W., McGuire, P.M., Wang, Y., Lu, S. & Woods, C.A. 2006. Phylogeny and biogeography of the *Petaurista philippensis* complex (Rodentia: Sciuridae), inter- and intraspecific relationships inferred from molecular and morphometric analysis. Mol Phylogen Evol 38:755-766. http://dx.doi.org/10.1016/j.ympev.2005.12.002
7. Choudhury, A. 2007. A new flying squirrel of the genus *Petaurista* link from Arunachal Pradesh in north-east India. The Newsletter & Journal of the Rhino Foundation for nat in NE India 7: 26-32.
8. Choudhury, A. 2009. One more new species of giant flying squirrel of the genus *Petaurista* Link, 1795 from Arunachal Pradesh in north-east India. The Newsletter & Journal of the Rhino Foundation for nat. in NE India 8: 26-34.
9. Nandini, R. 2001. The distribution and status of flying squirrels in Karnataka and Goa. Technical report. SACON, Coimbatore.
10. Nandini, R. 2001. The status and distribution of the small travancore flying squirrel (*Petinomys fuscocapillus fuscocapillus*) and the large brown flying squirrel (*Petaurista philippensis*) in the Western Ghats. Technical report. SACON, Coimbatore.
11. Umapathy, G. & Kumar, A. 2000. The occurrence of arboreal mammals in the rain forest fragments in the Anamalai Hills, south India. Biol Cons 92:311-319. http://dx.doi.org/10.1016/S0006-3207(99)00097-X

12. Prater, S.H. 2005. The book of Indian Animals. 5th ed. Bombay Natural History Society and Oxford University Press, p 194-197.

13. ● Kumara, H.N. & Singh, M. 2006. Distribution and relative abundance of giant squirrels and flying squirrels in Karnataka, India. Mammalia 70: 40-47. http://dx.doi.org/10.1515/MAMM.2006.006

14. Tehsin, R.H. 1980. Occurrence of the large brown flying squirrel and mouse deer near Udaipur, Rajasthan. J Bom Nat Hist Soc 77: 498.

15. Koli, V.K. 2012. Study on the occurrence, distribution and ethology of Indian giant flying squirrel (*Petaurista philippensis* Elliot) in southern Rajasthan. Ph. D. Thesis, Mohanlal Sukhadia University Udaipur, Rajasthan.

16. ● Nandini, R. & Parthasarathy, N. 2008. Food habits of the Indian giant flying squirrel (*Petaurista philippensis*) in a rain forest fragment, Western Ghats. J Mammal 89(6):1550-1556. http://dx.doi.org/10.1644/08-MAMM-A-063.1

17. Bhatnagar, C., Koli, V.K. & Sharma, S.K. 2010. Summer diet of Indian giant flying squirrel (*Petaurista philippensis* Elliot) in Sitamata wildlife sanctuary, Rajasthan, India. J Bom Nat Hist Soc 107:183-188.

18. ●● Koli, V.K., Bhatnagar, C. & Sharma, S.K. 2013. Food habits of indian giant flying squirrel (*Petaurista philippensis* Elliot) in tropical deciduous forest, Rajasthan, India. Mammal Study 38: 251-259. http://dx.doi.org/10.3106/041.038.0409

19. Bhatnagar, C., Sharma, S.K. & Koli, V.K. 2010. High day temperature and sleep out behaviour of Elliot's giant flying squirrel *Petaurista philippensis* (Elliot) in Sitamata Wildlife Sanctuary, Rajasthan, India. J Bom Nat Hist Soc 107: 245-246.

20. Koli, V.K., Bhatnagar, C. & Sharma, S.K. 2012. Sunbasking behaviour of Elliot's giant flying squirrel *Petaurista philippensis* (Elliot) in Sitamata Wildlife Sanctuary, Rajasthan, India. J Bom Nat Hist Soc 109:196-197.

21. ● Koli, V.K., Bhatnagar, C. & Mali, D. 2011. Gliding behaviour of indian giant flying squirrel (*Petaurista philippensis* Elliot). Curr Sci 100: 1563-1568.

22. Zacharias, V.J. & Bhardwaj, A. K. 1997. Observation on the flying squirrels *Petaurista petaurista philippensis* (Elliot) and *Petinomys fuscocapillus* (Jerdon) in Periyar tiger reserve and its neighbourhood. Indian Forester 123:973-974.

23. Walston, J., Duckworth, J.W. & Molur, S. 2008. *Petaurista philippensis*. In: IUCN 2008. IUCN Red List of Threatened Species. Downloaded on 10 February 2012.

24. Molur, S., Srinivasulu, C., Srinivasulu, B., Walker, S., Nameer, P.O. & Ravikumar, L. 2005. Status of non-volant small mammals: Conservation assessment and management plan (C.A.M.P.) workshop report. Zoo Outreach Organisation/CBSG-South Asia., Coimbatore, India.

25. Hutton, A.F. 1947. Nesting habits of the flying-squirrel (*Petaurista philippensis*). J Bom Nat Hist Soc 46:539-540.

26. Kuo, C.C. & Lee, L.L. 2003. Food availability and food habits of Indian giant flying squirrels (*Petaurista philippensis*) in Taiwan. J Mammal 84:1330-1340. http://dx.doi.org/10.1644/BOS-039

27. Kuo, C.C. & Lee, L.L. 2012. Home range and activity of the Indian giant flying squirrel (*Petaurista philippensis*) in Taiwan: influence of diet, temperature, and rainfall. Acta Theriol 57: 269-276. http://dx.doi.org/10.1007/s13364-012-0076-y

28. Fan, P. & Jiang, X. 2009. Predation on giant flying squirrels (*Petaurista philippensis*) by black

crested gibbons (*Nomascus concolor jingdongensis*) at Mt. Wuliang, Yunnan, China. Primates 50:45-49.
http://dx.doi.org/10.1007/s10329-008-0110-5

29. Thorington, R.W., Pitassy, D. & Jansa, S.A. 2002. Phylogenies of flying squirrels (Pteromyinae). J Mammal Evol 9:99-135.
http://dx.doi.org/10.1023/A:1021335912016

30. Champion, H.G. & Seth, S.K. 1968. A revised survey of forest types of India, Government of India, New Delhi, p 404.

31. Jamdar, N. 1985. The reflected glow from the eyes of the large red flying squirrel (*Petaurista petaurista*). J Bom Nat Hist Soc 82:397.

32. Buchanan, B.W. 1993. Effect of enhanced lighting on the behavior of nocturnal frogs. Anim Behav 45: 893-899.
http://dx.doi.org/10.1006/anbe.1993.1109

33. Lee, P.F., Progulske, D.R. & Lin, Y.S. 1993. Spotlight counts of giant flying squirrels (*Petaurista petaurista* and *P. alborufus*) in Taiwan. Bull Inst Zool, Acad Sinica 32: 54-61.

34. Lee, P.F., Progulske, D.R. & Lin, Y.S. 1986. Ecological studies on two sympatric *Petaurista* species in Taiwan. Bull Inst Zool, Acad Sinica 25: 113-124.

35. Ando, M. & Kuramochi, Y. 2008. Vocal communication of the Japanese giant flying squirrel *Petaurista leucogenys*. J Agricul Sci, Tokyo Uni Agricul 53: 176-183 (In Japanese with english abstract).

36. Murrant, M.N., Bowman, J., Garroway C.J., Prinzen, B., Mayberry H. & Faure P.A. 2013. Ultrasonic vocalizations emitted by flying squirrels. PLOS one 8:1-6.
http://dx.doi.org/10.1371/journal.pone.0073045

37. Eisenberg, J.F. 1978. The evolution of arboreal herbivores in the class Mammalia. In: Montgomery, G. G. (eds), the ecology of arboreal folivores, Smithsonian Institution Press, Washington, D. C., pp, 135-152.

38. Koli V.K. 2014. Morning territorial calls of male oriental magpie robin (*Copsychus saularis*). Taprobanica 6: 60-62.
http://dx.doi.org/10.4038/tapro.v6i1.7092

39. Kloubec, B. & Capek, M. 2000. Diurnal, nocturnal and seasonal patterns of singing activity in Marsh Warblers. Biologia 55:185-193.

40. Dunmak, A. & Sitasuwan, N. 2007. Song dialect of oriental magpie-robin (*Copsychus saularis*) in northern Thailand. The Nat Hist J Chulalongkorn Uni 7:145-153.

41. Dudley, R., Byrnes, G., Yanoviak, S.P., Borrell, B., Brown, R.M. & McGuire, J.A. 2007. Gliding and the functional origins of flight: biomechanical novelty or necessity? Ann Rev Ecol Evol Syst 38:179-201.
http://dx.doi.org/10.1146/annurev.ecolsys.37.091305.110014

42. Shattuk, M.R. & Williams, S.A. 2010. Arboreality has allowed for the evolution of increased longevity in mammals. PNAS 107: 4635-4639.
http://dx.doi.org/10.1073/pnas.0911439107

43. Garstang, M., Larom, D., Raspet, R. & Lindeque, M. 1995. Atmospheric controls on elephant communication. The J Exp Biol 198:939-951.

44. Babu, S. & Jayson, E.A. 2009. Anti-predator behaviour of large brown flying squirrel (*Petaurista philippensis*): is this an effective census method to survey the species. Curr Sci 96:772-773.

18

The Incremental Investment Value of Waterfowl Impoundments in the Upstate of South Carolina

R. Z. Clardy[1], W. M. Smathers, Jr.[2] & T. J. Straka[1,*]

[1] School of Agricultural, Forest, and Environmental Sciences, Clemson University, Clemson, SC 29634, USA.

[2] Department of Economics, Clemson University, Clemson, SC 29634, USA.

* Corresponding author address: School of Agricultural, Forest, and Environmental Sciences, P.O. Box 340310, Clemson University, Clemson, SC 29634-0310, USA; e-mail: tstraka@clemson.edu.

Keywords

Waterfowl;
Impoundment;
Financial analysis;
Discounted cash flow analysis;
GIS.

Abstract

Financial incentives and economic factors play a role in expansion and survival of waterfowl in the Upstate of South Carolina. Satisfying a basic biological need, waterfowl impoundments are an effective way to ensure that migrating ducks have a food source, cover, and water on their journey south during the winter months (duck hunting season), but also year-round to ensure these birds return annually. Development of the region's rural areas has resulted in a decrease in total area of flooded wetlands designated to waterfowl and a decline in waterfowl populations. At the same time, the willingness to pay to hunt ducks has escalated. South Carolinians drive across the United States in search for duck hunting when the same hunting experience could be created in their own backyard. The incremental value of creating a waterfowl impoundment was determined on five different properties in the Upstate of South Carolina (the study area). A discounted cost flow analysis approach was used that considered initial costs (a pond leveler, site preparation work, and blind building), recurring costs, and hunter revenue. The model also considered the opportunity cost of foregoing the harvest of planted crops on the land. Realistic examples are provided with corresponding net present values and geographic information system generated maps illustrate the layout of these impoundments. Waterfowl impoundments were found to produce significant financial return.

Introduction

The population of South Carolina has increased by nearly 10% in the last decade, contributing to the rapid depletion of wetlands and related waterfowl areas [1]. At the same time, the number of Upstate South Carolina duck hunters has increased, following a national trend of heightened popularity for the sport [2]. Decreasing waterfowl habitat and populations are creating a problem for sportsmen seeking hunting opportunities [3]. The popularity of waterfowl hunting means sportsmen are willing to pay for hunting opportunities and a significant revenue source is available to landowners with property that could be managed for waterfowl. United States Fish and Wildlife Service waterfowl experts believe, "in order to create a healthy population of waterfowl, we must produce two ducks tomorrow where one is produced today" [4]. Creative landowners now realize the demand for prime duck hunting habitat can be translated into reliable hunting revenue streams.

These same landowners are creating and managing waterfowl impoundments [5]. Waterfowl impoundment has various definitions [6, 7]; the one used here is a wetland area that is manipulated by providing a source of water and food to attract waterfowl by annual plantings and maintenance. Waterfowl impoundments become a financial investment to the landowner, with both cost and revenue streams. Choice hunting habitat attracts duck hunters that will pay top dollar for access privileges. In South Carolina, waterfowl impoundments are seldom primary sources of income; they usually provide incremental or extra income. One caveat involved with converting or altering wetlands is compliance with federal regulations (Clean Water Act), enforced by the U.S. Army Corps of Engineers, that deals with protection of wetland and flooded areas in South Carolina.

Waterfowl are known to be a very diverse wildlife species that require a specific, complex environment in order to survive [8]. In South Carolina the two most abundant and hunted waterfowl species are the mallard (*Anas platyrhynchos*) and the wood duck (*Aix sponsa*). These ducks are considered puddle ducks or "dabbling ducks" meaning they feed on the water surface or right beneath the surface [9, 10]. Waterfowl impoundments are the perfect habitat for these species.

When planting an impoundment for waterfowl in South Carolina, there are a few important considerations: the presence of water, a means to drain the water, methods to plant the area, and the ability to re-flood the area. Typically waterfowl habitat requires an impoundment be located away from major roads and highly populated urban areas. Topography is another major consideration: the land has to be floodable from a water source located on the property and also have the capacity to hold the water for substantial periods of time. When flooding an impoundment after planting, the optimum depth of the water is eight to twelve inches deep. This ensures that the seed falling to the bottom can still be utilized by ducks. When these factors come together, the landowner has an attractive incremental source of income.

The two most commonly planted crops for waterfowl in Upstate South Carolina are corn and Japanese millet. Both are planted when the land is drained and, after the crops are established and begin growing, the impoundment is slowly flooded to begin attracting waterfowl. Rice requires flooding after planting and is one of the most planted crops for waterfowl in the United States [11]. Soil conditions and topography in Upstate South Carolina are not conducive for rice production.

Net Present Value

Net present value (NPV), sometimes called discounted cash flow analysis, is commonly used to evaluate potential discounted value of an investment. It represents the value of the investment today (or at year 0) considering interest and the timing of cash flows. Waterfowl impoundments are long term investments with many associated costs and revenues. The NPV calculation involves subtracting the discounted value of all costs from the discounted value of all revenues [12]. NPV is a financial criterion; a higher NPV at a given interest rate or discount rate equates to a more attractive investment alternative.

Calculating NPV first involves determining the timing of the cash flows. Creating a waterfowl impoundment on a piece of land involves initial one-time costs, annual planting costs, and periodic maintenance costs. The initial cost of building the impoundment is a onetime cost that occurs at year zero. A second onetime cost is constructing blinds for hunters and then placing them on the impoundment. These costs have to occur once the impoundment is complete and hunting begins; we assumed the blinds construction also occurred at year 0 for all impoundments. Planting and fertilization occur in year 0 to establish the food source (corn and millet). The final onetime cost at year 0 is either the spillway or pond leveler alternative. Other costs are reoccurring, either annually or every few years. The impoundment must be replanted annually with corn and millet; this is a perpetual annual cost. Maintenance is assumed to occur every five years; this is a perpetual periodic cost.

The revenue from an impoundment is annual income hunters pay to use the resource (the fee for renting a blind). The cost of renting a blind for a duck season can vary from state to state and also year by year, and can be quite expensive. We used average Upstate South Carolina rates, but even within the state rental rates are highly variable. Single blinds in South Carolina have rented for fees as high as $10,000 per year (Ronald Fleming, personal communication, 4 April 2012). Obviously, the blind rental fee will have a major impact on the NPV calculation; we feel the blind fees used in this study are very conservative for the region.

Three NPV formulas were used in this study. The first discounts single sums from some future year to today [12]. This formula is used to discount the onetime costs and revenues, like construction, initial planting, spillways, and pond levelers. The formula reduces the net cash flow in any year to its present value today and is presented below:

$$\mathrm{NPV} = \sum_{t=0}^{N} \frac{R_t}{(1+i)^t}$$

where: R = discounted revenues minus discounted costs in dollars, t = time in years, and i = discount rate, expressed as a decimal.

The two additional equations were used to discount series of cash flows. The present value of a perpetual periodic cash flow series equation was used to discount cash flows that occur periodically throughout the management regime (specifically, in our case, the maintenance cost that occurs every five years). This equation is presented below:

$$V_0 = \frac{p}{(1+\mathrm{i})^{\mathrm{n}} - 1}$$

where: p = periodic cash flow payment or revenue in dollars, and n = number of years between payments.

The final equation was for the present value of perpetual annual cash flow series. Examples of this are the corn and millet that must be replanted each year and the hunting lease revenue that occurs annually. This equation is presented below:

$$V_0 = \frac{a}{i}$$

where: a = annual cash flow in dollars.

Both perpetual periodic and perpetual annual formulas are for cash flows that occur forever. The difference is one occurs each and every year and one occurs every few years. Single occur just once. Any cost that occurred at year 0 needed no discounting and was simply added to the sum at today's value.

The Discount Rate and Opportunity Cost

Interest is rent for the use of money and represents an opportunity cost. We used four discount rates in this analysis to account for market variation in these rates. We assumed costs and revenues increasing at about the inflation rate and use a real discount rate (net of inflation). Four percent discount should approximate current discount rates. We include 2% as relevant to the lower rates sometimes charged in the current stagnant economy and 7% and 10% discount rates are included for perspective on what to expect when interest rates rise. This range of discount rates presents an opportunity for a sensitivity analysis of scenarios for landowners with different discount rate expectations.

The general question of opportunity cost is also of interest. When a landowner agrees to a certain type of management regime for a property, there is always the underlying question, "What if I had done something else?" For these impoundment areas another common practice would be to plant corn or millet for harvest as crops for grain.

Corn is currently a booming agricultural crop because of its extensive use in the production of ethanol. Based on current average prices and yields, corn can yield revenues of $1,300 per ha. [16]. However, it is a row crop that is best harvested using a combine, making it difficult to manage in a potential impoundment area. Millet is a different type of crop commonly used for hay production and wildlife values. Millet is harvested by cutting it down at the stalk, and then drying and baling it for livestock feed. When used for wildlife, millet is knocked to the ground causing voluntary millet to re-sprout in future years. An average price for a bale of millet is $40 per round bale. Japanese millet is different because it is primarily used for wildlife values and the cost of harvest is usually overlooked because it is planted in unharvestable areas like wetlands.

Potential revenue streams generated by corn and millet production were less than those expected from a waterfowl impoundment. The small size and limited access characteristics of an impoundment area are generally not compatible with agricultural production and would reduce revenue streams. A waterfowl impoundment would have significant economic advantages over crop production in these low-lying areas.

Some might argue that an opportunity cost should be imposed for the use of the land. This would be equivalent to a land rental fee. If the landowner had a viable alternative use for the land, such a rental charge would be appropriate. However, in this case the land is assumed to be part of an estate with no other use beyond hunting or recreation.

Methods

Locating and mapping properties with impoundment development opportunities was the initial focus. The study areas were five wetlands in the ten-county area of Upstate South Carolina, chosen for economic and geographic feasibility using the ArcGIS program. First, wetland areas that had the potential to be flooded or were already flooded and had a viable water source were identified. ArcGIS has the ability to simultaneously consider multiple variables, like distance from roads, urban areas, and proximity to potential hunter populations (for the convenience factor of driving distance). The advanced GIS programs allowed for interaction of all variables on a map of South Carolina and highlighted areas of interest, allowing for a choice of five areas that appeared to be optimum impoundment areas.

A complete database used to promote Upstate South Carolina called "Ten at the Top" was used to evaluate potential impoundment areas for access potential, proximity to urban populations, and ownership patterns [13]. For example, proximity to a large city was a major assessment factor because people will pay more to duck hunt near their home or place of work. To consider this convenience factor, all of the selected properties are located within 15 miles of a large city. Table 1 describes the five properties identified as prime impoundment areas and Figure 1 shows outlines of the areas to be managed.

Table 1: Potential Upstate South Carolina Waterfowl Impoundments.

Impoundment	Size (ha)	County	Number of Blinds
1	3.14	Anderson	3
2	1.73	Anderson	2
3	9.03	Anderson	5
4	3.22	Greenville	3
5	1.94	Greenville	2

Recall that waterfowl impoundments involve initial, annual, and periodic costs. An initial cost is a cost that occurs at year zero or today and not again during the management regime. On all five properties in this study, water was already present meaning that water would not need to be pumped in from an outside source (i.e., a creek or river) to evaluate the capabilities of retaining water. The basic net present value formula recognizes that initial costs occur at year zero and need not be discounted. The initial cost of implementing an impoundment includes:

1. Bulldozing to clear of any trees and do grading,
2. Blind building,
3. Initially planting corn and millet, and a
4. Water leveler or spillway.

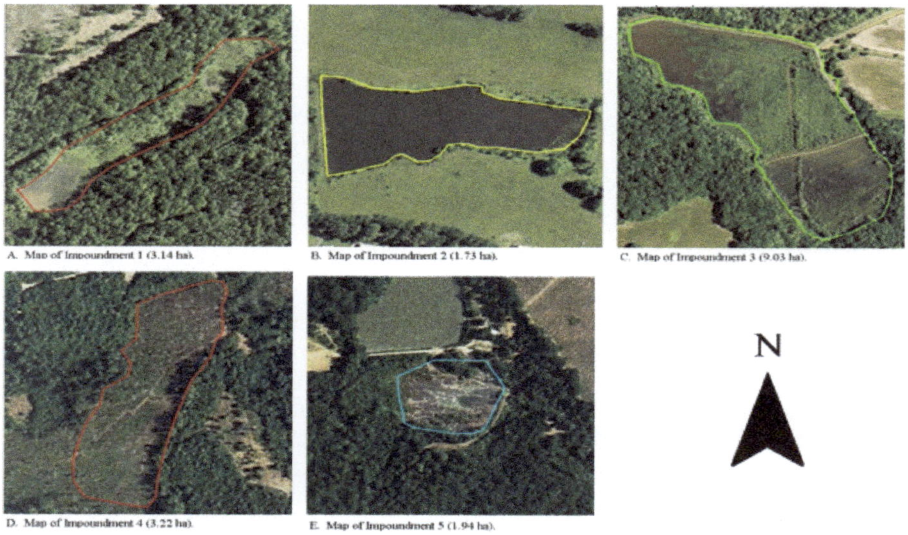

Fig. 1: Impoundment pond maps.

Table 2: Costs and revenue for Upstate South Carolina waterfowl impoundments.

	Impoundment 1	Impoundment 2	Impoundment 3	Impoundment 4	Impoundment 5
Area (ha)	3.14	1.73	9.03	3.22	1.94
			Costs at year 0		
Dozer ($/hr)	$80.00	$80.00	$80.00	$80.00	$80.00
Blind	$500.00	$500.00	$500.00	$500.00	$500.00
Millet ($/ha)	$210.00	$210.00	$210.00	$210.00	$210.00
Corn ($/ha)	$741.00	$741.00	$741.00	$741.00	$741.00
Pond Leveler	$350.00	0	0	$350.00	0
Spillway	0	$700.00	700.00	0	$700.00
Dozer (hours)	15	10	20	15	11
			Costs after year 0		
Millet ($/ha)	$210.00	$210.00	$210.00	$210.00	$210.00
Corn ($/ha)	$741.00	$741.00	$741.00	$741.00	$741.00
Maintenance	$1,000.00	$1,000.00	$1,000.00	$1,000.00	1,000.00
			Revenue		
Blinds ($ ea.)	$2,500.00	$2,500.00	$2,500.00	$2,500.00	$2,500.00
Blinds (no.)	3	2	5	3	2

Current rates for 2012 were used in the study [14]. The rate per hour for a bulldozer up to 250 horse power is $80.00/hour. A bulldozer will be required on each site in order to grade the impoundments and create water levels that are approximately eight inches to twelve inches in depth, along with removing any trees as specified by the land owner. Table 2 details technical requirements for each impoundment, like construction hours and umber of blinds. Blind building is essential for hunters to ensure a maximum chance in harvesting waterfowl and providing them maximum cover and a comfortable, safe hunt. The duck blinds used on each impoundment were 4x8 box type blind placed on stilts directly above the water. The average cost of building one of these blinds was approximately $500 (Ronald Fleming, personal communication, 4 April 2012). A conservative estimate of blind numbers for impoundments was used in Table 1 to keep financial aspects of the analysis reasonable.

Water levels are the most important step in the initial creation of a duck impoundment. If the water levels are too high, the crops planted will not be able to survive, and if they are not high enough, the impoundment will not hold as many ducks. The water leveler that was used is the "Clemson Beaver Pond Leveler" [3]. The Clemson University beaver pond leveler is a system of pipes that were inserted in a beaver dam at the lowest point of water flow that allows the water to be maintained at a specific level or for the entire area to be drained. The beaver pond leveler makes it hard for beavers to detect where the water is escaping from, meaning they will not dam up the flow. The average cost of installing a beaver pond leveler was $350 [15]. A spillway system, a second type of water leveler, was also used. In order to correctly install this system, a spillway was added to the lowest point in the dike. A pump located in the primary water source was added, along with a spillway, to drain existing water out of each property. The process involved placing a pipe in the currently existing dike located on the properties. The pipe must be adequate size to support the amount of water flow. In all the areas used for this study a gate blocking the pipe ran at the lowest part of the dam was installed. This ensured after the fall flooding of the impoundment that the water would not escape. The cost of this was approximately $700.00 (J&M Construction, personal communication, 12 April 2012). Three of the areas used the Clemson University beaver pond leveler and two used the spillway system.

Annual costs reoccur every year for the duration of the impoundment and must be discounted back to year zero. It is essential that each year the impoundment become replanted to ensure that the waterfowl have an annual food source. Thus, planting and fertilizing costs also occur annually. The two food sources that will be planted in all five impoundments are corn and millet. Corn is a row crop that is best planted with a row planter for maximum growth. Planting corn costs approximately $741.00 per ha with fertilizer. Millet can be planted using a grain drill or by broadcasting the seed in the areas of interest. Millet can be broadcast using a spreader on a tractor at a cost of approximately $210.00 per ha (Ronald Fleming, personal communication, 4 April 2012). We did not include liability insurance as an annual expense. A landowner might want to include this expense. Also, land opportunity cost was not included as an annual expense and is discussed later.

There were six costs represented at year zero and two annual costs that occurred every year in perpetuity. A maintenance cost was also charged every five years for repairs and any extra work that may need to be done on the impoundments. Table 3 presents cash flows generated from the parameters detailed in Table 2. NPV was determined using four different real discount rates (2%, 4%, 7% and 10%). Current market conditions are likely best-represented by a 4% discount rate; the higher discount rates will allow the reader to interpret results for higher discount rates likely to develop with improving economic conditions. NPV estimates are presented in Table 4 and are broken down into four columns using the four discount rates so landowners can better approximate their own discount rate situation.

Table 3: Cash flows generated for Upstate South Carolina waterfowl impoundments.

	Impoundment 1	Impoundment 2	Impoundment 3	Impoundment 4	Impoundment 5
			Costs at year 0		
Dozer	$1,200.00	$800.00	$1,600.00	$1.200.00	$880.00
Blind	$1,500	$1,000	$2,500	$1,500	$1,000
Millet	$659.40	$363.30	$1,896.30	$676.20	$407.40
Corn	$2,326.74	$1,281.93	$6,691.23	$2,386.02	$1,437.54
Pond Leveler	$350.00	0	0	$350.00	0
Spillway	0	$700.00	700.00	0	$700.00
			Costs after year 0		
Millet	$659.40	$363.30	$1,896.30	$676.20	$407.40
Corn	$2,326.74	$1,281.93	$6,691.23	$2,386.02	$1,437.54
Maintenance	$1,000.00	$1,000.00	$1,000.00	$1,000.00	1,000.00
			Revenue		
Blinds	$7,500.00	$5,000.00	$12,500.00	$7,500.00	$5,000.00

Table 4: Net present values per ha for five impoundment areas, at various percent discount rates.

	NPV/ha			
Impoundment	2%	4%	7%	10%
1	$210,048.94	$102,194.68	$55,963.42	$37,464.48
2	$153,985.35	$75,808.29	$41,295.90	$35,658.68
3	$172,627.55	$79,808.54	$40,020.60	$24,099.09
4	$206,168.86	$100,216.60	$54,800.48	$36,627.60
5	$143,720.14	$69,835.88	$38,163.19	$25,487.68

Results

Table 4 presents NPV per ha for each impoundment option. As expected, as the discount rate increases, the NPV of the impoundment decreases. The impoundments have realistic revenue and cost structures that varied due to the natural factors one would expect in Upstate South Carolina. ArcGIS insured appropriate and representative impoundment opportunities were evaluated. In all cases, even at the highest discount rates, all of the impoundments produced positive NPVs. Figures 1A to 1F produced by ArcGIS, and Table 4 will allow landowners to easily compare their impoundment opportunities to other possible land use alternatives.

Note that all NPVs were positive, even at 10% interest. That indicates that the rate of return on all impoundments exceeded 10%. Rate of return is defined as the point where NPV becomes zero and turns negative. All five impoundments have a relatively large NPV remaining at 10% interest, so the rate of return would be much higher than 10%.

Current demand for duck hunting is very high and duck blind leasing opportunities are limited, ensuring for the immediate future that duck blinds will not go unleased (Ronald Fleming, personal communication, 4 April 2012). The additional revenue can be an excellent source of income for impoundment owners. While reviewing the benefits of producing a waterfowl impoundment, a few key results stood out. Using the 4% discount rate, the most likely relevant one in today's economy, the NPV of impoundment 3 did not produce the highest NPV. Given its large size, one would expect it to produce the highest NPV. This shows the general positive financial trends these impoundment produce, but also emphasizes that the various technical and financial variables interact to produce unexpected NPVs. The data do suggest that there is an optimal size for an impoundment and NPV is likely to decrease when that size is exceeded. Net revenue seems to increase with impoundment size to some optimum size and then decreases as the impoundment becomes too large to support hunting opportunities. These were simplified examples of real impoundments, but the variance in NPVs should alert landowners that assumptions on number of blinds per impoundment and revenue per blind can greatly impact results. It should be noted all assumptions like planting costs and maintenance are reasonable and based on real world estimates from experts. But actual on-the ground situations will require each landowner to make his or her own assumptions.

Achieving maximum revenue on each impoundment is essential in a landowner's decision to create a waterfowl impoundment. Each impoundment has a different number of blinds located on it that is suited for the size and layout of each impoundment. Overcrowding a duck impoundment is historically a large problem. It is essential to provide the ducks that utilize the impoundments a safe place to acquire the three essential components of these areas (water, food and cover). Blinds should not be placed too close together and a safe zone should be present in each one. If every duck that utilizes the impoundment gets shot at by hunters, they will not be as likely to return to the impoundment from year to year. The problem with this is that landowners want to make the highest revenue possible, causing them to place as many blinds as they can, as close as possible, with still providing a safe hunting environment. While this may seem like a good idea, in the long run the impoundment will not obtain the maximum number of waterfowl possible because of the lack of space for them to safely land.

While building duck impoundments can be a costly practice, there are ways to save some money and cut back the initial costs of implying them. As noted in Table 2, the cost of constructing a duck blind and having it installed is approximately \$500 per blind. This cost is represented for having someone else do the labor for the landowner. A duck blind can be made very easily and the landowner can save money by simply doing it himself. The key to success is providing a comfortable blind for hunters to hunt out of and this can be created at a much lower cost than represented in this study. The \$1,000 maintenance cost is a fixed cost that will actually vary from year to year, with \$1,000 being an average. The largest key to keeping maintenance cost to a minimum is to not cut corners in the initial building stage. If the impoundments are built correctly from the start, the maintenance cost incurred in the future will be much lower than they would be on an impoundment that was not constructed correctly.

Conclusions

Wetlands designated for waterfowl are becoming scarcer as population increases cause Upstate South Carolina to develop. This high demand for the sport of duck hunting has presented landowners with an opportunity for an excellent incremental revenue source from suitable wetland areas. Waterfowl impoundments are meant to provide food, water and cover for migrating waterfowl that pass through South Carolina to and from the northern United States. Waterfowl impoundments offer a financially attractive opportunity for increase waterfowl habitat and provide hunting opportunities. We've shown the levels of profitability that can be expected and some of the practical limitations. The value of a waterfowl impoundment planted cannot be surpassed by anyone who has a passion for the conservation of wildlife and also additional revenue. All it takes is a water source, the right topography, and the drive to see the land make a difference.

References

Five "key references", selected by the authors, are marked below (Three recommended (●) and two highly recommended (●●) papers).

1. U.S. Census Bureau. 2012. State and County QuickFacts. http://quickfacts.census.gov/qfd/states/45000.html. Cited 12 May 2012.
2. Hefner, J.M. & Brown, J.D. 1984. Wetland trends in the southeastern United States. Wetlands 4(1):1-11.
DOI: 10.1007/BF03160482
3. Clemson University Cooperative Extension Service. 1994. The Clemson beaver pond leveler. AFW Leaflet 1. Clemson University Cooperative Extension Service, Department of Aquaculture, Fisheries, and Wildlife, Clemson, SC, USA.
4. Bellrose, F.C. & Low, J.B. 1978. Advances in waterfowl management research. Wildlife Soc. B. 6(2):63-72.
5. McHugh, K.E. & Mings, R.C. 1991. On the road again: Seasonal migration to a sunbelt metropolis. Urban Geogr. 12(1):1-18.
DOI: 10.2747/0272-3638.12.1.1
6. ● Weller, M.W. 1989. Waterfowl management techniques for wetland enhancement, restoration and creation useful in mitigation procedures. In: Kusler, J.A. & Kentula, M.E. (eds.), Wetland Creation and Restoration, The State of the Science. Island Press, Washington, DC, USA, pp. 517-528.
7. Cross, D. & Vohs, P. (eds.). 1988. Waterfowl Management Handbook. U.S. Fish and Wildlife Service, Fort Collins, CO, USA.
8. ● Frederickson, L.H. & Reid, F.A. 1988. Waterfowl use of wetland complexes. In: Cross, D. & Vohs, P. (eds.), Waterfowl Management Handbook. U.S. Fish and Wildlife Service, Fort Collins, CO, USA, Ch. 13.2.1.
9. Jorde, D.G., Krapu, G.L. & Crawford, R.D. 1983. Feeding ecology of mallards wintering in Nebraska. J. Wildl. Manage.47(4):1044-1053.
DOI: 10.2307/3808163
10. ●● USDA Natural Resources Conservation Service. 2001. Shallow Water Management for Dabbling Waterfowl (Fish and Wildlife Habitat Management Guidesheet). http://www.mn.nrcs.usda.gov/technical/ecs/wild/waterfowl.pdf. Cited 12 May 2012.

11. ●● Ringleman, J.K. 1988. Managing agricultural foods for waterfowl. In: Cross, D. & Vohs, P. (eds.), Waterfowl Management Handbook. U.S. Fish and Wildlife Service, Fort Collins, CO, USA, Ch. 13.4.3.

12. ● Bullard, S.H. & Straka, T.J. 1998. Basic Concepts in Forest Valuation and Investment Analysis. Forestry Suppliers, Inc., Jackson, MS, USA.

13. Ten at the Top. 2010. Ten at the Top: Fostering Regional Collaboration in South Carolina's Upstate, Greenville, SC, USA. http://www.tenatthetop.org. Cited 12 May 2012.

14. FEMA. 2012. Schedule of Equipment Rates. Department of Home Land Security, Department of Federal Emergency Management Agency, Washington, DC, USA. http://www.fema.gov/government/grant/pa/eqrates.shtm. Cited 12 May 2012.

15. Wiest, R.A. 1998. A Landowner's Guide to Building Forest Access Roads (NA-TR-06-98). USDA Forest Service, Northeastern Area, State and Private Forestry, Rador, PA, USA.

16. CME Group. 2012. Agricultural Commodities Products. Chicago Mercantile Exchange, Chicago, IL, USA. http://www.cmegroup.com/trading/agricultural/. Cited 12 May 2012.

19

Evaluating the Effect of Short-Term Capture and Handling on Fecal Glucocorticoid Metabolite Levels in Mourning Doves

D.S. Jachowski[1,2,*], T.W. Mong[1], B.E. Washburn[1,3], J.H. Schulz[1,4] & J.J. Millspaugh[1]

[1] Department of Fisheries and Wildlife Sciences, Univ. of Missouri, 302 ABNR Building, Columbia, MO 65211, USA.

[2] Department of Fish and Wildlife Conservation, Virginia Tech, 106 Cheatham Hall, Blacksburg, VA 24061, USA.

[3] USDA, Wildlife Services, National Wildlife Research Center, 6100 Columbus Avenue, Sandusky, OH 44870, USA.

[4] Missouri Department of Conservation, Conservation Research Center, 1110 South College Avenue, Columbia, MO 65201, USA.

* Corresponding author e-mail: djachowski@gmail.com.

Keywords
Stress;
Physiology;
Handling;
Captivity;
Mourning dove.

Abstract
The measurement of stress hormones (i.e., glucocorticoids) has greatly advanced animal conservation. Fecal glucocorticoid metabolite assays are valuable because they are noninvasive, but their ability to detect responses to short-term (<30 min) stressors in a way similar to blood serum assays is comparatively less well understood. We evaluated whether fecal glucocorticoid metabolites (FGMs) increased in captive wild mourning doves (*Zenaida macroura*) exposed to either a brief (<2 min) capture, handling and release (CHR) or capture stress protocol (CSP; i.e., capture, hold for 30 min, release) treatment. Previous studies have shown that mourning doves exhibit elevated FGMs within 2-3 hrs of experimental challenges. Therefore, we attempted to collect feces every hour for 24 h pre-treatment and 36 h post-treatment. We did not detect a consistent increase in FGMs in response to CHR or CSP treatments. Though additional research is needed, FGM levels were lower the longer birds were held in captivity and we did not observe sex-based or seasonal differences in FGM responses. For mourning doves, and likely other species, plasma corticosterone analysis is better suited to assess responses to short-term stressors. Alternatively, FGMs are ideal for research focused on longer-term patterns in physiological state because they are not sensitive to exposure to temporary, acute stressors.

Introduction

Measurement of stress hormone production in wildlife has dramatically improved our understanding of their physiology, behavior, and conservation [1, 2, 3, 4]. For avian species, in particular, non-lethal means of assessing stress hormone concentrations through collection of blood samples has revolutionized our understanding of physiological mechanisms involved with migration [5, 6], social interactions [7], foraging behavior [8, 9, 10], reproductive timing [11], and parasitism [12]. Of particular interest to conservation managers, elevated glucocorticoid levels are associated with individual fitness or health that can, in turn, be used to assess broader population- or species-level responses to environmental change [2, 13, 14, 15].

Monitoring changes in fecal glucocorticoid metabolite (FGM) levels provides a noninvasive alternative to monitoring glucocorticoid concentration in blood plasma [16, 17, 18, 19]. FGM analyses have been widely used, particularly in wild settings, because samples can be collected frequently and non-invasively from focal individuals [20, 21] and FGMs might provide a broader integrated measure of the stress response than blood glucocorticoid levels [22]. Moreover, studies using blood glucocorticoid sampling have found that capture and handling might compromise an accurate assessment of glucocorticoid levels [23, 24]. The ability of FGM assays to detect responses to short-term stressors in wild animals, often of interest in field studies, is less well understood. For example, when FGM techniques are combined with behavioral studies involving capture, handling, and restraint to collect biological data (e.g., attach leg bands or radio transmitters), it is important to determine whether or not fecal glucocorticoid analyses are capable of detecting associated short-term stress responses if and when they occur [25, 26].

In this study we investigated if FGM assays were able to detect exposure to short-term stressors by continuously sampling mourning doves (*Zenaida macroura*) in a captive setting. Mourning doves provide an ideal study subject because FGM assays have been validated, and elevations in FGMs are known to occur within 2-3 hrs following exposure to acute stressors such as experimental challenge with adrenocorticotrophin (ACTH) [27]. We built on this basic experimental understanding of FGM responses in mourning doves by conducting trials on captive wild doves exposed to two different stress regimes; a short-term (<2 min) stressor mimicking capture and banding, and a longer term (<30 min) stressor simulating the amount of time needed to capture, handle, and attach a radio transmitter to a bird.

Methods

We captured wild mourning doves near Columbia, Missouri (38°54'N, 92°18'W) using modified Kniffin traps [28] during December 2000 to June 2001. Upon capture, we banded each bird with an individually numbered metal leg band, and assigned age and gender based on plumage characteristics [29, 30]. We transported mourning doves to the T.S. Baskett Wildlife Research and Education Area, located 15 km south of Columbia, Missouri (38°45'N, 92°11'W). During captivity, doves were kept individually in outdoor 1.8 m x 1.8 m x 1.8 m wooden framed pens raised 0.6 m above ground [31]. A wild-bird seed mixture (sorghum grains, wheat, crimped corn, oats, and sunflower seed) and water were provided *ad libitum*.

Due to logistical constraints and our need to continuously sample dove glucocorticoid metabolites to address our study objectives, our sampling was necessarily focused on intensively monitoring a small number of individuals (n = 6). We administered 3 different treatment protocols over the course of 18 months (Table 1). During the first experiment in winter (January) 2001, we administered a capture stress protocol (CSP) treatment [1] to two doves that had been in captivity for 6 weeks (Table 1). These birds were captured at 1200 and immediately placed into a breathable cotton sack. We placed the birds in a sternal position in the sack for 30 min, at which time we released them back into their respective pens (total disturbance time of 30 min 47 s +/- 16 SE s).

Table 1: Summary of experimental treatments used to evaluate short-term stress events in wild mourning doves held in captivity.

Treatment	Season	Year	Gender M^1,F^2	Start Time	Captivity[3]
CHR [4]	Summer	2001	0, 2	0800	12 weeks
Control 1	Summer	2001	1, 1	N / A	12 weeks
CSP [5]	Winter	2001	0, 2	1200	6 weeks
CSP	Winter	2002	1, 1	1600	20 weeks
Control 2	Winter	2002	1, 1	N / A	20 weeks

[1] Number of male mourning doves in experiment.
[2] Number of female mourning doves in experiment.
[3] Length of time wild mourning doves were held in captivity prior to the beginning of the experiments.
[4] Capture, handling, and restraint treatment. Birds assigned to this treatment were captured using a mesh net, removed from the net, held in our hands for 50 s, and released into their respective pen. Total disturbance time was 72 +/-3 SE s. This same procedure was repeated with each bird 30 min later.
[5] Capture stress protocol (Wingfield et al. 1992). Birds assigned to the stress protocol group were captured using a net, removed, and immediately placed into a breathable cotton sack. We placed the birds in a sternal position in the sack for 30 min, at which time we released them back into their respective pens. Total disturbance time of 30 min 47 s +/- 16 SE s.

A second experiment was conducted in summer (July) 2001, when we administered brief (<2 min) capture, handling, and release (CHR) treatments to 2 doves and simultaneously monitored 2 additional doves that were not handled and where only droppings were collected (serving as a control). Individual birds were randomly assigned to treatments. We attempted to assign one male and one female to each treatment, but errors in assigning gender [30] precluded equal numbers of males and females in each treatment. Control birds were not captured, handled, or restrained. The 2 doves assigned to the CHR treatment group were captured using a mesh net (at 0800), removed from the net, held in our hands for 50 s, and released into their respective pen (total disturbance time of 72 +/-3 SE s). This same procedure was repeated with each bird 30 min later to extend the length of adrenocortical stimulation.

We conducted a third experiment in winter (February) 2002, when we again administered a CSP treatment to 2 doves (Table 1). We followed methods similar to winter 2001, with three key differences: (1) this group of doves had been in captivity for 20 weeks, (2) we administered treatment at 1600 (as opposed to 1200) to try to determine if response differed by time of day [e.g., 32, 33], and (3) we simultaneously sampled 2 control doves (Table 1).

We collected droppings (feces plus urine) from each individual bird every hour, when available, beginning 24 h pre-treatment and continuing for 36 h post-treatment (except for the winter 2001 experiments when sample collection continued until 30 h post-treatment). While feces and urine often contain differing FGM levels and thus sample-to-sample variation in feces:urine ratio could influence our ability to detect trends in FGMs [17], we were logistically unable to separate feces from urine for this study. Droppings were frozen at –20°C within 10 min of collection. We placed frozen droppings into a lyophilizer (Freeze-dry Specialties, Inc., Osseo, MN, USA) for 24 h. Once freeze-dried, we sifted samples through a stainless steel mesh to remove large particles; each sample was then thoroughly mixed. We extracted fecal glucocorticoid metabolites from feces using a modification of Schwarzenberger et al. [34]. We placed dried feces (0.016 to 0.120 g) in a test tube with 2.0 mL of 90% methanol and vortexed the sample at high speed in a multi-tube vortexer for 30 min. We centrifuged samples at 500 x g for 20 min, saved the supernatant, and stored it at –84° C until assayed.

We used I^{125} corticosterone radioimmunoassay (RIA) kits (ICN #07-120103, ICN Biomedicals, Solon, OH, USA) previously validated for use in mourning doves [27] to quantify glucocorticoid metabolite concentrations. Droppings were analyzed in 10 assays, with each assay including all the samples from an individual bird. We followed the ICN protocol for the I^{125} corticosterone RIA, except that we halved the volume of all reagents. Inter-assay variation for four assays was 5.3% and average intra-assay variation was 1.3%.

We calculated summary statistics of mean, standard deviation and coefficient of variation for each individual dove prior to and following treatment. We also calculated baseline FGM levels for each individual during each experiment using methods outlined by Brown et al. [35] that involved an iterative process where we excluded FGM values that exceeded the mean plus 1.5 SD and recalculated a mean value until no values exceeded the mean plus 1.5 SD. The final mean FGM value for each individual dove in each experiment was termed the baseline for that individual.

Results

We collected and assayed an average of 53.5 samples/bird (SE = 2.33, Range = 47–60, $n = 10$) during this study (total of 535 droppings). Mourning doves defecated at a rate of 0.9 samples/bird/hour during our experiments (SE = 0.04, Range = 0.8–1, $n = 10$).

We did not observe consistent short-term elevations in FGMs following exposure to acute CHR and CSP stressors. Baseline FGM values for individuals in both treatment and control groups were in a similar range of 25-32 ng/g, and we observed no discernible trend in FGMs following our summer 2001 CHR treatment (Fig. 1). Further, the CHR treatment and control groups exhibited similar amounts of variability and similar FGM values over the course of the study (Table 2).

During the winter 2001 CSP treatment, pre-treatment (i.e., baseline) FGM levels were two times higher than FGM levels of all other birds during the pre-treatment period (Fig. 2). Post-treatment, in both doves we observed peaks in FGMs that were two times baseline FGMs in the first 6 hrs after treatment and then again 23–29 hrs post-treatment (Fig. 2). However, between these peaks FGMs were not consistently elevated nor did variation in FGMs increase substantially compared to pre-treatment levels (Table 2).

In contrast to the winter 2001 CSP treatment, we observed no discernible increase in FGMs following CSP treatment in winter 2002 (Fig. 3). The doves used in the winter 2002 CSP treatment had been maintained in captivity for 20 weeks and were treated at 1600 (cf. the winter 2001 CSP treatment in Table 1) and both treatment and control groups exhibited lower baseline FGMs in comparison to doves in the two previous experiments (Fig. 3). Post-treatment, we did not observe an increase in FGMs for

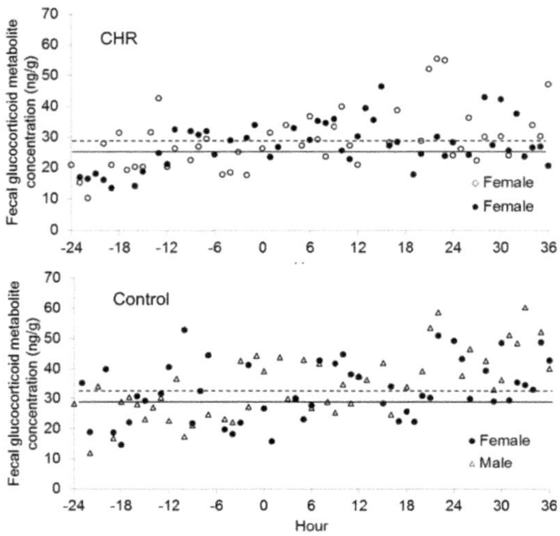

Figure 1: Fecal glucocorticoid concentrations (ng/g) of two mourning doves assigned to the capture, handling, and release (CHR) treatment group ($n = 2$ females) in summer 2001 over 60 h (top panel). Study began at 08:00 h. Time 0 represents the time when capture, handle and restraint (CHR) treatments were administered. No treatments were administered to the control group (n = 1 male, n = 1 female) (lower panel). Lines represent baseline fecal glucocorticoid concentrations (ng/g) for each individual (dashed line corresponds with hollow points, solid line with solid points).

Table 2: Summary of experimental treatments used to evaluate short-term stress events in wild mourning doves held in captivity.

Treatment	Gender (M, F)	Mean (SD) fecal glucocorticoid metabolite concentration		Coefficient of Variation	
		Pre-treatment	Post-treatment	Pre-treatment	Post-treatment
CHR	F	23.3 (7.0)	33.2 (9.7)	0.3002	0.2925
	F	23.6 (7.3)	29.7 (7.0)	0.3086	0.2360
Control 1	M	27.4 (8.3)	39.3 (9.8)	0.3042	0.2491
	F	29.5 (10.6)	34.7 (9.2)	0.3607	0.2642
CSP	F	44.4 (14.0)	50.6 (21.3)	0.3155	0.4202
	F	56.5 (16.2)	76.6 (16.0)	0.2872	0.2093
CSP	M	16.7 (2.8)	17.3 (4.0)	0.1694	0.2317
	F	24.4 (5.8)	29.6 (7.9)	0.2379	0.2689
Control 2	M	10.1 (2.1)	12.8 (2.8)	0.2085	0.2177
	F	14.6 (4.0)	16.1 (3.5)	0.2762	0.2180

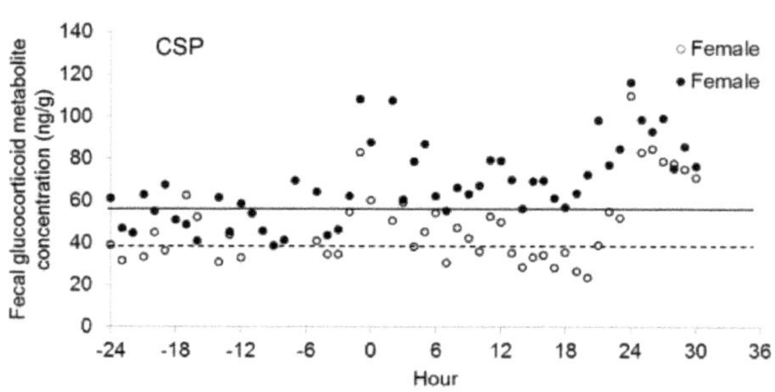

Figure 2: Fecal glucocorticoid concentrations (ng/g) of two mourning doves assigned to the stress protocol treatment group in winter 2001 (n = 2 females) over 54 h. Time 0 represents the time when capture stress protocol (CSP; Wingfield et al. 1992) treatments were administered. Study began at 12:00 h. Lines represent baseline fecal glucocorticoid concentrations (ng/g) for each individual (dashed line corresponds with hollow points, solid line with solid points).

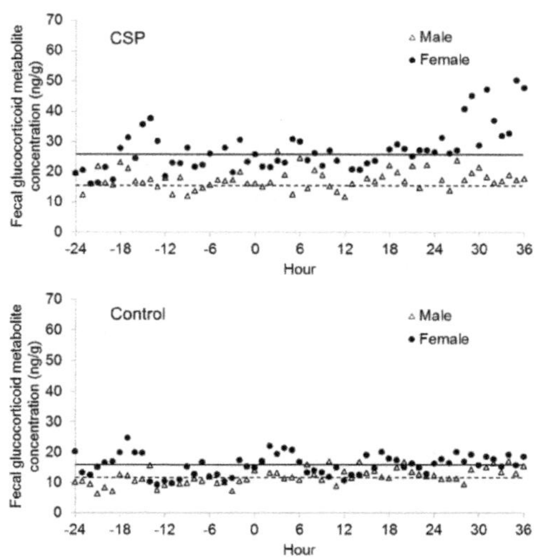

Figure 3: Fecal glucocorticoid concentrations (ng/g) of two mourning doves assigned to the capture stress protocol (CSP; Wingfield et al. 1992) treatment group (n = 1 male, n = 1 female) in winter 2002. Time 0 represents the time when treatments were administered. Study began at 16:00 h. No treatments were administered to the control group (n = 1 male, n = 1 female) (lower panel). Lines represent baseline fecal glucocorticoid concentrations (ng/g) for each individual (dashed line corresponds with hollow points, solid line with solid points).

the male dove, and only observed an increase in FGMs for the female dove after 24 hrs (Fig. 3). Across our entire study, this was the only evidence to suggest sex-based differences in FGM responses by mourning doves to our treatments. Finally, coefficient of variation values were similar pre- and post-treatment across both groups (Table 2).

Discussion

Although our study was restricted to a small number of individual birds, failure to observe consistent elevations in FGMs following CHR or CSP treatments suggests that, compared to blood, fecal sampling is not sensitive to elevations in glucocorticoid concentrations in response to short-term, acute stressors in mourning doves. Previous studies with captive mourning doves indicated that a significant increase in plasma corticosterone occurred in response to the CHR protocol used in this study [36, 37]. Therefore, although plasma corticosterone levels were likely elevated in response to CHR during our study, no elevation was evident in the fecal glucocorticoid metabolite profiles. Because FGMs reflect adrenocortical activity over an extended time period [22], and any acute response is lost because it is integrated, we think that this resulted in a loss of resolution of the stress response in feces.

Stressor-specific differences associated with the duration of a stressor might have contributed to the lack of elevations in FGM following capture and handling. In contrast to our relatively short term (<30 min) handling stress protocols, elevations in FGMs have been observed following exposure to other types of stressors. For example, dickcissels (*Spiza americana*) that were captured and had radio-transmitter backpacks attached to them elicited a peak in FGMs for 3 hrs, between 9-12 hrs following transmitter attachment [25]. We did not observe a similar acute stress response in our study despite knowing that it only takes 2-3 hrs for glucocorticoids to be detected in feces in mourning doves following ACTH challenge [27]. The absence of short-term elevations in FGMs during this study suggests that stressors ⩾30 min (e.g., a permanently attached transmitter) might result in an acute or long-term elevation in FGMs, but that activities associated with handling <30 min do not elevate FGMs in mourning doves. Therefore, our results provide some assurance for managers and researchers that individual doves will not demonstrate long-term physiological effects resulting from capture and marking lasting <30 min as is frequently performed during national banding activities [38, 39, 40].

Captivity can influence endocrine functional responses in avian species [41] and significantly affect corticosterone levels [5, 42]. Following capture, handling and radio-transmitter attachment in wild dickcissels that were acclimated to captivity for at least 2 weeks, Seudkamp Wells et al. [25] observed elevations in FGMs in birds receiving transmitter attachments, but no response in control birds. Similarly, our findings support previous reports that detecting stress responses in FGMs is likely dependent on acclimation time. Evidence from our 2 winter CSP treatments suggest that length of time mourning doves were held in captivity was associated with decreased inter-sample variability and overall lower FGM levels. Further, decreased variability in FGMs over the course of our study suggests that by week 20, either physiological acclimatization had occurred (i.e., doves no longer considered captivity to be a stressor) or that chronic stress diminished the overall ability of the birds to mount a physiological stress response to stressors [43, 44, 45]. Therefore, similar to blood-based corticosterone assays, the effects of captivity on FGMs is an important, potentially confounding factor that should be accounted for when trying to extend findings from the laboratory to the field, and vice-versa [41, 42].

Further research is needed to differentiate the potential effect of time of day or season on mourning dove FGMs. Evidence from blood sampling of circulating corticosterone levels in avian species suggests that diel rhythms occur; baseline corticosterone levels are lowest mid-day and individuals that are most sensitive to stressors exhibit greatest increases in corticosterone levels early in the morning after waking [32, 33]. In contrast, the limited evidence we observed for elevation in FGMs following treatment was during 1200 (Fig. 2) and not during the earlier 0800 treatment. Further, comparison of controls between experiments conducted at 0800 and 1600 (although during differing seasons) supports the overall trend in that mourning dove baseline FGMs were not lowest in the early morning hours after waking. In addition to fine-scale diel rhythms, seasonal trends in corticosterone and adrenal responsiveness have been observed in a diversity of avian species [46, 47, 48, 49, 50]. For example, Romero et al. [51] found that plasma corticosterone levels and responses to CSP differed by season in White-crowned Sparrows, and that levels and responsiveness were lowest during fall migration and in winter, and highest during active breeding in summer. Although we observed some of the highest FGM values for doves sampled during the winter 2001 period, doves sampled during winter 2002 that were exposed to a similar CSP treatment protocol exhibited the lowest observed FGM values over the course of the study (Fig. 3). Collectively, our findings suggest that time of day and seasonal differences on FGMs were inconsistent, and likely of secondary importance compared to other environmental factors, most notably time in captivity.

Although our results suggest that FGM analyses are not appropriate for detecting acute physiological responses to short-term stressors in mourning doves and likely other avian species, these methods apparently provide a better-integrated measure of adrenocortical responses in long-term field studies. The main utility of FGM monitoring techniques likely involves long-term studies where frequently repeated sample collection from focal individuals is required [27, 52]. Fecal glucocorticoid metabolite analyses might provide a better-integrated measure of adrenocortical responses because they are less sensitive to short-term stressors or disturbances (e.g., predator walking by captive bird colony) that could otherwise bias interpretation of assay results [22, 53]. Further, noninvasive FGM monitoring, used concurrently with the collection of other physiological (e.g., body condition) or demographic parameters (e.g., recruitment), might be effective for examining the effects of potential chronic stressors [54, 55]. We encourage researchers to consider stressor-specific differences, ease of sampling, and the physiological question of interest (i.e. baseline conditions or response to short-term, acute stressors) when determining which sampling technique is most appropriate.

Acknowledgments

Financial and logistical support for this project was provided by the University of Missouri (MU) Department of Fisheries and Wildlife Sciences, a MU Life Science Mission Enhancement Postdoctoral Fellowship, a MU Research Board Grant, the Missouri Department of Conservation (Federal Aid in Wildlife Restoration Project W-13-R), and the 2001 Webless Migratory Game Bird Research Program (United State Fish and Wildlife Service and the United States Geological Survey – Biological Resources Division). Fecal glucocorticoid assays were conducted in the Wildlife Stress Physiology Laboratory in the Department of Fisheries and Wildlife Sciences at the University of Missouri-Columbia. We thank B. Crampton, S. Kistner, B. Hoenes, T. Bonnot, R. Woeck, and C. Rittenhouse for their assistance in pen construction and/or feces collection. This research was approved by the University of Missouri Animal Care and Use Committee (Protocol #3581).

References

Five "key references", selected by the authors, are marked below (Three recommended (●) and two highly recommended (●●) papers).

1. Wingfield, J.C., Vleck, C.M., & Moore, M.C. 1992. Seasonal changes in the adrenocortical response to stress in birds of the Sonoran Desert. J Exp Zool 264:419–428. http://dx.doi.org/10.1002/jez.1402640407
2. Bonier, F., Moore, I.T., Martin, P.R., & Robertson, R.J. 2009. The relationship between fitness and baseline glucocorticoids in a passerine bird. Gen Comp Endocr 163:208-213. http://dx.doi.org/10.1016/j.ygcen.2008.12.013
3. Ouyang, J.Q., Sharp, P.J., Dawson, A., Quetting, M., & Hau, M. 2011. Hormone levels predict individual differences in reproductive success in a passerine bird. P Roy Soc B 278: 2537-2545. http://dx.doi.org/10.1098/rspb.2010.2490
4. Schoech, S.J., Rensel, M.A., & Heiss, R.S. 2011. Short-and long-term effects of developmental corticosterone exposure on avian physiology, behavioral phenotype, cognition, and fitness: a

review. Curr Zool 57:514-530.

5. Piersma, T., & Ramenofsky, M. 1998. Long-term decreases of corticosterone in captive migrant shorebirds that maintain seasonal mass and moult cycles. J Avian Biol 29:97–104. http://dx.doi.org/10.2307/3677186

6. Piersma, T., Reneerkens, J., & Ramenofsky, M. 2000. Baseline corticosterone peaks in shorebirds with maximal energy stores for migration: a general preparatory mechanism for rapid behavioral and metabolic transitions? Gen Comp Endocr 120:118-126. http://dx.doi.org/10.1006/gcen.2000.7543

7. Carere, C., Groothuis, T.G.G., Möstl, E., Daan, S., & Koolhaas, J.M. 2003. Fecal corticosteroids in a territorial bird selected for different personalities: daily rhythm and the response to social stress. Horm Behav 43:540-548. http://dx.doi.org/10.1016/S0018-506X(03)00065-5

8. Pravosudov, V.V. 2003. Long-term moderate elevation of corticosterone facilitates avian food-caching behaviour and enhances spatial memory. P Roy Soc B 270: 2599-2604. http://dx.doi.org/10.1098/rspb.2003.2551

9. Lohmus, M., Sundström, L.F., & Moore, F.R. 2006. Non-invasive corticosterone treatment changes foraging intensity in red-eyed vireos *Vireo olivaceus*. J Avian Biol 37: 523-526. http://dx.doi.org/10.1111/j.0908-8857.2006.03733.x

10. Crossin, G.T., Trathan, P.N., Phillips, R.A., Gorman, K.B., Dawson, A., Sakamoto, K.Q. & Williams, T.D. 2012. Corticosterone predicts foraging behavior and parental care in macaroni penguins. Am Nat 180:E31-E41. http://dx.doi.org/10.1086/666001

11. Schoech, S .J., Rensel, M.A., Bridge, E.S., Boughton, R.K., & Wilcoxen, T.E. 2009. Environment, glucocorticoids, and the timing of reproduction. Gen Comp Endocr 163:201-207. http://dx.doi.org/10.1016/j.ygcen.2008.09.009

12. Raouf, S.A., Smith, L.C., Brown, M.B., Wingfield, J.C., & Brown, C.R. 2006. Glucocorticoid hormone levels increase with group size and parasite load in cliff swallows. Anim Behav 71:39-48. http://dx.doi.org/10.1016/j.anbehav.2005.03.027

13. Marra, P.P., & Holberton, R.L. 1998. Corticosterone levels as indicators of habitat quality: effects of habitat segregation in a migratory bird during the non-breeding season. Oecologia 116:284-292. http://dx.doi.org/10.1007/s004420050590

14. ● Walker, B.G., Boersma, P.D., & Wingfield, J.C. 2005. Field endocrinology and conservation biology. Integr Comp Biol 45:12-18. http://dx.doi.org/10.1093/icb/45.1.12

15. Arlettaz, R., Patthey, P., Baltic, M., Leu, T., Schaub, M., Palme, R., & Jenni-Eiermann, S. 2007. Spreading free-riding snow sports represent a novel serious threat for wildlife. P Roy Soc B 274:1219-1224. http://dx.doi.org/10.1098/rspb.2006.0434

16. Wasser, S.K., Hunt, K.E., Brown, J.L., Cooper, K., Crockett, C.M., Bechert, U., Millspaugh, J.J., Larson, S., & Monfort, S.L. 2000. A generalized fecal glucocorticoid assay for use in a diverse array of non-domestic mammalian and avian species. Gen Comp Endocrinol 120:260–275. http://dx.doi.org/10.1006/gcen.2000.7557

17. ●● Millspaugh, J.J., & Washburn, B.E. 2004. Use of fecal glucocorticoid metabolite measures in conservation biology research: considerations for application and interpretation. Gen Comp Endocr 138:189-199. http://dx.doi.org/10.1016/j.ygcen.2004.07.002

18. Smith, J.E., Monclús, R., Wantuck, D., Florant, G.L. & Blumstein, D.T. 2012. Fecal glucocorticoid metabolites in wild yellow-bellied marmots: Experimental validation, individual differences and ecological correlates. Gen Comp Endocr 178:417-426. http://dx.doi.org/10.1016/j.ygcen.2012.06.015

19. Narayan, E.J., Webster, K., Nicolson, V., Mucci, A. & Hero, J.M. 2013. Non-invasive evaluation of physiological stress in an iconic Australian marsupial: the koala (*Phascolarctos cinereus*). Gen Comp Endocr 187: 39-47. http://dx.doi.org/10.1016/j.ygcen.2013.03.021

20. Sheriff, M.J., Krebs, C.J., & Boonstra, R. 2010. Assessing stress in animal populations: Do fecal and plasma glucocorticoids tell the same story? Gen Comp Endocr 166:614–619. http://dx.doi.org/10.1016/j.ygcen.2009.12.017

21. Sheriff, M.J., Dantzer, B., Delehanty, B., Palme, R. & Boonstra, R. 2011. Measuring stress in wildlife: techniques for quantifying glucocorticoids. Oecologia 166: 869-887. http://dx.doi.org/10.1007/s00442-011-1943-y

22. Harper, J.M., & Austad, S.N. 2000. Fecal glucocorticoids: a noninvasive method of measuring adrenal activity in wild and captive rodents. Physiol Biochem Zool 73:12–22. http://dx.doi.org/10.1086/316721

23. Le Maho, Y., Karmann, H., Briot, D., Handrich, Y., Robin, J.P., Mioskowski, E., Cherel, Y., & Farni, J. 1992. Stress in birds due to routine handling and a technique to avoid it. Am J Physiol 263:R775–R781.

24. Romero, L.M., & Reed, J.M. 2005. Collecting baseline corticosterone samples in the field: is under 3 min good enough? Comp Biochem Physiol A 140:73-79. http://dx.doi.org/10.1016/j.cbpb.2004.11.004

25. ●● Seudkamp Wells, K.M., Washburn, B.E., Millspaugh, J.J., Ryan, M.R., & Hubbard, M.W. 2003. Effects of radio-transmitters on fecal glucocorticoid levels in captive dickcissels. Condor 105:805-810. http://dx.doi.org/10.1650/7174

26. Pereira, R.J.G., Granzinolli, M.A.M., Barros, F.M., & Duarte, J.M.B. 2009. Influence of radiotransmitters on fecal glucocorticoid levels of free-ranging male American kestrels. J Wildlife Manage 73:772-778. http://dx.doi.org/10.2193/2008-184

27. ● Washburn BE, Millspaugh JJ, Schulz JH, Jones SB, Mong T. 2003. Using fecal glucocorticoids for stress assessment in mourning doves. Condor 105: 696-706. http://dx.doi.org/10.1650/7216

28. Reeves, H.M., Geis, A.D., & Kniffin, F.C. 1968. Mourning dove capture and banding. Special Scientific Report-Wildlife 117, U.S. Fish and Wildlife Service, Washington, D.C.

29. Mirarchi, R.E. 1993. Sexing, aging, and miscellaneous research techniques. In: Baskett,T.S., Sayre, M.W., Tomlinson, R.E. & Mirarchi, R.E. (eds.). Ecology and management of the mourning dove. Wildlife Management Institute, Washington, D.C., pp 399–408.

30. Schulz, J.H., Sheriff, S.L., He, Z., Braun, C.E., Drobney, R.D., Tomlinson, R.E., Dolton, D.D., & Montgomery, R.A. 1995. Accuracy of techniques used to assign mourning dove age and gender. J Wildl Manage 59:759–765. http://dx.doi.org/10.2307/3801953

31. Mong, T.W., Schulz, J.H., & Millspaugh, J.J. 2002. Design and construction of mourning dove research pens. T Missouri Acad Sci 36:21-24.

32. ● Breuner, C.W., Wingfield, J.C. & Romero, L.M. 1999. Diel rhythms of basal and stress-induced corticosterone in a wild, seasonal vertebrate, Gambel's white-crowned sparrow. J Exp Zool 284:334–342.

33. Tarlow, E.M., Hau, M., Anderson, D.J., & Wikelski, M. 2003. Diel changes in plasma melatonin and corticosterone concentrations in tropical Nazca boobies (*Sula granti*) in relation to moon phase and age. Gen Comp Endocr 133: 297-304.
http://dx.doi.org/10.1016/S0016-6480(03)00192-8
34. Schwarzenberger, F., Möstl, E., Bamberg, E., Pammer, J. & Schmehlik, O. 1991. Concentrations of progestagens and oestrogens in the faeces of pregnant lipizzan, trotter and thoroughbred mares. J Reprod Fertil Suppl 44:489–499.
35. Brown, J.L., Bellem, A.C., Fouraker, M., Wildt, D.E., & Roth, T.L. 2001. Comparative analysis of gonadal and adrenal activity in the black and white rhinoceros in North America by noninvasive endocrine monitoring. Zoo Biol 20:463-486.
http://dx.doi.org/10.1002/zoo.10028
36. Roy, C., & Woolf, A. 2001. Effects of hunting and hunting-hour extension on Mourning Dove foraging and physiology. J Wildl Manage 65:808–815.
http://dx.doi.org/10.2307/3803030
37. Washburn, B.E., Morris, D.L., Millspaugh, J.J., Faaborg, J. & Schulz, J.H. 2002. Using a commercially available radioimmunoassay to quantify corticosterone in avian plasma. Condor 104:558–563.
http://dx.doi.org/10.1650/0010-5422(2002)104[0558:UACART]2.0.CO;2
38. Otis, D.L. & White, G.C. 2002. Re-analysis of a banding study to test the effects of an experimental increase in bag limits of mourning doves. J App Stat 29:479–495.
http://dx.doi.org/10.1080/02664760120108539
39. Otis, D.L. 2006. Mourning dove hunting regulation strategy based on annual harvest statistics and banding data. J Wildl Manage 70:1302–1307.
http://dx.doi.org/10.2193/0022-541X(2006)70[1302:MDHRSB]2.0.CO;2
40. Otis, D.L., Schulz, J.H. & Scott, D.P. 2008. Mourning dove (*Zenaida macroura*) harvest and population parameters derived from a national banding study. U.S. Department of Interior, Fish and Wildlife Service, Biological Technical Publication FWS/BTP-R3010-2008, Washington, DC, USA.
41. Romero, L.M. & Wingfield, J.C. 1999. Alterations in hypothalamic-pituitary-adrenal function associated with captivity in Gambel's white-crowned sparrows (*Zonotrichia leucophrys gambelii*). Comp Biochem Physiol 122B:13–20.
http://dx.doi.org/10.1016/S0305-0491(98)10161-X
42. Marra, P.P., Lampe, K.T., & Tedford, B.L. 1995. Plasma corticosterone levels in two species of Zonotrichia sparrows under captive and free-ranging conditions. Wilson Bull 107:296–305.
43. Romero, L.M. 2004. Physiological stress in ecology: lessons from biomedical research. Trends Ecol Evol 19:249-255.
http://dx.doi.org/10.1016/j.tree.2004.03.008
44. Cyr, N.E., & Romero, L.M. 2007. Chronic stress in free-living European starlings reduces corticosterone concentrations and reproductive success. Gen Comp Endocr 151:82-89.
http://dx.doi.org/10.1016/j.ygcen.2006.12.003
45. Dickens, M.J., Earle, K.A., & Romero, L.M. 2009. Initial transference of wild birds to captivity alters stress physiology. Gen Comp Endocr 160:76-83.
http://dx.doi.org/10.1016/j.ygcen.2008.10.023
46. Romero, L.M., Soma, K.K. & Wingfield, J.C. 1998a. Changes in pituitary and adrenal sensitivities allow the Snow Bunting (*Plectrophenax nivalis*), an Arctic-breeding songbird, to modulate corticosterone release seasonally. J Comp Physiol B 168:353–358.
http://dx.doi.org/10.1007/s003600050154

47. Romero, L.M., Soma, K.K. & Wingfield, J.C. 1998b. The hypothalamus and adrenal regulate modulation of coricosterone release in redpolls (*Carduelis flammea* - an Arctic-breeding song bird). Gen Comp Endocr 109:347–355. http://dx.doi.org/10.1006/gcen.1997.7048

48. Romero, L.M., & Remage-Healey, L. 2000. Daily and seasonal variation in response to stress in captive Starlings (*Sturnus vulgaris*): corticosterone. Gen Comp Endocr 119:52–59. http://dx.doi.org/10.1006/gcen.2000.7491

49. Romero, L.M. 2002. Seasonal changes in plasma glucocorticoid concentrations in free-living vertebrates. Gen Comp Endocr 128:1-24. http://dx.doi.org/10.1016/S0016-6480(02)00064-3

50. Landys, M.M., Goymann, W., Schwabl, I., Trapschuh, M., & Slagsvold, T. 2010. Impact of season and social challenge on testosterone and corticosterone levels in a year-round territorial bird. Horm Behav 58: 317-325. http://dx.doi.org/10.1016/j.yhbeh.2010.02.013

51. Romero, L.M., Ramenofsky, M., & Wingfield, J.C. 1997. Season and migration alters the corticosterone response to capture and handling in an Arctic migrant, the White-crowned sparrow (*Zonotrichia leucophrys gambelii*). Comp Biochem Physiol 116C:171–177.

52. Jachowski, D.S., Slotow, R., & Millspaugh, J.J. 2013. Delayed physiological acclimatization by African elephants following reintroduction. Anim Conserv 16:575-583. http://dx.doi.org/10.1111/acv.12031

53. Ninnes, C.E., Waas, J.R., Ling, N., Nakagawa, S., Banks, J.C., Bell, D.G., Bright, A., Carey, P.W., Chandler, J., Hudson, Q.J., et al. 2010. Comparing plasma and faecal measures of steroid hormones in Adelie penguins *Pygoscelis adeliae*. J Comp Phys B 180: 83-94. http://dx.doi.org/10.1007/s00360-009-0390-0

54. Ellenberg, U., Setiawan, A.N., Cree, A., Houston, D.M., & Seddon, P.J. 2007. Elevated hormonal stress response and reduced reproductive output in yellow-eyed penguins exposed to unregulated tourism. Gen Comp Endocr 152: 54-63. http://dx.doi.org/10.1016/j.ygcen.2007.02.022

55. Cry, N.E., & Romero, L.M. 2008. Fecal glucocorticoid metabolites of experimentally stressed captive and free-living starlings: Implications for conservation research. Gen Comp Endocr 158:20-28. http://dx.doi.org/10.1016/j.ygcen.2008.05.001

20

Population Density Estimation of the European Wildcat (*Felis silvestris silvestris*) in Sicily Using Camera Trapping

S. Anile[1,*], C. Amico[2] & B. Ragni[3]

[1] Dipartimento di Biologia Animale "Marcello La Greca", Università di Catania, Via Androne 81, 95124 Catania, Italy.

[2] Ripartizione Faunistico Venatoria di Catania, Via San Giuseppe alla Rena 30/b, 95121 Catania, Italy.

[3] Dipartimento di Biologia Cellulare ed Ambientale, Università degli Studi di Perugia, Via Elce di Sotto, 06123 Perugia, Italy.

* Corresponding author e-mail: stefanoanile@yahoo.it.

Keywords

Capture-recapture analyses; Etna Volcano; *Felis silvestris silvestris*; Marking system.

Abstract

The wildcat is an elusive species that is threatened with extinction in many areas of its European distribution. In Sicily the wildcat lives in a wide range of habitats; this study was done on Mount Etna. Previous camera trap monitoring was conducted in 2006 (pilot study) and in 2007 (first estimation of wildcat population size by using camera trapping with capture-recapture analyses) in the same study area. In 2009 digital camera traps in pairs were used at each station with the aim of determining the density of the wildcat's population by using capture–recapture analyses. The coat-colour and markings system was used to determine both the taxonomical status of the photographed cat and the individual identification.

Two trap-lines adjacent to each other were run in two consecutive data collection periods. Camera traps worked together for 1080 trap-days and we obtained 42 pictures of wildcats from 32 events of photographic capture, from which 10 individuals (excluding four kittens) were identified. The history capture of each individual was constructed and the software CAPTURE (model heterogeneity) was used to generate an estimation of the population density (0.28 ± 0.1 wildcat/1 Km^2) by using the Full Mean Maximum Distance Moved for wildcats caught more than once to calculate the effective sampled area.

The wildcat's population density on Mount Etna falls within the medium/high range of those found throughout Europe. Recaptures of the same individual and reproductions events have been recorded across the years. Comparison with previous studies suggested that wildcat's density in our study area might be stable.

Introduction

Assessing accurate estimates of population density is one of the major goals for wildlife management, especially for those species that are difficult to observe directly owing to their behaviour and the ecological conditions of their habitat. Sicily is characterized by the presence of one of the most important insular populations [1] of the threatened European wildcat (*Felis silvestris silvestris* Schreber, 1777) which is not the result of human introduction [2].

Currently the wildcat is affected by several threat factors such as illegal shooting, hybridization with domestic cat (reported mainly from Scotland and Hungary), loss of habitat and road killing (mainly in Germany)[3-5]: despite this only one study has been carried out in the past on this taxon on Sicily [6].

Wildcat populations are scarce and fragmented [1], therefore conservation measures should focus on the best ecological information, particularly in an accurate assessment of population numbers; especially for those inhabited islands, where immigration from other populations is impossible [7-9].

Wildcats typically live at low densities [10], with the activity rhythm that mainly occurs during night-time [11], on the contrary spending most of day-time in dense vegetation habitat [12]. Given that direct visual observation as a way to detect the wildcat's presence is quite difficult [10] and time-consuming. Assessment of wildcat population numbers have been obtained through questionnaire surveys [13,14], visual observations [15,16], indirect counts of tracks [17] or non invasive DNA techniques [18].

Camera trapping is a non-invasive technique that can be successfully applied to monitor rare, nocturnal and forest species like many felids [19, 20], but so far it was rarely used to assess the wildcat density.

In 2006 a pilot study with camera traps was started in the southwest part of Etna Regional Park [21] and in 2007 it was conducted our first study that, by using camera trapping and capture-recapture analyses, aimed to obtain an estimation of the wildcat's population size [22]: the minimum number of individuals identified, the inter-trap distances and the capture success rate from these previous studies represented the background knowledge to develop the protocol for this project conducted during 2009.

We increased of about 60% the sampling effort reached in the previous study (11 stations and 671 trap-days) by adding more camera stations, that were equipped by a pair of digital camera traps instead of the single film camera trap previously used.

The aim of the present study was to assess the population size of the wildcat through established procedures for capture-recapture analyses of a closed population [19,23,24], by using camera-trapping in place of traps and the individual variation in coat coloration and markings of the wildcat to recognize 'recaptures' in photographs.

Results were compared with those of the previous studies [21,22] performed in the same area; recaptures and reproductions were detected across few years, enlarging the application fields of this technique even to this endangered small felid [25,26].

Materials and methods

Study Area

The study area is located on the southwest side of Mount Etna (Sicily) and ranges from 900 to 2.000 m a.s.l., corresponding to the southernmost location of the European wildcat's distribution [1] (Fig. 1). The Etna Regional Park (59.000 ha) represents one of the largest suitable area for wildcats in Sicily (25.711 Km^2), therefore playing a crucial role for wildcat's conservation in this island [6].

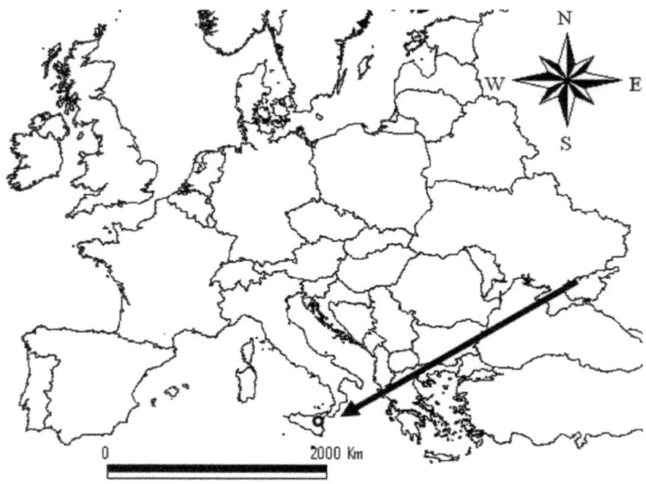

Fig. 1: Location of the Volcano Etna in Sicily.

The landscape is made up of quite recent large lava flows and volcanic inactive secondary cones of different ages, intermixed with wide patches of different trees (*Pinus laricio*, *Quercus pubescens*, *Quercus ilex*, *Castanea sativa*, *Populus tremulus* and *Genista etniensis*).

The resulting habitat structure is quite heterogeneous: woods are interrupted by open fields and widespread refuges are provided by the many cavities of the volcanic soil.

In addition to the European wildcat, the study site is inhabited by the red fox (*Vulpes vulpes*), pine marten (*Martes martes*), brown hare (*Lepus corsicanus*), wood mouse (*Apodemus sylvaticus*), Savi's pine vole (*Microtus savii*) and an abundant population of rabbit (*Oryctolagus cuniculus*).

Human activity is rather low and vehicles access (only under permission of the park authorities) into the study area is provided by country roads travelled for the management of woodland, sheep farming and tourist trekking.

The climate is typically Mediterranean, but during some winters the snow cover may become relevant. Rainfalls are less abundant from the end of spring to the end of summer.

Camera-trapping Protocol

The protocol followed throughout the present research represents the result of an experimental methodological process started in 2006, in the same study area, on the same target species and with the general purpose to assess important aspects of wildcat spatial organization [21,22].

A preliminary survey of the study area, based on the collection of cat signs (e.g. scats, footprints, trails, scratching, prey remains) was conducted to find the best places to set camera traps. Eighteen digital camera traps (DFV® equipped with Sony® DSC-W55) with passive infrared motion/heat sensors were arranged in nine trap stations, in pairs, in order to obtain photographs of both sides of the wildcats, allowing better identification of each individual. Each camera was accommodated in an iron box, locked with a padlock and then tied to a tree (at 50 ± 10 cm from the ground) with a chain. Camera traps were set with a delay time of 10 min. between successive photos. We used a 4 GB Memory Stick Duo that, in combination with a 7 Mega-Pixels resolution, could store up to 572 photographs. The cameras were set to allow the shooting of up to three photographs if the target remained in the detection zone of the sensor. The whole camera trapping period lasted continuously from 14 May to 11 September, but it was arranged in two consecutive sessions, each one lasting 60 days and based on nine trap stations; the trap stations were checked at least once a week.

Depending on the topography and the accessibility of the study area, the trap stations were distributed in order to obtain an inter-trap-station distance of 1351 ± 790m (Fig. 2); this ensured that at least one camera station was located inside the smallest home range size, occupied by this felid in Mediterranean habitat [10,27,11,28] and that no holes, large enough to host a wildcat home range, were left inside the study area [19,23,24].

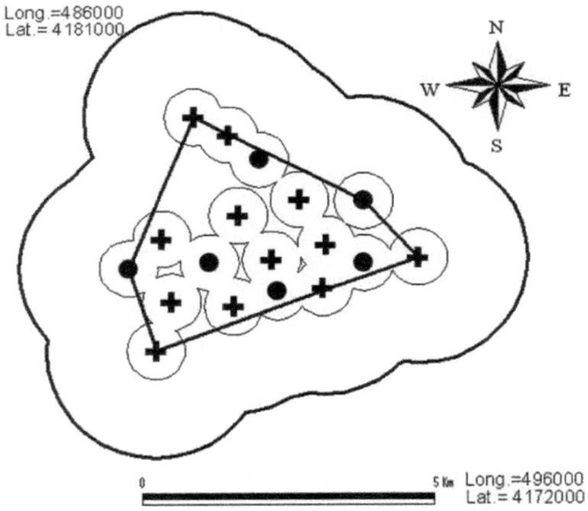

Fig. 2: Trapping station distribution during the study (n=18). Crosses: stations that "captured" wildcats; points: stations that didn't "capture" wildcats; buffers around each stations have 500 m radius; Minimum Convex Polygon (M.C.P): encompasses the trapping stations area (1090 ha); out-boundary area: $W_{F.M.M.D.M.}$ area (4558 ha) created by using the F.M.M.D.M. (1870 m) as a buffer around each stations. Coordinates refer to the vertices of the figure.

Finally, no attractants or baits were used in order to avoid differential responses according to sex, age and status [19,23].

To calculate the effective sampled area *W* we used the Full Mean Maximum Distance Moved (F.M.M.D.M), calculated for wildcats photographed from at least two camera trap stations, because recent camera trapping studies conducted in combination with radiotracking on Ocelot (*Leopardus pardalis*) [29] and Jaguar (*Panthera onca*) [30] show that the F.M.M.D.M. could better approximate the home range size.

Data Analysis

The first step of the data analysis procedure was to define the taxonomic status of "captured" cats, between free-ranging domestic cat (*Felis silvestris catus*), European wildcat (*F. s. silvestris*) and hybrids between them. This was done by using the system of coat coloration and markings proposed by Ragni and Possenti [31], which were used for "a priori" classification in several genetic studies [4,32,33,34,35]: all of them showed an high congruence between this set of diagnostic morphological traits and the genetic distinction between domestic cat, wildcat and their hybrids.

Furthermore, in order to distinguish individuals, we considered number, shape, dimension and position of stripes, bands and spots on the trunk and limbs, number and shape of the rings on the tail, as well as the dimension of its black tip; the observation of unequivocal body signs such as scars on face, lips and ears was also useful in individuals identification (Fig. 3).

Fig. 3: Example of individual identification in wildcats through comparison of coat pattern of the same region in different wildcats.

In order to assess all the above mentioned criteria, all wildcat's photographs (unequivocally classified by station, camera, and data) were independently examined by all authors following principles of conservativeness and accuracy. Only concordant identifications of the individuals were then considered and included in the capture history.

Furthermore, photographs of wildcats showing the same body region were compared with those from a complete data set from previous studies conducted during 2006 (24 pictures) [21], 2007 (27 pictures) [22] and 2008 (8 pictures) [36]: the procedure aimed to establish the minimum number of individuals photographed [29].

A matrix with the capture history of each photographed specimen in the sampling intervals (occasions) was generated: data from the two trapping lines were treated as if they had run simultaneously [20,37] (each occasion was of 12 days long), resulting in 10 sampling occasions.

The software program CAPTURE was used to generate population estimate based on capture–recapture models under the assumption of a closed population where immigration, emigration, births and deaths did not affect the estimate [38,39,40].

Closure test in CAPTURE [39] and the more robust closure test of Stanley and Burnham [41] were performed to test the assumption of a closed population.

Capture-recaptures models differ in their assumption about capture probabilities, in particular individually heterogeneity (Mh) assumes that each individual had its own probability of being captured independently of time and behaviour, time model (Mt) accounts for variation in capture probabilities across occasions, behaviour model (Mb) considers a differential response (trap-happy or trap-avoid) if the individual has been previously captured and finally, capture probabilities are assumed to be constant (M0) respect to time and individuals. In addition, CAPTURE allows estimation under 4 more models that combine the sources of variation cited above (Mbh, Mth, Mtb, Mtbh).

Finally CAPTURE, using an overall model selection, compares the relative fit and the Goodness-of-Fit across all these models [20], assigning a score (from 0 to 1) that indicates which one is the best model (score 1).

Results

Camera traps worked continuously for a total of 1080 trap days. Thirty two wildcat events produced 42 photographs (indeed in 31.2% of wildcat events the pairs of camera traps located for each station snapped photographs of both sides of the wildcat) from 12 trapping stations (Table 1).

Table 1: *X and Y*: U.T.M coordinates reference system WGS84; *Starting and Ending*: monitoring periods; *Pictures*: wildcat photograph recorded for each station and simultaneous shot of both cameras at the same event in brackets.

Stations	X	Y	Starting	Ending	Pictures
1	493127	4176420	14/5/2009	13/7/2009	1
2	492201	4176312	14/5/2009	13/7/2009	
3	491505	4175875	14/5/2009	13/7/2009	6 (3)
4	490719	4175804	14/5/2009	13/7/2009	
5	490619	4176374	14/5/2009	13/7/2009	4 (4)
6	489569	4176316	14/5/2009	13/7/2009	
7	488916	4175628	14/5/2009	13/7/2009	2 (1)
8	490055	4177155	14/5/2009	13/7/2009	5 (2)
9	491098	4177436	14/5/2009	13/7/2009	1
10	489974	4175547	13/7/2009	11/9/2009	2
11	488185	4176216	13/7/2009	11/9/2009	
12	488750	4176737	13/7/2009	11/9/2009	1
13	489291	4178907	13/7/2009	11/9/2009	2
14	489883	4178595	13/7/2009	11/9/2009	3
15	490410	4178149	13/7/2009	11/9/2009	
16	492195	4177410	13/7/2009	11/9/2009	
17	491561	4176643	13/7/2009	11/9/2009	2
18	488647	4174766	13/7/2009	11/9/2009	3

From the whole pool of 42 photographs we were able to select 40 photographs useful for the individual identification.

Ten different individuals were identified, all belonging to *Felis silvestris silvestris* subspecies, from which we were able to sex two males and two females and in addition the presence of four kittens with their mother was recorded: the rate of capture success was 2.9 captures/100 trap-days.

Two captures (ID=5 and ID=6) were excluded from the capture history (Table 2) because the same individuals were captured twice in the same occasion (respectively seventh and tenth occasion).

Table 2: Capture history of the ten individually identified wildcats; each of the 10 trapping occasions comprises two periods of six days each; 1: at least one "capture" of an identified individual; 0: no capture (see the text for further specifications).

Wildcats	Occasions									
1	0	0	0	0	1	0	0	0	0	0
2	1	0	1	0	0	1	0	0	0	0
3 ♂	0	1	0	0	0	0	0	0	1	0
4 ♂	1	0	0	1	1	0	0	0	1	0
5	0	0	0	0	0	0	1	0	0	0
6 ♀	1	1	0	1	0	0	0	0	0	1
7	0	1	1	0	0	0	0	0	0	1
8	0	0	0	0	0	0	0	0	1	0
9	0	0	0	0	0	0	0	0	1	0
10 ♀	0	0	0	0	0	0	0	0	0	1

The overall model selection test (Table 3) indicates that the best fitting model of our data was M(b) where behaviour is considered as a source of variation. However, considering that 1) the captures of two wildcats (ID= 2 and ID=7) on both trapping lines produced a trap-happy bias in capture history (capture events have been assigned to the corresponding occasion, so distorting the real occurrence of such events across the time) that would favour M(b); 2) the ratio of recaptures/captures (13/23) suggested no evidence for trap-shyness; 3) no attractors were used. We therefore decided to use the heterogeneity model M(h) that reflects the biological aspects and the social organization of most of the felids [23,42].

The closure test in CAPTURE and the closure test of Stanley and Burnham provided no evidence for a violation of closure assumption.

Table 3: Result of program CAPTURE. The asterisk indicates the model that we decided to use to estimate the population density of wildcats. Estimators for the more complex models (bh, th, tb, tbh) are not yet available.

Models	M(0)	M(h)	M(b)	M(t)	M(bh)	M(th)	M(tb)	M(tbh)
Value selection score	0.33	0.30*	1.00	0.61	0.00	0.75	0.72	0.59
Test for Closure Capture	z = 0.093, P = 0.53							
Test for Closure Stanley and Burnham	X^2 = 9.860, df 7, P = 0.196							
Population estimate *N*, with SE	13, 2.36							
95% Confidence Interval	8.38 – 17.62							
Profile likelihood	11 to 21							
Probability of capture *p*	0.18							

Recaptures for wildcats 'caught' more than once (n = 4) resulted in a F.M.M.D.M = 1870 m and therefore $W_{F.M.M.D.M}$ (the merged area of all of the circular buffers with radius = F.M.M.D.M. created for each trapping station) was 4588 ha (Fig. 2).

Finally the population density $D = N/W$ was 0.28 ± 0.1 wildcat/1 km^2.

Discussion

Our studies show that camera trapping can be used to accurately estimate species abundance and population density [43], and so it represents an essential tool for wildcat conservation.

Compared to our previous studies, the rate of capture success was equal (2.9 captures/100 trap-days) to that obtained in 2006 [21], whilst it was slightly lower than the one of 2007 [22] (4 captures/100 trap/days). The percentage of usable photographs for the present study was 95%, whilst the percentage of usable photographs in the previous studies was 59% in 2007 and 37.5% in 2006: presumably the use of two cameras at each trap station increased the probability to obtain valuable photographs for the individual identification [20].

Six out of the 10 identified wildcats have been already photographed during the previous studies and most of the recaptures have been obtained from the same locations: these findings could be due to a stable occupancy of the home ranges over the time [27] in adult wildcats, enhancing the role of the camera trapping even for medium-long term studies [25].

The F.M.M.D.M for the present study was greater than that calculated in 2007 (1080 m); it seems reasonable that the increased effort in term of area surveyed during 2009 allowed to detect recaptures at longer distances [44].

The dimension of individual home ranges of some felid species fits better with F.M.M.D.M than with H.M.M.D.M estimators, so the 2007 population density (0.96 wildcat/1 km^2) was recalculated [22] using F.M.M.D.M., which resulted in a population density of 0.46 ± 0.13wildcat /1 km^2.

The confidence intervals of the two monitorings conducted in 2007 and 2009 slightly overlapped suggesting that the Mount Etna population of wildcats might be stable.

Usually more than one wildcat was detected from a single station so wildcats probably share trails that occur in the study area, as in other felid species [19]; additionaly, radiotracking studies on wildcat shown that overlap between home range could occur [10,11,27,28].

Other important aspects to be highlighted are that wildcats ranging in our study area didn't appear to be intimidated by the camera trap, neither did they avoid the camera traps after the first shot or after we checked them. Furthermore, it should be considered both the high numbers of recaptures and the wildcats with only one capture, photographed by the peripheral stations according to the principle that animals having their home range on the edge of the trap layout have less probability to be photographed [45]: that is, no evidences of the trap-shyness behaviour [46] have been obtained during this survey.

The density estimate for 2009 is in the medium/high range when compared to studies where the wildcat's population density is available, even if calculated with other methods: Central Appennines (0.2–0.3/km^2) [17], Voijvodina (0.164–0.449/km^2) [15], 'optimal forest habitats' in Western Europe (0.3–0.5/km^2) [47], the Stromberg of Baden-Wuettemburg (SW Germany) (0.17–0.25/km^2) [16], the Polish Carpathian Mountains (0.1–0.13/km^2) [13], the Basel Southwest Mountains (Switzerland) (0.35/km^2) [18], most recent density available for Hungary (0.11/ km^2) [14].

The population density of the Sicilian wildcat on Mount Etna could be favorably influenced by some reasonable factors: the presence of an abundant population of European wild rabbit, the "staple" prey for the wildcat where available [48]; the heterogeneous habitat of Etna provides optimal features in term of hunting areas and refuges [12,28,49,50]; the Mediterranean climate ensures, even when snow cover could be abundant, that it persists only for at most several weeks [51].

Finally, the protected status of Etna (Natural Regional Park) might be a favorable factor, even if wildcat conservation actions have never been undertaken.

The goodness of camera trapping to estimate the population density of the wildcat is reasserted, so we support the use of this non-invasive methodology as a standard-tool for wildcat conservation. All the camera trapping procedures such as displacing, field setting, checking and data download were conducted by the same person and this is a particular cue when funding opportunities are scarce and more over, the camera trapping equipment could be useful for future surveys [52].

Our next step will be to obtain DNA by systematic scat and hair collection during camera trapping surveys monitoring in order to obtain an independent estimate of the wildcat's population.

Acknowledgements

Ripartizione Faunistico Venatoria of Catania allowed to use all the camera trapping equipments.
Stefano Anile thanks: his family for economic and moral support; E. Centofanti for camera trap assistance.
The authors thank four anonymous referees that greatly improved the first version of the manuscript.

References

Five "key references", selected by the authors, are marked below (Three recommended (●) and two highly recommended (●●) papers).

1. Nowell, K. & Jackson, P. 1996. Wild Cats: Status survey and conservation action plan. International Union for Nature Conservation (IUCN) /Cat Specialist Group, Gland, Switzerland. pp. 110-113. The Burlington Press, Cambridge, U.K.
2. Kitchener, A. C. & Rees, E. E. 2009. Modelling the dynamic biogeography of the wildcat: implications for taxonomy and conservation. J. Zool. 279: 144-155. DOI:10.1111/j.1469-7998.2009.00599.x
3. Krone, O., Guminsky, O., Meinig, H., Herrmann, M., Trinzen, M. & Wibbelt, G. 2008. Endoparasite spectrum of wild cats (*Felis silvestris* Schreber, 1777) and domestic cats (*Felis catus* L.) from the Eifel, Pfalz region and Saarland, Germany. Eur. J. Wildl. Res. 54: 95-100. DOI:10.1007/s10344-007-0116-0
4. Oliveira, R., Godinho, R., Randi, E. & Alves, P.C. 2008. Hybridization versus conservation: are domestic cats threatening the genetic integrity of wildcats (*Felis silvestris silvestris*) in Iberian Peninsula? Philos. Trans. R. Soc.: Biol. Sci. 363: 2953-2961. DOI:10.1098/rstb.2008.0052
5. Klar, N., Herrmann, M. & Kramer-Schadt, S. 2009. Effects and mitigation of road impacts on individual movement behavior of wildcats. J. Wildl. Manag. 73 (5): 631-638. DOI:10.2193/2007-574
6. Ragni, B. & Seminara, S. 1987. Il gatto selvatico. [The wildcat]. A. R. T. A. Sicilia, Fondo mondiale per la Natura, sezione di Polizzi Generosa, Palermo.

7. Daniels, M.J., Balharry, D., Hirst, D., Kitchener A.C. & Aspinall, R.J. 1998. Morphological and pelage characteristics of wild living cats in Scotland: implications for defining the `wildcat'. J. Zool. 244: 231-247.
8. Daniels, M.J., Beaumont, M.A., Johnson, P.J., Balharry, D., Macdonald, D.W. & Barratt, E. 2001. Ecology and genetics of wild-living cats in the north-east of Scotland and the implications for the conservation of the wildcat. J. Appl. Ecol. 38: 146-161.
9. Kitchener, A.C., Yamaguchi, N., Ward, J.M. & Macdonald, D.W. 2005. A diagnosis for the Scottish wildcat (*Felis silvestris*): a tool for conservation action for a critically-endangered felid. Anim. Conserv. 8: 223-237. DOI:10.1017/S1367943005002301
10. Monterroso, P., Brito, J.C., Ferreras, P. & Alves, P.C. 2009. Spatial ecology of the European wildcat in a Mediterranean ecosystem: dealing with small radio-tracking datasets in species conservation. J. Zool. 279: 27-35. DOI:10.1111/j.1469-7998.2009.00585.x
11. Germain, E., Benhamou, S. & Poulle, M. L. 2008. Spatio-temporal sharing between the European wildcat, the domestic cat and their hybrids. J. Zool. 276: 195-203. DOI:10.1111/j.1469-7998.2008.00479.x
12. Jerosch, S., Gotz, M., Klar, N. & Mechthild, R. 2010. Characteristics of diurnal resting sites of the endangered European wildcat (*Felis silvestris silvestris*): Implications for its conservation. J. Nat. Conserv. 18: 45-54. DOI:10.1016/j.jnc.2009.02.005
13. Okarma, H., Sniezko, S. & Olszanska, A. 2002. Occurence of wildcat in the Polish Carpathian Mountains. Acta Theriol. 47 (4): 499-504.
14. Heltai, M., Biro', Z. & Szemethy, L. 2006. The changes of distribution and population density of wildcats *Felis silvestris* Schreber, 1775 in Hungary between 1987-2001. Nat. Conserv. 62: 37-42.
15. Dimitrijevic, S. 1980. Ecomorphological characteristics of the wildcat (*Felis silvestris* Schreber, 1777) on the territory of Voijvodina. Arhiv. biol. Nauka Beograd 58 (3a): 219-282
16. Heller, M. 1992. Status and conservation of the wildcat (*Felis silvestris*) in the region Baden-Wurttemberg, Southwest Germany. In Environmental encounters (eds). Seminary on the biology and conservation of the wildcat (*Felis silvestris*). Proceedings of the Convention on the Conservation of the European Wildlife and Natural Habitats; Strasbourg. N° 16, p 42-45.
17. Ragni, B. 2006. Il gatto selvatico. [The wildcat]. In: Perdisa, A.(Eds.), Salvati dall'arca. [Saved by the ark]. Bologna, pp. 35-36.
18. Weber, D., Stoeckle, T.M. & Roth, T. 2008. Entwicklung und Anwendung einer neuen Wilkatzen-Nachweismethode. [Developing and applying a new method for wildcat monitoring]. Internal report. Hintermann & Weber AG Rodersdorf 29.
19. ● Karanth, K.U. 1995. Estimating tiger populations from camera-trap data using capture-recapture models. Biol. Conserv. 71: 333-338.
20. Jackson, R.M., Roe, J.D., Wangchuk, R. & Hunter, D.O. 2005. Surveying Snow Leopard Populations with Emphasis on Camera Trapping: A Handbook. The Snow Leopard Conservancy Sonoma, California, 73. Available via www.SnowLeopardConservancy.org.
21. Anile, S., Bizzarri, L. & Ragni, B. 2009. Camera trapping the European wildcat (*Felis silvestris silvestris*) in Sicily (Southern Italy): preliminary results. Hystrix Ital. J. Mammal. (n.s) 20(1): 55-60.
22. Anile, S., Bizzarri, L. & Ragni, B. 2010. Estimation of European wildcat population size in Sicily (Italy) using camera-trapping and capture-recapture analyses. Ital. J. Zool. 77 (2): 241-

246.
DOI:10.1080/11250000903419731
23. Karanth, K.U. & Nichols, J.D. 1998. Estimation of tiger densities in India using photographic captures and recaptures. Ecol. 79 (8): 2852-2862.
24. Karanth, K.U., Chundawat, R.,S., Nichols, J.D. & Samba Kumar, N.S. 2004. Estimation of tiger densities in the tropical dry forests of Panna, Central India, using photographic capture–recapture sampling. Anim. Conserv. 7: 285-290.
DOI:10.1017/S1367943004001477
25. Karanth, K.U., Nichols, J.D., Samba Kumar, N.S. & Hines, J.E. 2006. Assessing tiger population dynamics using photographic capture-recapture sampling. Ecol. 87 (11): 2925-2937.
26. Rowcliffe, M.J. & Carbone, C. 2008. Surveys using camera traps: are we looking to a brighter future? Anim. Conserv. 11: 185-186.
DOI:10.1111/j.1469-1795.2008.00180.x
27. Birò, Z., Szemethy, L. & Heltai, M. 2004. Home range sizes of wildcats (*Felis silvestris*) and feral domestic cats (*Felis silvestris f.catus*) in a hilly region of Hungary. Mamm. Biol. 69: 302-310.
28. Sarmento, P., Cruz, J., Tarroso, P. & Fonseca, C. 2006. Space and habitat selection by female european wildcats (*Felis silvestris silvestris*). Wildl. Biol. Pract. 2 (2): 79-89.
DOI:10.2461/wbp.2006.2.10
29. ●● Soisalo, M.K. & Cavalcanti, S.M.C. 2006. Estimating the density of a jaguar population in the Brazilian Pantanal using camera-traps and capture–recapture sampling in combination with GPS radio-telemetry. Biol. Conserv. 129: 487-496.
DOI:10.1016/j.biocon.2005.11.023
30. ● Dillon, A. & Kelly, M.J. 2008. Ocelot home range, overlap and density: comparing radio telemetry with camera trapping. J. Zool. 275: 391-398.
DOI:10.1111/j.1469-7998.2008.00452.x
31. Ragni, B. & Possenti, M. 1996. Variability of coat-colour and markings system in *Felis silvestris*. Ital. J. Zool. 63: 285-292.
32. Randi, E., Pierpaoli, M., Beaumont, M., Ragni, B. & Sforzi, A. 2001. Genetic identification of wild and domestic cats (*Felis silvestris*) and their hybrids using bayesian clustering methods. Mol. Biol. Evol. 8 (9): 1679-1693.
33. Pierpaoli, M., Birò, Z. S., Herrmann, M., Hupe, K., Fernandes, M., Ragni, B., Szemethy, L. & Randi, E. 2003. Genetic distinction of wildcat (*Felis silvestris*) populations in Europe, and hybridization with domestic cats in Hungary. Mol. Ecol.12: 2585-2598.
DOI:10.1046/j.1365-294X.2003.01939.x
34. Lecis, R., Pierpaoli, M., Birò, Z., Szemethy, L., Ragni, B., Vercillo, F. & Randi, E. 2006. Bayesian analyses of admixture in wild and domestic cats (*Felis silvestris*) using linked microsatellite loci. Mol. Ecol. 15: 119-131.
DOI:10.1111/j.1365-294X.2005.02812.x
35. Oliveira, R., Godinho, R., Randi, E., Ferrand, N. & Alves, P. C. 2007. Molecular analysis of hybridisation between wild and domestic cats (*Felis silvestris*) in Portugal: implications for conservation. Conserv. Genet.
DOI:10.1007/s10592-007-9297-z
36. Anile, S. 2008. Ecologia, biologia, genetica e conservazione del gatto selvatico (*Felis silvestris silvestris* Schreber, 1777) nel Parco dell'Etna. [Ecology, biology, genetic and conservation of the wildcat (*Felis silvestris silvestris* Schreber, 1777) in the Etna Park]. PhD Thesis, University of Catania

37. Trolle, M. & Kery, M. 2003. Estimation of ocelot density in the Pantanal using capture-recapture analysis of camera-trapping data. J. Mammal. 84 (2): 607-614.

38. Otis, D.L., Burnham, K.P., White, G.C. & Anderson, D.R. 1978. Statistical inference from capture data on closed populations. Wildl. Monogr. 62: 1-135.

39. White, G.C., Anderson, D.R., Burnham, K.P. & Otis, D.L. 1982. Capture-recapture and Removal Methods for Sampling Closed Populations. Internal report. Los Alamos National Laboratory, Los Alamos, USA.

40. Rexstad, E. & Burnham, K.P. 1991. User's Guide for Interactive Program CAPTURE. Abundance Estimation of Closed Populations. Internal report. Colorado State University Fort Collins, Colorado, USA.

41. Stanley, T.R. & Burnham, K.P. 1999. A closure test for time-specific capture-recapture data. Environ. and Ecol. Stat. 6 (2): 197-209.

42. Harmsen, B.J., Foster, R.J. & Doncaster, C.P. 2010. Heterogeneous capture rates in low density populations and consequences for capture-recapture analysis of camera-trap data. Popul. Ecol. DOI:10.1007/s10144-010-0211-z

43. ● Rowcliffe, M.J. & Carbone, C. 2008. Surveys using camera traps: are we looking to a brighter future? Anim. Conserv. 11: 185-186.
DOI:10.1111/j.1469-1795.2008.00180.x

44. ●● Dillon, A. & Kelly, M.J. 2007. Ocelot *Leopardus pardalis* in Belize: the impact of trap spacing and distance moved on density estimates. Oryx 41 (4): 1-9.
DOI:10.1017/S0030605307000518

45. Royle, J.A., Nichols, J.D., Karanth, K.U. & Gopalaswamy, A.M. 2009. A hierarchical model for estimating density in camera-trap studies. J. Appl. Ecol. 46: 118-127.
DOI:10.1111/j.1365-2664.2008.01578.x

46. Wegge, P., Pokheral, C.P. & Jnawali, S.R. 2004. Effects of trapping effort and trap shyness on estimates of tiger abundance from camera trap studies. Anim. Conserv. 7: 251-256.

47. Stahl P. & Leger F. 1992. Le chat sauvage (*Felis silvestris*, Schreber, 1777). [The European wildcat (*Felis silvestris* Schreber, 1777)]. In: Artois M. & Maurin H. (eds). Encyclopédie des Carnivores de France. [Encyclopedia of carnivores from France]. Société Française pour l'Etude et la Protection des Mammifères (S.F.E.P.M.), Bohallard, Puceul, France. (17), p 50.

48. Lozano, J., Moleon, M. & Virgos, E. 2006. Biogeographical patterns in the diet of the wildcat, *Felis silvestris* Schreber, in Eurasia: factors affecting the trophic diversity. J. Biogeogr. 33: 1076-1085.
DOI:10.1111/j.1365-2699.2006.01474.x

49. Klar, N., Néstor, F., Kramer-Schadt, S., Herrmann, M., Trinzen, M., Buttner, I. & Niemitz, C. 2008. Habitat selection models for European wildcat conservation. Biol. Conserv. 141: 308-319.
DOI:10.1016/j.biocon.2007.10.004

50. Lozano, J. 2010. Habitat use by European wildcats (*Felis silvestris*) in central Spain: what is the relative importance of forest variables? Anim. Biodivers. Conserv. 33 (2): 143-150.

51. Mermod, C.P. & Liberek, M. 2002. The role of snowcover for European wildcat in Switzerland. Zool. Jagdwiss Suppl. 48: 17-24.

52. Rovero, F. & Marshall, A.R. 2009. Camera trapping photographic rate as an index of density in forest ungulates. J. Appl. Ecol. 46: 1011-1017.
DOI:10.1111/j.1365-2664.2009.01705.x

21

Preliminary Data Used to Assess the Accuracy of Estimating Female White-Tailed Deer Diel Birthing-Season Home Ranges Using Only Daytime Locations

S. Barber-Meyer[1,*] & L. D. Mech[1,2]

[1] U.S. Geological Survey, Northern Prairie Wildlife Research Center, 8711 – 37th St., SE, Jamestown, ND 58401-7317, USA.

[2] Postal address: The Raptor Center, 1920 Fitch Avenue, University of Minnesota, St. Paul, 55108, USA.

* Corresponding author: Postal address: USGS, 1393 Hwy 169, Ely, MN, USA 55731;
e-mail: sbarber-meyer@usgs.gov.

Keywords

Core use;
Fixed kernel;
Global Positioning System;
Habitat use;
Minimum convex polygon;
Odocoileus virginianus;
Radiotelemetry;
Very high frequency.

Abstract

Because many white-tailed deer (*Odocoileus virginianus*) home-range and habitat-use studies rely only on daytime radio-tracking data, we were interested in whether diurnal data sufficiently represented diel home ranges. We analyzed home-range and core-use size and overlap of 8 adult-female Global-Positioning-System-collared deer during May and June 2001 and 2002 in the Superior National Forest, Minnesota, USA. We used 2 traditional means of analysis: minimum-convex polygons (MCP) and fixed kernels (95% FK, home range and 50% FK, core use) and two methods to partition day and night location data: (1) daytime = 0800-2000 h versus nighttime = 2000-0800 h and (2) sunup versus sundown. We found no statistical difference in size of home-range and core-use areas across day and night comparisons; however, in terms of spatial overlap, approximately 30% of night-range areas on average were not accounted for using daytime locations, with even greater differences between core-use areas (on average approximately 50%). We conclude that diurnal data do not adequately describe diel adult-female-deer, May-June home-ranges due to differences in spatial overlap (location). We suggest research to determine (1) if our findings hold in other circumstances (e.g., exclusive of the parturition period, other age classes, etc.), (2) if our conclusions generalize under other conditions (e.g., across deer range, varying seasons, etc.), (3) if habitat-use conclusions are affected by the incomplete overlap between diurnal and diel data, (4) how many nocturnal locations must be included to generate sufficient overlap, and (5) the influence of using other kernel sizes (e.g., 75%, 90%).

Introduction

VHF (very high frequency) radio locations have been used since the 1960's to investigate animal movements and related ecology [1-2]. Traditionally, due largely to logistics (e.g. aerial telemetry), such data were primarily restricted to diurnal or crepuscular periods and often used to assess home ranges [3-4]. Many white-tailed deer (*Odocoileus virginianus*) home-range and habitat-use studies have traditionally comprised only daytime or crepuscular locations (e.g. [5-7]). Depending on how well diurnal data represent diel home ranges, certain subjects such as behavior, habitat use, habitat selection, etc. may not be definitively addressed using diurnal-only data. Thus we were interested in the degree to which diurnal data accurately represented diel home ranges.

Whereas this subject has been assessed for wolves (*Canis lupus*, [8]), it has yet to be examined in white-tailed deer during May-June (e.g., this period encompasses the typical birthing season in our study area, [9]). One study investigated whether year-around diurnal data were appropriate to estimate diel habitat-use patterns in white-tailed deer but did not address home-range comparisons [10]. Another compared global positioning system (GPS) and VHF diurnal and nocturnal home ranges of white-tailed deer during winter [11]. That study determined that whereas diurnal and nocturnal winter home ranges were similar, their "differences may have important implications for studies focused on deer use of space, habitat, and resources at a finer scale" [11, page 779]. We assessed: (1) whether May-June diurnal deer data adequately represented corresponding nocturnal data and (2) whether different partitioning methods of day and night affected this comparison.

Methods

Study area

Our study area was comprised of 2,060-km^2 in the Superior National Forest, Minnesota, USA during May-June 2001 and 2002. (See [9] for detailed description). During 2002, prefawning deer densities throughout our study area were 4-14/10 km^2 (M. H. Dexter, Minnesota Department of Natural Resources, unpublished report). The primary predator of deer there is the wolf [9,12-13], and mean wolf density was approximately 25/1000 km^2 during our study [14].

Methods

During March 2001 and April 2002, we live-trapped, anesthetized, examined and fitted with GPS radiocollars [Advanced Telemetry Systems, Inc. (ATS), Isanti, MN] female deer [15]. We programmed GPS collars to acquire locations every 15 or 30 min, 24 h/day during May-June although some collars had varying schedules. We expected locations to be within 5 m and 30 m of the true location 50% and 95% of the time, respectively [16-17].

Because we posed similar questions for deer as Demma and Mech did for wolves [8] we generally followed the methods used by those researchers. We plotted all data in ESRI® ArcMap™ 10.0 (2010) and used Geospatial Modeling Environment version

0.5.5 Beta (2011) to generate home ranges and calculate centroids, polygon area and overlapping polygon area. We excluded locations if a deer had not yet completed her spring migration, or if the deer was dispersing or moving to a novel home range as determined by examination of the plotted locations.

Because we were interested in whether methods traditionally used by researchers to estimate home ranges based on diurnal data were comparable to results based on nocturnal data, we used 2 traditionally applied [3] home-range-estimation methods (minimum convex polygon; MCP [18] and fixed kernel; FK [19]) to estimate May-June deer home ranges. May-June data were the only locations available to us for these tests. To generate our kernel-density estimates we used fixed smoothing with least-squares cross validation and home and core-range areas delineated at 95% (95% FK) and 50% (50% FK) probability contours, respectively. We generated day and night home ranges based on: (1) daytime (2000-0800h) vs. nighttime (0800-2000h) and (2) sunup vs. sundown (National Oceanic and Atmospheric Administration solar calculator: http://www.esrl.noaa.gov/gmd/grad/solcalc/) [8]. To remove potential bias in pairwise comparisons of day-night home ranges, we generated equal sample sizes of day and night data by randomly excluding data from the larger sample in each comparison [8]. We assessed proportion of overlap between day and night MCPs and FKs and used 2-tailed, paired t-tests in Excel (version 14.0.7106.5003, Microsoft® Office Professional Plus 2010) to examine area differences [8]. We similarly assessed partitioning method by evaluating overlap and area differences between daytime and sunup MCPs and FKs [8].

Results

During March 2001 and April 2002, we fitted 8 female deer aged 1 to 13-years old with GPS radiocollars, and we analyzed GPS locations during May-June of each deer's capture year (Table 1). [Note deer ages increase by 1 yr in May/June (Table 1) relative to their spring capture ages.] We excluded the first 2 of 416 locations for Deer 8000 and the first of 1,616 locations for Deer 8044 during that period because these locations indicated the deer were still migrating to their May-June ranges. We also excluded the last 118 of 674 locations for Deer 8014 because these locations indicated the deer dispersed from its May-June home range. These exclusions were done before the random data exclusions. After all exclusions, the mean number of locations available was 456.4 (SE = 119.3) for daytime vs. nighttime comparisons and 452.8 (SE = 115.0) for sunup vs. sundown comparisons (Table 1).

Mean areas of daytime MCPs and nighttime MCPs were not significantly different nor were mean areas of sunup and sundown MCPs (Table 1; $t_7 = -1.50$, $P = 0.18$ and $t_7 = 1.23$, $P = 0.26$, respectively). Overlap averaged 72% (SE = 0.04, range = 0.50-0.81) for daytime vs. nighttime MCPs and averaged 70% (SE = 0.03, range = 0.53-0.85) for sunup vs. sundown MCPs (Table 1).

Mean areas of daytime 95% FKs and mean nighttime areas were not significantly different nor were mean areas of sunup and sundown 95% FKs (Table 1; $t_7 = -0.73$, $P = 0.49$ and $t_7 = 0.70$, $P = 0.51$, respectively). Overlap averaged 68% (SE = 0.04, range = 0.49-0.83) for daytime vs. nighttime 95% FKs and averaged 66% (SE = 0.04, range = 0.49-0.82) for sunup vs. sundown 95% FKs (Table 1).

Table 1: Comparisons of areas (ha) of minimum convex polygons (MCP) and fixed kernels (FK) of day and night locations of female deer by Global-Positioning-System (GPS) collars during May-June 2001 and 2002, Superior National Forest, northeastern Minnesota, USA. The number of locations refers to those available to select from when generating each polygon.

MCP			Daytime vs. Nighttime					Sunup vs. Sundown				
Deer no.	Age (Yr)	Dates of GPS collar locations used	No. locations	Daytime area	Nighttime area	Ratio daytime to nighttime area	Overlap	No. locations	Sunup area	Sundown area	Ratio sunup to sundown area	Overlap
8000	2	5/7/2001-6/30/2001	153	76.1	84.4	0.90	0.81	187	78.1	82.0	0.95	0.85
8002	2	5/15/2001-6/28/2001	625	114.8	135.7	0.85	0.70	518	136.9	92.5	1.48	0.64
8004	14	5/4/2001-6/30/2001	408	143.6	177.9	0.81	0.65	406	159.2	164.4	0.97	0.68
8012	3	5/1/2001-6/27/2001	49	387.8	411.5	0.94	0.81	68	393.7	415.8	0.95	0.81
8014	2	5/3/2001-6/7/2001	247	122.4	96.7	1.27	0.75	259	124.4	92.6	1.34	0.71
8044	6	5/13/2002-6/30/2002	572	81.4	81.6	1.00	0.50	748	99.5	64.6	1.54	0.53
8050	2	5/1/2002-6/30/2002	469	76.1	86.7	0.88	0.72	364	74.6	81.5	0.92	0.67
8058	2	5/7/2002-6/28/2002	1128	92.0	96.8	0.95	0.81	1072	90.5	79.8	1.13	0.71
		Mean	456.4	136.8	146.4	0.95	0.72	452.8	144.6	134.2	1.16	0.70
		SE	119.3	36.9	39.6	0.05	0.04	115.0	37.1	41.6	0.09	0.03

FK (95% probability contour)			Daytime vs. Nighttime					Sunup vs. Sundown				
Deer no.	Age (Yr)	Dates of GPS collar locations used	No. locations	Daytime area	Nighttime area	Ratio daytime to nighttime area	Overlap	No. locations	Sunup area	Sundown area	Ratio sunup to sundown area	Overlap
8000	2	5/7/2001-6/30/2001	153	80.0	117.8	0.68	0.63	187	87.9	111.9	0.79	0.49
8002	2	5/15/2001-6/28/2001	625	67.5	68.2	0.99	0.63	518	73.1	56.2	1.30	0.60
8004	14	5/4/2001-6/30/2001	408	132.0	156.9	0.84	0.74	406	123.8	155.8	0.79	0.76
8012	3	5/1/2001-6/27/2001	49	729.8	726.5	1.00	0.66	68	716.2	659.0	1.09	0.65
8014	2	5/3/2001-6/7/2001	247	119.0	127.2	0.94	0.83	259	123.9	126.1	0.98	0.82
8044	6	5/13/2002-6/30/2002	572	64.6	44.8	1.44	0.49	748	60.9	40.8	1.49	0.49
8050	2	5/1/2002-6/30/2002	469	73.7	70.2	1.05	0.71	364	75.7	68.9	1.10	0.74
8058	2	5/7/2002-6/28/2002	1128	74.1	67.4	1.10	0.75	1072	73.8	62.1	1.19	0.72
		Mean	456.4	167.6	172.4	1.01	0.68	452.8	166.9	160.1	1.09	0.66
		SE	119.3	80.8	80.3	0.08	0.04	115.0	78.9	72.6	0.09	0.04

FK (50% probability contour)			Daytime vs. Nighttime					Sunup vs. Sundown				
Deer no.	Age (Yr)	Dates of GPS collar locations used	No. locations	Daytime area	Nighttime area	Ratio daytime to nighttime area	Overlap	No. locations	Sunup area	Sundown area	Ratio sunup to sundown area	Overlap
8000	2	5/7/2001-6/30/2001	153	15.9	29.8	0.53	0.35	187	18.7	28.6	0.66	0.34
8002	2	5/15/2001-6/28/2001	625	13.7	12.7	1.08	0.68	518	12.9	12.3	1.05	0.69
8004	14	5/4/2001-6/30/2001	408	18.5	21.3	0.87	0.67	406	16.8	22.5	0.74	0.68
8012	3	5/1/2001-6/27/2001	49	161.2	135.1	1.19	0.52	68	157.8	118.8	1.33	0.39
8014	2	5/3/2001-6/7/2001	247	30.2	33.4	0.91	0.63	259	30.7	31.6	0.97	0.66
8044	6	5/13/2002-6/30/2002	572	17.9	5.9	3.03	0.33	748	14.9	5.2	2.87	0.35
8050	2	5/1/2002-6/30/2002	469	19.3	20.8	0.93	0.28	364	19.5	20.9	0.93	0.28
8058	2	5/7/2002-6/28/2002	1128	19.8	15.2	1.30	0.67	1072	18.8	15.1	1.24	0.59
		Mean	456.4	37.1	34.3	1.23	0.52	452.8	36.3	31.9	1.22	0.50
		SE	119.3	17.8	14.7	0.27	0.06	115.0	17.5	12.8	0.25	0.06

Mean areas of daytime 50% FKs (core-use area) and sunup 50% FKs were not significantly different from nighttime and sundown areas (Table 1; $t_7 = 0.66$, $P = 0.53$ and $t_7 = 0.81$, $P = 0.44$, respectively). Overlap averaged 52% (SE = 0.06, range = 0.28-0.68) for daytime vs. nighttime 50% FKs and averaged 50% (SE = 0.06, range = 0.28-0.69) for sunup vs. sundown 50% FKs (Table 1).

Mean overlap for daytime vs. sunup MCPs was 92% (SE = 0.03, range = 0.82-0.99), 90% (SE = 0.01, range = 0.86-0.95) for 95% FKs and 86% (SE = 0.01, range = 0.80-0.91) for 50% FKs.

Discussion

Because our preliminary data are biased toward females (8 of 8), 2-yr olds (5 of 8) and the birthing season (which can alter and restrict home-range use [20-22]), our conclusions should not be extended to other age classes and genders.

While we found no significant differences in home-range and core-use area sizes between day and night for both partitioning techniques, overlap measures indicated approximately 30% of night ranges on average were not accounted for using day locations, with even greater differences between core use areas. In some cases ~70% of the core-use area did not overlap. We conclude that diurnal data do not adequately describe diel adult female deer May-June home-ranges (due to differences in spatial overlap) using the MCP or 95% FK methods, and stress that core-use estimation (50% FK) especially may miss potentially important nocturnal locations. Indeed, for biological questions, management or conservation purposes, the absolute size of a home range may be less important than its location and shape. Our May-June findings echo the conclusions of Kochanny et al. [11] that although adult female deer diurnal and nocturnal winter home ranges were similar, important differences existed on a finer scale. We suggest additional research to determine if our findings hold under other circumstances (e.g., exclusive of the parturition period, other age classes), to determine if our conclusions generalize across deer range, seasons and varying factors that determine home-range size and location [23] and to determine how the degree of non-overlap between diurnal and diel ranges translates into biologically meaningful metrics such as varying habitat use (see [24] as it relates to roe deer, *Capreolus capreolus*).

Both partitioning techniques (daytime vs. nighttime or sunup vs. sundown) revealed similar day results as evidenced by relatively high overlap values for each of the home-range and core-use estimation methods. Thus, we conclude that either partitioning technique is valid in areas similar to our study area during May-June. Partitioning technique may be more relevant during winter when temperature fluctuations due to sunup would be more significant. However, this remains to be tested.

Although it was beyond the scope of this study to detail the strengths and weaknesses of traditional home range estimation methods, important critiques of the methods exist (e.g., overestimation of home range, sensitivity to small samples, etc.) and the method selected can alter results considerably [25]. While our study instead focused on the application of these traditional methods, we nevertheless recommend additional research investigating alternate kernel sizes (e.g., 75%, 90%), determining how many nocturnal locations are required to generate adequately shaped (i.e., sufficient overlap) 24-hr home ranges, and assessing how white-tailed deer behavior, habitat use and selection can be influenced by location sampling regimes [25].

Because much traditional research derived from VHF diurnal data exists on female-deer home ranges, our findings suggest that, in areas similar to our study area during

May and June, diurnal home ranges and location data for adult does are not sufficient for management decisions requiring diel ranges except for coarse-location needs. Although GPS-collaring systems are relatively more expensive, usually of shorter duration and often deployed on fewer animals, we recommend these for studies of diel deer behavior, habitat use and selection if diel VHF telemetry locations are not feasible.

Acknowledgements

This study was supported by the US Geological Survey (USGS). We thank numerous USGS volunteer technicians who live-trapped deer for this study. We thank Dominic Demma (Alaska Department of Fish and Game) for analysis advice and Chris Kochanny (Sirtrack Wildlife Telemetry Consultant) and 5 anonymous reviewers for reviewing earlier drafts of this manuscript. Any use of trade, firm, or product names is for descriptive use only and does not imply endorsement by the U. S. Government.

References

Five "key references", selected by the authors, are marked below (Three recommended (●) and two highly recommended (●●) papers).

1. Cochran, W.W., Lord, R.D. 1963. A radio system for wild animals. Journal of Wildlife Management 27: 9–24.
http://dx.doi.org/10.2307/3797775
2. Fuller, M.R., Millspaugh, J.J., Church, K.E., Kenward, R.E. 2005. Wildlife radio telemetry. In Braun (ed.). Techniques for wildlife investigations and management. The Wildlife Society. Bethesda, Maryland, USA. pp. 377-417.
3. ●● Laver, P.N., Kelly, M.J. 2008. A critical review of home range studies. Journal of Wildlife Management 72: 290–298.
http://dx.doi.org/10.2193/2005-589
4. ●● Powell, R.A. 2012. Movements, home ranges, activity and dispersal. In Boitani & Powell (eds.). Carnivore ecology and conservation: a handbook of techniques. Oxford University Press, United Kingdom. pp. 188-217.
5. Montgomery, G.G. 1963. Nocturnal movements and activity rhythms of white-tailed deer. Journal of Wildlife Management 27: 422-427.
http://dx.doi.org/10.2307/3798516
6. Kammermeyer, K.E., Marchinton, R.L. 1977. Seasonal changes in circadian activity of radio-monitored deer. Journal of Wildlife Management 41: 315-317.
http://dx.doi.org/10.2307/3800612
7. ● Beier, P., McCullough, D.R. 1990. Factors influencing white-tailed deer activity patterns and habitat use. Wildlife Monographs 109: 3-51.
8. Demma, D.J., Mech, L.D. 2011. Accuracy of estimating wolf summer territories by daytime locations. American Midland Naturalist 165: 436–445.
http://dx.doi.org/10.1674/0003-0031-165.2.436
9. Nelson, M.E., Mech, L.D. 1981. Deer social organization and wolf depredation in northeastern Minnesota. Wildlife Monographs 77: 1–53.
10. Kernohan, B.J., Jenks, J.A., Naugle, D.E., Millspaugh, J.J. 1996. Estimating 24-h habitat use patterns of white-tailed deer from diurnal use. Journal of Environmental Management 48: 299-303.
http://dx.doi.org/10.1006/jema.1996.0079

11. ● Kochanny, C.O., DelGiudice, G.D., Fieberg, F. 2009. Comparing Global Positioning System and Very High Frequency telemetry home ranges of white-tailed deer. Journal of Wildlife Management 73: 779-787.
http://dx.doi.org/10.2193/2008-394

12. Frenzel, L.D. 1974. Occurrence of moose in food of wolves as revealed by scat analysis: a review of North American studies. Le Naturaliste Canadien 101: 467–479.

13. Nelson, M.E., Mech, L.D. 1986. Mortality of white-tailed deer in northeastern Minnesota. Journal of Wildlife Management 50: 691–698.
http://dx.doi.org/10.2307/3800983

14. Mech, LD. 2009. Long-term research on wolves in the Superior National Forest. In Wydeven, VanDeelen, & Heske (eds.). Recovery of gray wolf in the Great Lakes region of the United States: An endangered species success story. Springer, New York, New York, USA. pp. 15-34.

15. Merril, S.B., Adams, L.G., Nelson, M.E., Mech, L. D. 1998. Testing releasable GPS radiocollars on wolves and white-tailed deer. Wildlife Society Bulletin 26: 830-835.

16. Moen, R., Pastor, J., Cohen, Y. 1997. Accuracy of GPS telemetry collar locations with differential correction. Journal of Wildlife Management 61: 530–539.
http://dx.doi.org/10.2307/3802612

17. Dussault, C., Courtois, R., Ouellet, J.-P., Huot, J. 2001. Influence of satellite geometry and differential correction on GPS location accuracy. Wildlife Society Bulletin 29: 171-179.

18. Mohr, C.O. 1947. Table of equivalent populations of North American small mammals. American Midland Naturalist 37: 223–249.
http://dx.doi.org/10.2307/2421652

19. Seaman, D.E., Powell, R.A. 1996. An evaluation of the accuracy of kernel density estimators for home range analysis. Ecology 77: 2075–2085.
http://dx.doi.org/10.2307/2265701

20. Lesage, L., Crête, M., Huot, J., Dumont, A., Ouellet, J.P. 2000. Seasonal home range size and philopatry in two northern white-tailed deer populations. Canadian Journal of Zoology 78 (11): 1930-1940.
http://dx.doi.org/10.1139/z00-117

21. Bertrand, M.R, DeNicola, A.J., Beissinger, S.R., Swihart, R.K. 1996. Effects of parturition on home ranges and social affiliations of female white-tailed deer. Journal of Wildlife Management 60: 899-909.
http://dx.doi.org/10.2307/3802391

22. D'Angelo, G.J., Comer, C.E., Kilgo, J.C., Drennana, C.D., Osborn, D.A., Miller, K.V. 2004. Daily movements of female white-tailed deer relative to parturition and breeding. Proceedings of the Southeastern Association of Fish and Wildlife Agencies 58: 292-301.

23. Stewart, K.M., Bower, R.T., Weisberg, P.J. 2011. Spatial use of landscapes. In Hewitt (ed.). Biology and management of white-tailed deer. CRC Press, New York, New York, USA. pp. 181-217.

24. Bonnot, N., Morellet, N., Verheyden, H., Cargnelutti, B., Lourtet, B., Klein, F., Hewison, A. J. M. 2013. Habitat use under predation risk: hunting, roads and human dwellings influence the spatial behaviour of roe deer. European Journal of Wildlife Research 59(2): 185-193.
http://dx.doi.org/10.1007/s10344-012-0665-8

25. ● Börger, L., Franconi, N., De Michele, G., Gantz, A., Meschi, F., Manica, A., Lovari, S., Coulson, T. 2006. Effects of sampling regime on the mean and variance of home range size estimates. Journal of Animal Ecology 75(6): 1393-1405.
http://dx.doi.org/10.1111/j.1365-2656.2006.01164.x

22

Hunting Issues in Contemporary Society: Perspectives of Resident Hunters in Montana (USA)

S. L. Eliason[1]

[1] Department of Social Sciences and Cultural Studies, Montana State University, Billings 1500 University Dr, Billings, Mt 59101, USA.

* Corresponding author email: seliason@msubillings.edu.

Keywords

Hunters;
Hunting Issues;
Montana;
Perceptions;
Social Change.

Abstract

Hunting is a popular recreational activity in many rural regions of the United States. There is a need for information about individuals who hunt, including their attitudes and perceptions of the activity. The objective of this exploratory research was to identify issues affecting hunters in modern society. This study took a qualitative approach to data collection and examined hunting issues in Montana. Data were gathered from a mail survey that used open-ended questions. The survey was sent to a random sample of resident elk hunters. Responses revealed five central concerns of hunters in Montana. These included the presence of wolves on the landscape, the commercialization of wildlife, nonresident hunters, high license prices, and motorized vehicles and the hunting experience. The findings of this study enhance our understanding of hunting issues in contemporary society.

Introduction

Although the number of individuals who participate in hunting has declined over time [1-10], hunting persists as a popular recreational activity in many rural regions of the United States [11-13]. Individuals participate in hunting for a variety of reasons that include recreation, tradition, nature, to get a trophy, or to obtain meat [12, 14-20]. Given the abundance of deer and elk in the Rocky Mountain region of the U.S., big game hunting is a time honored tradition in many western states [15, 21-23].

Hunting in the twenty-first century takes place in an era of significant social change [1, 7, 10, 12, 15, 24-30]. Wildlife agencies are aware of the declining number of young people who are participating in hunting, and these organizations understand that the successful recruitment and retention of hunters is essential for the hunting tradition to continue and for hunting to remain a viable tool for wildlife management [6, 8, 9, 21, 22, 31]. For this reason, many states have instituted special “youth hunts” for young individuals in an effort to get them interested in hunting at a young age [3]. These recruitment programs allow youth to hunt during special seasons that are open only for them in order to give them an opportunity to have a good hunting experience without all the crowds and competition present during the general hunting seasons.

The forces of development and population change are transforming both the physical and cultural landscapes of rural areas in the western U.S. [32-35]. Because of low profit margins and loss of interest in ranching combined with increasing land values in the western U.S., many ranchers are selling their land to real estate developers who are turning these tranquil and bucolic agricultural areas into subdivisions and ranchettes [35]. Population change is also occurring in rural locations in the eastern U.S. Describing their study in Maine, Teisl, Boyle, and Record [36] stated "... population growth had negative effects on the sale of resident hunting licenses and may be the consequence of suburbanization and countryside sprawl that reduces or limits access to hunting on private land" (p. 15). In addition to reducing hunting opportunities, these practices result to the loss of wildlife habitat, which is essential for healthy wildlife populations [37, 38].

Demographic change also includes an influx of newcomers to rural communities in a number of western states [33, 35, 39, 40]. In some instances, these individuals are purchasing ranches and not allowing open access to the property for hunting [32, 39]. All of these practices result in fewer places for individuals to hunt and contribute to the concentration of hunters in other areas, especially public land.

Paradoxically, the need for more hunters comes at a time when in some instances wildlife agencies have found it necessary to limit the number of hunters for some popular big-game species. Given the effects of a growing population, some states in the western U.S. have begun placing restrictions on the number of resident big-game hunting licenses. For example, Utah allowed unregulated sales of deer hunting licenses to resident hunters for decades. In 1994, the total number of deer hunting licenses was capped at less than 100,000. These restrictions help to conserve the resource as well as to alleviate hunter overcrowding in the field.

Change to traditional hunting practices has also come in the form of new wildlife management policies on private property. In particular, wildlife agencies in some western states have enacted private property programs that generate substantial revenue for landowners and private enterprise, with limited benefits to public hunters [40-43]. These include Colorado's "Ranching for Wildlife" program and Utah's "Cooperative Wildlife Management Unit" or CWMU program [41, 42, p. 140]. These programs have institutionalized the commercialization of wildlife in these states.

Theoretical Perspective

This study used a symbolic interaction theoretical framework [44-47]. The symbolic interactionist perspective is concerned with the meanings of things for individuals [44]. This theory has utility for the study of leisure and recreation because it helps us understand how individuals interpret phenomena and situations they encounter in their lives [45]. In a description of this perspective, Colton [45] stated:

> Symbolic interactionism emphasizes that human society uses symbols and meanings, and behavior is based on these meanings. Meanings are derived through social interaction but can be handled and modified by an individual in dealing with things encountered. ...Meanings form bases for action in leisure, recreation, and tourism. Wide ranges of meanings can exist for activities and

> roles. Meanings can also be placed on situtations (sic) where action takes place. (pp. 346, 351)

According to Garst, Williams, and Roggenbuck [48], "Individuals and groups socially construct reality and shared experiences, language, and meanings that create a basis for knowledge and understanding" (p. 92). Describing the importance of meanings, Jordan et al. [49] stated that "The concept of meanings is...important because meanings are generally related to actions" (p. 19).

The words of individuals are imbued with meaning. In order to understand the viewpoints of individuals, it is essential to obtain their perspective in their own words. According to Henderson [50]:

> The meanings of any symbol (e.g., leisure) have their origins in interactions, which are defined and changed by individuals according to the meanings that are held. The individual studied is the expert and the attempt is to describe their vocabularies, ways of looking, and sense of the important and the unimportant. (p. 48).

Because of this emphasis, the symbolic interaction perspective provides a framework that helps us understand the meanings of things for individuals and how they view the world [44]. Scholars using qualitative research methods have used this perspective to gain an in-depth perspective of hunters [12, 23]. In a qualitative study of Idaho elk hunters, Tynon [23] reported that "hunters' responses were in their own words, reflecting their own meanings, as opposed to lists of phrases representative of researcher-generated constructions" (p. 40). Dizard [12] successfully used interviews with hunters to provide "insight into why people hunt—what it means to them" (p. 46).

Study Site

Geographically, Montana is the fourth largest state in the U.S. From a population standpoint, it is one of the smallest with a population of just over one million residents. Montana is a rural state that contains a variety of terrains including mountains in the west and plains in the eastern part of the state. Hunting is a popular recreational activity for residents, and thousands of nonresidents also come to the state each year to hunt a variety of wildlife species including deer, elk, antelope, and upland game birds [15, 21, 22, 51]. Wildlife issues such as bison and wolf management are important politically and receive considerable media attention at the local and national levels [52-58]. According to Gude et al. [21], "Declining hunter participation in Montana could have cultural, political, economic, and wildlife management and conservation implications" (p. 472).

One particularly timely topic that warrants attention is the identification of hunting issues in the contemporary U.S. from the perspective of hunters themselves. Given current trends associated with declining hunter numbers [1, 8, 9], the critical need for the successful recruitment and retention of hunters [3-6, 21, 22, 28], and social change that is occurring with respect to hunting [1, 12, 24], it is important for wildlife agencies and managers to be aware of factors affecting the hunting experience of those who participate in this recreational activity. The purpose of the present research was to identify, from the perspective of resident hunters, issues associated with hunting in Montana and why those issues matter to them.

Methods

This study took a qualitative approach to data collection. Data were obtained from a mail survey. The survey contained mostly open-ended questions about hunting [59, 60]. The study population consisted of Montana resident elk hunters. A random sample of 1,000 resident elk hunters in 2004 was obtained from Montana Fish, Wildlife and Parks, the state agency responsible for managing fish and wildlife in the state. A survey was sent to these individuals in 2005. A total of two-hundred and fifty-five completed surveys were returned, for a response rate of 26%.

Qualitative studies have the advantage of providing in-depth information that is rich in detail. Neuman [60] stated "To learn how a respondent thinks and discover what is important to him or her...open questions are best. ...Open-ended questions are especially valuable in early or exploratory stages of research" (pp. 324-325).

Given the exploratory nature of the study, it was important to obtain data from respondents in their own words [50, 60]. The survey contained a question that allowed hunters to provide information about their perceptions of hunting in the state: "Do you have any other comments about hunting in Montana?" Respondents were asked to provide answers in their own words so that issues important to Montana hunters could be obtained in rich detail that captured the essence of their feelings and perceptions.

For data analysis, written comments from each survey were systematically examined by the author with the intent of identifying common themes concerning the hunting experience in Montana. A variety of comments were provided by hunters, including some positive comments about hunting in the state. Hunter comments also revealed a number of concerns or issues pertaining to hunting in the state, which formed the basis for this paper. Common themes that emerged with respect to hunting issues were placed into topical categories developed by the author. Representative statements from hunters that were illustrative of each category were selected for inclusion in the paper. Summarization and interpretation of the responses is provided by the author.

Results and Discussion

Written responses to the question "Do you have any other comments about hunting in Montana?"indicated that Montana hunters who completed the survey were concerned about the sport. Data analysis revealed five central issues or concerns expressed by these hunters. These issues were perceived by hunters as threats to the hunting experience and included the presence of wolves on the landscape, the commercialization of wildlife, nonresident hunters, high license prices, and motorized vehicles and the hunting experience. Each of these issues will be discussed in greater detail in this section.

The Presence of Wolves on the Landscape

One of the major issues affecting hunters in Montana is the presence of wolves on the landscape. Wolves were reintroduced in Yellowstone National Park in 1995 [56, 57, 61]. As the wolf population rapidly expanded, many of these animals migrated outside of the Park boundaries and they currently occupy much of western and central Montana. In some areas of the state including around Yellowstone National Park and in western Montana, elk populations have experienced significant declines in recent years.

Because wolves are predators that consume elk and other big game animals, wolves are often blamed by hunters for declining wildlife populations. As the following comments suggest, hunters in the study tended to believe that wolf reintroduction was a mistake. This is because they perceived wolves were having a deleterious effect on elk herds in some parts of the state:

> *Fishing, trapping and hunting have been a big part of my life. The wolves were a big mistake. If anyone can't figure [out] why the Yellowstone elk are disappearing, well they aren't very smart. I do a lot of hunting around Belfry and Red Lodge. It's real sad what was done to that elk herd 2 years ago. The wolves are part of that problem, I feel also. They kept them down in the ranchers' fields.*

> *Its too bad that the elk numbers and migration patterns have been affected so dramatically by the wolf problem. We have seen wolves and their tracks in the mountains since forever. They had found a balance in nature-the smart folks who upset that balance didn't do any Montana residents any favors.*

> *I am concerned that the lack of state control on wolves is leading us down a road of less and less hunting opportunities. These critters need to be managed, and if the sportsmen can help do it, it will become a much more positively accepted effort.*

> *Wolves, out of state landowners, and anti-hunting terrorists will be the downfall of our Montana hunting (& fishing) heritage. Until we once again eradicate all three, hunting will only suffer and become harder to fill our hunting tags.*

> *The Montana Fish Wildlife and Parks is supposed to be a sportsman organization and the best way to pressure a species is to hunt it! Wolf and grizzly bear preservation is not in the interest of the sportsman-it could be.*

> *The introduction of the "wolf" was the end of Montana's future for big game hunting.*

> *I hunt elk, so the introduction of wolves was a huge mistake by the fed's.*

> *Get rid of the wolves that have killed off the elk population.*

At the time of the survey in 2005, wolves were protected because of their status as an endangered species and wolf hunting was not allowed in Montana. In an effort to help control the rapidly expanding wolf population, Montana Fish, Wildlife and Parks established a wolf hunt and now issues unlimited tags to residents and nonresidents. There is, however, a quota on how many wolves can be killed each season and once the quota has been reached the hunt ends. Wolf hunts were held in 2009, 2011, 2012, and 2013.

The Commercialization of Wildlife

The ability of the public to enter private property for the purpose of accessing natural resources and hunting has been identified as a significant issue in the U.S. [3, 62-64]. Another pressing issue evident in the comments of hunters was the trend toward the commercialization of wildlife in Montana [42, 65, 66]. This most commonly involves outfitters who lease private property (ranches) from landowners for the exclusive use of their paying clientele, which results in the exclusion of public hunters from these properties [32, 40, 42]. This has become an especially contentious issue in Montana because in the past many ranchers would allow resident hunters to access their land for hunting [32, 67-69].

While there is a substantial amount of public land in Montana, the majority of land is in private ownership. According to official statistics, 59% of Montana land is privately owned, 35% is in the public sector, and the remaining property (4%) is Indian trust land [70]. Loss of access to private land for hunting purposes is a significant issue for Montana hunters because many of the prime hunting locations in the state are found on private land [43]. The commercialization of wildlife has also contributed to the poaching problem in Montana, and state game wardens contend that a substantial amount of poaching occurs on private land behind locked gates [43].

The following comments reflect hunter concerns toward the commercialization of wildlife in Montana and suggest that it has resulted in strained relations among hunters, outfitters, and landowners because it is associated with the loss of access to private land for resident hunters:

> *Hunting in Montana has been great to me during my era in life but I don't think it will stay that way. Every year more and more good hunting areas on private land are getting shut off to the average Montanan. Money has spoiled hunting all over the state. More and more land is either being bought up and locked up by wealthy out-of-staters or it is being leased out by outfitters which cater to the rich out of state hunters. My dad summed it up once by saying "What used to be a poor man's way of life has become a rich man's pleasure." With all of the private land being tied up by the wealthy the local people are forced to hunt public land. This over runs all of the good public areas with too many hunters and most of the game is pushed onto private land that is closed to the public.*

> *I want hunting to be geared for the guy who loves the outdoors and wildlife and not geared for the people who want to profit from hunting. For example, outfitters leasing land and excluding the public.Land owners profiting from wildlife.*

Expanding the amount of private land open to hunting – and making knowledge of and access to such land easier – is now the #1 issue in sustaining hunting into the future. ... We must also turn back the privatization of wildlife by ranchers, outfitters, and lease holders.

It is getting more and more difficult to get access to property to hunt on. Hunting is getting to be a business which in my mind is a detriment to the sport. As more and more land owners lease their land to "rich" hunters the sport suffers. Hunting clubs, hunting leases, [and] hunting for pay should all be illegal.

I fear that it may, or is, becoming a sport for the very wealthy. Private land and public resource is a huge concern of mine especially in central and eastern Montana.

I resent the SOB's from out-of-state buying up and/or leasing up Montana and locking out schmucks like me.

I hope that FWP will work on land access for hunters.

Access to hunting land is a serious concern of hunters that shouldn't be taken lightly by wildlife agencies. In their study of Illinois hunters, Miller and Vaske [4] reported that "hunters who felt they did not have access to hunting lands were 106 times more likely to have decreased participation than hunters who had land available for hunting" (p. 268).

Nonresident Hunters

Another concern voiced by survey respondents was out of state (nonresident) hunters. While general deer and elk licenses are available for Montana resident hunters to purchase on an unrestricted basis, the number of these licenses available for sale to nonresident hunters is strictly limited by the state. In addition, nonresident hunters pay license fees that are substantially higher than those charged to resident hunters. It should be noted that nonresident hunters also contribute a substantial amount of money to the economies of communities across the state via their expenditures on goods and services such as food, lodging, gas, outfitters, and taxidermists [51, 71-74].

Despite the positive economic contributions made by nonresident hunter to the Montana economy, many of hunters in the study held the perception that the state issued too many nonresident hunting licenses and as a result the number of nonresident hunters was far too high. These comments were characterized by territorial and possessive tones that were indicative of an insider/outsider perspective, and suggested that the large number of nonresident hunters was unfair to Montana residents, who they felt should receive priority to the states wildlife resources since they reside there:

Montana is hunted by more out of state hunters than any other state (not including Alaska). We obviously have some wonderful country to hunt. But it is important to treat that land with respect, so our hunting heritage will be passed on to the next generation.

We need to limit our out of state licenses and permits so that locals have a better chance. I don't think out of staters abide by our laws nor do they respect them.

Its almost like they are taking over.

There are getting to be a lot of out of state hunters and as a Montana resident we have to compete with them and I don't think that it is fair.

[The state] should limit more the number of out of state hunters. Hunting should be preserved for the future generations of Montana children.

Put Montana hunters first instead of the out-of-state just because they have to pay more.

If out of state permits have sold out you are not charging enough.

Too many out of staters.

High License Prices

Another issue that emerged from the analysis was the perceived high cost of resident hunting licenses. Because wildlife agencies receive significant funding from license sales, this issue is of concern since "...hunting and fishing licenses compete with other goods and services that consumers may buy" (p. 3) [36]. In a description of the possible effects of increases in hunting license fees on hunters, Robison and Ridenour [5] stated "Hunting is increasingly expensive and time-constraining, and license fee hikes may have an impact on hunting (Teisl, Boyle, & Record, 1999). Hunting fees have increased almost every single year since the early 1970s, averaging about 6% a year" (p. 422).

Many individuals in the present study held the perception that hunting license fees were simply too expensive for Montana resident hunters. These individuals based their claims on the premise that Montana wages were low in comparison with other states, and argued that rising license costs represented a significant hardship for many Montana families who like to hunt:

I think the price of big game licenses are getting much too high for the local people. Montana does not have the high wages some of the other states have and the parents of young people have to help the kids with buying their licenses, and that is too bad.

If the license fees continue to rise, and access to public lands is becoming harder to come by then local people that live and work here won't be able to afford "hunting."

Residents of Montana should not be "fee'd out" of hunting in Montana. The

> *out of state hunters should be paying more, and the in state hunters should be paying less.*
>
> *The price of tags are going up. A family with 4 or 5 people in it will spend a lot of money on a sportsman license.*
>
> *Keep it basic, inexpensive. Limit it as much as possible to people who live in Montana.*
>
> *The cost is getting too high!*

While there is some evidence that Montana wages are low relative to other states [75, 76], the cost of resident hunting licenses appears to be quite reasonable in comparison to fees charged to resident hunters in other western states. To put the situation into perspective, it is useful to provide a comparison of Montana with another western state, Utah. In 2013 adult resident hunters in Utah paid $50 for a general season elk permit plus $26 for a prerequisite hunting license [77]. Adult residents of Montana paid $20 for a general elk license and $10 in prerequisite fees that include $8 for a conservation license and $2 for a hunting access enhancement fee [78].

For limited entry bull elk permits that are issued through a drawing, Utah residents paid $285 plus a $10 application fee [77], while Montana residents paid $4 for a permit plus a $5 application fee (in addition to possessing a general elk license [$20] and the prerequisite conservation license [$8] and hunting access enhancement fee [$2]) [78].

Given that Montana hunters have been accustomed to spending very little money for hunting licenses over a span of decades, any fee increases are likely to be viewed as burdensome by resident hunters. In a recent study, Schorr et al. [22] analyzed deer and elk hunting license sales in Montana during the time period of 2002-2011 and stated "Our analysis suggests that as license price increased retention and recruitment rates decreased in Montana" (p. 950). Describing the impact of license fee increases on the sale of hunting licenses in Alabama, Mehmood et al. [31] stated:

> In 1980, license fees were raised to $10.25 (from $5.00) and total sales dropped by over 40,000. Again in 1990, license fees were increased to $15.00 and sales dropped by 20,000. In each of these cases, sales never returned to their preincrease level. (p. 244).

Motorized Vehicles and the Hunting Experience

A final issue reported by Montana hunters had to do with the use of motorized vehicles as they affect the hunting experience. Responses revealed several concerns related to the use of motorized vehicles, all of which when taken together suggest the need to strike a balance in terms of providing adequate public access to hunting areas while at the same time protecting the resource and the quality and integrity of the hunting experience.

All Terrain Vehicles

In the past three decades there has been a dramatic increase in all terrain vehicle (ATV) use across the U.S. [79]. The proliferation of ATVs has made it easy for hunters to get into some formerly difficult-to-access areas that used to require a strenuous hike. Some of the respondents felt ATVs were a problem and indicated that the state needed to enact regulations to limit ATV use so the quality of hunting in Montana would not decline:

> *I think something should be done with the four wheelers. They run posted roads, drive all over parks, and ruin your hunt when you walk in.*
>
> *Stricter restrictions should be placed on ATV's. They are being misused and are ruining good hunting areas in Montana.*
>
> *ATV use is out of control and must be reduced or game and hunters will suffer.*

A Need for Walk In Areas

Wildlife species need habitat that provides breeding and escape cover. Motorized vehicles tend to disturb wildlife, and places that have extensive vehicle access generally contain less abundant game than remote areas that are inaccessible to vehicles. Also, excessive vehicle access can contribute to hunter overcrowding, which generally results in a less satisfying experience for hunters. Hunter responses stressed the need to walk in to hunting areas to preserve the quality of the hunting experience:

> *I tend to be frustrated by those who hunt from vehicles, pursue game with vehicles. My preference is to spend the majority of my hunting time quietly sitting and observing. That's my choice. To witness poor methods and ethics causes me to understand why many "hunters" do not harvest game and are not allowed on private property. The landowners that accommodate access for hunters become frustrated by poor methods, ethics, and failure to follow landowner instructions.*
>
> *Avoid road hunt situations. Someday we will all be too old to cover the kind of ground we could at thirty years of age. That is the nature of life. Do not make special rules for people to use motorized vehicles instead of walking.*

Fewer Road Closures

Big game animals such as deer and elk are large, and retrieving them from the field can pose a formidable challenge for hunters if they are a significant distance from a vehicle. Some hunters felt that there were too many existing road closures that hampered their ability to access hunting locations and made it difficult to retrieve game animals. These individuals claimed it was especially important to be able to use motorized vehicles to retrieve game animals that were harvested:

> *Access is a real issue. Too many road closures. Roads that have been traveled for years are now closed and basically the same as posted ground since the*

average hunter can't get to areas to hunt. Horse hunters and guides love it. Even if they would let you drive in to retrieve game it would help.

We need less road closures because it hampers the retrieval of game and access to areas.

If [the] BLM stops us from driving off road to get our game, I'm done.

Conclusion

Prior studies on hunters and hunting have focused on topics such as hunter satisfaction [16, 17, 20], declining hunter numbers [1, 8, 9], and hunter recruitment and retention [3-6, 21, 22, 28]. The findings from the present study enhance our understanding of the perceived challenges Montana hunters are facing in contemporary society. Some of these challenges are unique to Montana, while others are occurring across the nation to varying degrees. Of the issues identified in the present study, perhaps the most important is the commercialization of wildlife. When a monetary value is placed on wildlife and it becomes commoditized [42, 65, 66], it creates an incentive for landowners to lease their land to outfitters for the exclusive use of their paying clientele. This practice results in the loss of access to private property for resident hunters, and is perceived to be a major problem by hunters in the state.

Montana hunters in the study also held the perception that wolves were responsible for the decline of elk populations in some locations of the state. It is believed that wolves have also had a negative impact on big game herds in certain areas of other western states, including parts of Idaho and Wyoming. As with Montana, Idaho has established a wolf hunting season to control the wolf population. Although many environmentalists abhor the idea of killing wolves, the creation of wolf hunting seasons has most likely tempered the attitudes of sportsmen toward wolves in these states. As noted earlier, a respondent in the present study stated "...These critters need to be managed, and if the sportsmen can help do it, it will become a much more positively accepted effort."

Although it is facing a number of challenges [1, 12, 24, 26], hunting is a tradition that continues to make important recreational, economic, and cultural contributions to rural communities across America [2, 5, 8, 9, 15, 28]. As population growth and urbanization continue in modern society, hunters in the U.S. will continue to grapple with challenges to the hunting tradition [9, 28].

It is likely that a variety of trends and conditions contribute to declining interest in hunting by "pushing" hunters out of the sport. Depending on the particular location, these could include variables such as short hunting seasons, lack of game, cost of hunting licenses, hunter crowding on public lands, and lack of access to private land [12, 22, 27, 28, 31]. Other factors serve to "pull" hunters away from the activity such as new hobbies and interests that are the result of technology, e.g.,video games and electronic entertainment [5]. These factors also affect hunter recruitment, and serve to reduce the number of young hunters becoming involved in the activity.

There is a need for additional research dealing with the attitudes and perceptions of hunters toward hunting issues. In order to discern possible regional variations in hunter attitudes and perceptions, qualitative studies of hunters should be carried out in other states and countries to identify factors that affect the hunting experience. Future research should examine hunting issues by age category of hunters [21]. In their study of hunter participation, recruitment, and retention trends among deer and elk hunters in Montana, Gude et al. [21] stressed the importance of using the "age class" of hunters when analyzing these activities (pp. 472-473). In regard to hunting issues, it is possible that middle aged hunters may have different concerns about hunting than young hunters. For this reason, it would be useful for wildlife agencies to be aware of the concerns of hunters at different stages of the lifecycle so that programs could be developed and directed toward the improvement of the hunting experience for different segments of the hunting population.

The present research provides in-depth information about hunting issues in a single state. Even though the sample was randomly selected, a limitation of the study is that the qualitative nature of the methodology does not lend itself to generalizability of the findings. Nonetheless, it is hoped that the findings from the present study will be useful to researchers, agency personnel, and others who are interested in understanding issues of concern to hunters regarding the hunting experience in modern society.

References

Five "key references", selected by the authors, are marked below (Three recommended (●) and two highly recommended (●●) papers).

1. ● Heberlein, T.A. 1991. Changing attitudes and funding for wildlife—Preserving the sport hunter. Wildlife Society Bulletin 19(4):528-534.
2. Kendall, B.J., Guynn, D.C., Jr., Straka, T.J. & Yarrow, G.K. 2013. Hunter and landowner needs and expectations and the economic impact of hunting on rural South Carolina counties (USA). Wildlife Biology in Practice 9(2):76-90.
3. Larson, L.R., Stedman, R.C., Decker, D.J., Siemer, W.F. & Baumer, M.S. 2014. Exploring the social habitat for hunting: Toward a comprehensive framework for understanding hunter recruitment and retention. Human Dimensions of Wildlife 19(2):105-122. http://dx.doi.org/10.1080/10871209.2014.850126
4. Miller, C.A. & Vaske, J.J. 2003. Individual and situational influences on declining hunter effort in Illinois. Human Dimensions of Wildlife 8(4):263-276. http://dx.doi.org/10.1080/716100421
5. Robison, K.K. & Ridenour, D. 2012. Wither the love of hunting? Explaining the decline of a major form of rural recreation as a consequence of the rise of virtual entertainment and urbanism. Human Dimensions of Wildlife 17(6):418-436. http://dx.doi.org/10.1080/10871209.2012.680174
6. Ryan, E.L. & Shaw, B. 2011. Improving hunter recruitment and retention. Human Dimensions of Wildlife 16(5):311-317. http://dx.doi.org/10.1080/10871209.2011.559530
7. Schulz, J.H., Millspaugh, J.J., Zekor, D.T. & Washburn, B.E. 2003. Enhancing sport-hunting opportunities for urbanites. Wildlife Society Bulletin 31(2):565-573.
8. U.S. Department of the Interior. 2006a. 2006 national survey of fishing, hunting, and wildlife-associated recreation. U.S. Department of the Interior, Fish and Wildlife Service, and U.S. Department of Commerce, U.S. Census Bureau. Retrieved on January 6, 2010. Available: http://wsfrprograms.fws.gov/Subpages/NationalSurvey/nat_survey2006_final.pdf

9. Winkler, R. & Warnke, K. 2013. The future of hunting: An age-period-cohort analysis of deer hunter decline. Population and Environment 34(4):460-480. http://dx.doi.org/10.1007/s11111-012-0172-6

10. Zinn, H.C. 2003. Hunting and sociodemographic trends: Older hunters from Pennsylvania and Colorado. Wildlife Society Bulletin 31(4):1004-1014.

11. Boglioli, M. 2009. A Matter of Life and Death: Hunting in Contemporary Vermont. University of Massachusetts Press, Amherst, MA, USA.

12. ●● Dizard, J.E. 2003. Mortal Stakes: Hunters and Hunting in Contemporary America. University of Massachusetts Press, Amherst, MA, USA.

13. Stedman, R.C. & Heberlein, T.A. 2001. Hunting and rural socialization: Contingent effects of the rural setting on hunting participation. Rural Sociology 66(4):599-617. http://dx.doi.org/10.1111/j.1549-0831.2001.tb00086.x

14. Cohen, E. 2014. Recreational hunting: Ethics, experiences and commoditization. Tourism Recreation Research 39(1):3-18.

15. Eliason, S.L. 2008. A statewide examination of hunting and trophy nonhuman animals: Perspectives of Montana hunters. Society and Animals 16(3):256-278. http://dx.doi.org/10.1163/156853008X323402

16. Gigliotti, L.M. 2000. A classification scheme to better understand satisfaction of Black Hills deer hunters: The role of harvest success. Human Dimensions of Wildlife 5(1):32-51. http://dx.doi.org/10.1080/10871200009359171

17. Heberlein, T.A. & Kuentzel, W.F. 2002. Too many hunters or not enough deer? Human and biological determinants of hunter satisfaction and quality. Human Dimensions of Wildlife 7(4):229-250. http://dx.doi.org/10.1080/10871200214753

18. Kalof, L. & Fitzgerald, A. 2003. Reading the trophy: Exploring the display of dead animals in hunting magazines. Visual Studies 18(2):112-122. http://dx.doi.org/10.1080/14725860310001631985

19. Kellert, S.R. 1996. The Value of Life. Island Press, Washington, DC, USA.

20. Manfredo, M.J., Fix, P.J., Teel, T.L., Smeltzer, J. & and Kahn, R. 2004. Assessing demand for big-game hunting opportunities: Applying the multiple-satisfaction concept. Wildlife Society Bulletin 32(4):1147-1155. http://dx.doi.org/10.2193/0091-7648(2004)032[1147:ADFBHO]2.0.CO;2

21. Gude, J.A., Cunningham, J.A., Herbert, J.T. & Baumeister, T. 2012. Deer and elk hunter recruitment, retention, and participation trends in Montana. Journal of Wildlife Management 76(3):471-479. http://dx.doi.org/10.1002/jwmg.272

22. ● Schorr, R.A., Lukacs, P.M. & Gude, J.A. 2014. The Montana deer and elk hunting population: The importance of cohort group, license price, and population demographics on hunter retention, recruitment, and population change. Journal of Wildlife Management 78(5): 944-952. http://dx.doi.org/10.1002/jwmg.732

23. Tynon, J.F. 1997. Quality hunting experiences: A qualitative inquiry. Human Dimensions of Wildlife 2(1):32-46. http://dx.doi.org/10.1080/10871209709359085

24. Bronner, S.J. 2008. Killing Tradition: Inside Hunting and Animal Rights Controversies. The University Press of Kentucky, Lexington, KY, USA. http://dx.doi.org/10.5810/kentucky/9780813125282.001.0001

25. Goldfine, R. & Goldfine, O. 2003. Hunters and healers: Social change and cultural conflict in rural Maine. Visual Studies 18(2):96-111. http://dx.doi.org/10.1080/14725860310001631976

26. ●● Heffelfinger, J.R., Geist, V. & Wishart, W. 2013. The role of hunting in North American wildlife conservation. International Journal of Environmental Studies 70(3):399-413. http://dx.doi.org/10.1080/00207233.2013.800383

27. Mahoney, S.P., & Jackson, J.J., III. 2013. Enshrining hunting as a foundation for conservation – the North American model. International Journal of Environmental Studies 70(3):448-459. http://dx.doi.org/10.1080/00207233.2013.801178

28. Montgomery, R. & Blalock, M.G. 2010. The impact of access, cost, demographics, and individual constraints, on hunting frequency and future participation. Academy of Marketing Studies Journal 14(2):115-131.

29. Peggs, K. 2012. Animals and Sociology. Palgrave Macmillan, Basingstoke. http://dx.doi.org/10.1057/9780230377271

30. Peterson, M.N., Hansen, H.P., Peterson, M.J. & Peterson, T.R. 2010. How hunting strengthens social awareness of coupled human-natural systems. Wildlife Biology in Practice 6(2):127-143.

31. Mehmood, S., Zhang, D. & Armstrong, J. 2003. Factors associated with declining hunting license sales in Alabama. Human Dimensions of Wildlife 8(4):243-262. http://dx.doi.org/10.1080/716100423

32. Haggerty, J.H. & Travis, W.R. 2006. Out of administrative control: Absentee owners, resident elk and the shifting nature of wildlife management in Southwestern Montana. Geoforum 37(5):816-830. http://dx.doi.org/10.1016/j.geoforum.2005.12.004

33. Hines, J.D. 2007. The persistent frontier & the rural gentrification of the Rocky Mountain West. Journal of the West 46(1):63-73.

34. Peterson, M.N., Mertig, A.G. & Liu, J. 2007. Influence of urban immigrants on outdoor recreation and land use in Teton Valley. Journal of Park and Recreation Administration 25(4): 25-38.

35. Robbins, P., Meehan, K., Gosnell, H. & Gilbertz, S.J. 2009. Writing the new West: A critical review. Rural Sociology 74(3):356-382. http://dx.doi.org/10.1526/003601109789037240

36. Teisl, M.F., Boyle, K.J. & Record, R.E., Jr. 1999. License-sales revenues: Understanding angler and hunter reactions to changes in license prices. Human Dimensions of Wildlife 4(4): 1-17. http://dx.doi.org/10.1080/10871209909359162

37. Freyfogle, E.T. & Goble, D.D. 2009. Wildlife Law: A Primer. Island Press, Washington, DC, USA.

38. Knezevic, I. 2009. Hunting and environmentalism: Conflict or misperceptions. Human Dimensions of Wildlife 14(1):12-20. http://dx.doi.org/10.1080/10871200802562372

39. Robbins, P. 2006. The politics of barstool biology: Environmental knowledge and power in Greater Northern Yellowstone. Geoforum 37(2):185-199. http://dx.doi.org/10.1016/j.geoforum.2004.11.011

40. Robbins, P. & Luginbuhl, A. 2005. The last enclosure: Resisting privatization of wildlife in the Western United States. Capitalism Nature Socialism 16(1):45-61. http://dx.doi.org/10.1080/1045575052000335366

41. Eliason, S.L. 2000. Some polemical observations on Utah's cooperative wildlife management unit program. Wildlife Society Bulletin 28(1):264-267.

42. Eliason, S. 2012a. From the king's deer to a capitalist commodity: A social historical analysis of the poaching law. International Journal of Comparative and Applied Criminal Justice 36(2): 133-148.
http://dx.doi.org/10.1080/01924036.2012.669912

43. Eliason, S.L. 2012b. Trophy poaching: A routine activities perspective. Deviant Behavior 33(1):72-87.
http://dx.doi.org/10.1080/01639625.2010.548289

44. Blumer, H. 1969. Symbolic Interactionism: Perspective and Method. University of California Press, Berkeley, CA, USA.

45. Colton, C.W. 1987. Leisure, recreation, tourism: A symbolic interactionism view. Annals of Tourism Research 14(3):345-360.
http://dx.doi.org/10.1016/0160-7383(87)90107-1

46. Noe, F.P., Hammitt, W.E. & Bixler, R.D. 1997. Park user perceptions of resource and use impacts under varied situations in three national parks. Journal of Environmental Management 49(3):323-336.
http://dx.doi.org/10.1006/jema.1995.0093

47. Samdahl, D.M. 1988. A symbolic interactionist model of leisure: Theory and empirical support. Leisure Sciences 10(1):27-39.
http://dx.doi.org/10.1080/01490408809512174

48. Garst, B.A., Williams, D.R. & Roggenbuck, J.W. 2010. Exploring early twenty-first century developed forest camping experiences and meanings. Leisure Sciences 32(1):90-107.
http://dx.doi.org/10.1080/01490400903430905

49. Jordan, D.J., Cox, A., Thompson, T., Jeon, J., Palacios, I., Patterson, A., Peel, J. & Henderson, K.A. 2009. An exploration of the meanings of parks in Oklahoma. Journal of Park and Recreation Administration 27(2):17-32.

50. Henderson, K.A. 2006. Dimensions of Choice: Qualitative Approaches to Parks, Recreation, Tourism, and Leisure Research. Venture Publishing, State College, PA, USA.

51. U.S. Department of the Interior. 2006b. 2006 national survey of fishing, hunting, and wildlife-associated recreation: Montana. U.S. Department of the Interior, Fish and Wildlife Service, and U.S. Department of Commerce, U.S. Census Bureau. Retrieved on January 6, 2010. Available: http://www.census.gov/prod/2008pubs/fhw06-mt.pdf

52. Bidwell, D. 2010. Bison, boundaries, and brucellosis: Risk perception and political ecology at Yellowstone. Society and Natural Resources 23(1):14-30.
http://dx.doi.org/10.1080/08941920802132575

53. Brownell, J.L. 1987. The genesis of wildlife conservation in Montana. Master's Thesis, Montana State University, Bozeman, MT.

54. Kelley, J.M. 2001. Implications of a Montana voter initiative that reduces chronic wasting disease risk, bans canned shooting, and protects a public trust. Great Plains Natural Resources Journal 6:89-109.

55. Nesheim, D.A. 2012. Profit, preservation, and shifting definitions of bison in America. Environmental History 17(3):547-577.
http://dx.doi.org/10.1093/envhis/ems048

56. Scarce, R. 1998. What do wolves mean? Conflicting social constructions of *Canis lupus* in "bordertown." Human Dimensions of Wildlife 3(3):26-45.
http://dx.doi.org/10.1080/10871209809359130

57. Scarce, R. 2005. More than mere wolves at the door: Reconstructing community amidst a wildlife controversy. In: Herda-Rapp, A. & Goedeke, T.L. (eds.), Mad About Wildlife. Brill, Leiden, The Netherlands, pp. 123-146.

58. Shanahan, E.A., McBeth, M.K., Tigert, L.E. & Hathaway, P.L. 2010. From protests to litigation to YouTube: A longitudinal case study of strategic lobby tactic choice for the Buffalo Field Campaign. Social Science Journal 47(1):137-150.
http://dx.doi.org/10.1016/j.soscij.2009.10.002

59. Fowler, F.J., Jr. 1993. Survey Research Methods. Sage, Newbury Park, CA, USA.

60. Neuman, W.L. 2011. Social Research Methods: Qualitative and Quantitative Approaches. Allyn & Bacon, Boston, MA, USA.

61. Wilson, M.A. 1997. The wolf in Yellowstone: Science, symbol, or politics? Deconstructing the conflict between environmentalism and wise use. Society and Natural Resources 10(5): 453-468.
http://dx.doi.org/10.1080/08941929709381044

62. Ginger, C., Emery, M.R., Baumflek, M.J. & Putnam, D.E. 2012. Access to natural resources on private property: Factors beyond right of entry. Society and Natural Resources 25(7):700-715.
http://dx.doi.org/10.1080/08941920.2011.633596

63. Leopold, A. 1919. The National Forests: The last free hunting grounds of the nation. Journal of Forestry 17(2):150-153.

64. Stedman, R.C., Bhandari, P., Luloff, A.E., Diefenbach, D.R. & Finley, J.C. 2008. Deer hunting on Pennsylvania's public and private lands: A two-tiered system of hunters? Human Dimensions of Wildlife 13(4):222-233.
http://dx.doi.org/10.1080/10871200802010513

65. Posewitz, J. 1999. Inherit the Hunt: A Journey into the Heart of American Hunting. Falcon, Helena, MT, USA.

66. ● Posewitz, J. 2004. Rifle in Hand: How Wild America was Saved. Riverbend, Helena, MT, USA.

67. Swenson, J.E. 1983. Free public hunting and the conservation of public wildlife resources. Wildlife Society Bulletin 11(3):300-303.

68. Swensson, E.J. & Knight, J.E. 1998. Identifying Montana hunter/rancher problems and solutions. Journal of Range Management 51(4):423-427.
http://dx.doi.org/10.2307/4003327

69. Swensson, E.J. & Knight, J.E. 2001. Hunter management strategies used by Montana ranchers. Wildlife Society Bulletin 29(1):306-310.

70. Montana's Official State Website. 2010. Land ownership in Montana counties. Montana's Official State Website, Natural Resource Information System. Retrieved on August 3, 2010. Available: http://nris.mt.gov/montanafacts/county_own.asp

71. Eliason, S.L. 2011. Motivations for becoming an outfitter in big sky country. Human Dimensions of Wildlife 16(5):299-310.
http://dx.doi.org/10.1080/10871209.2011.553667

72. Nickerson, N.P. & Dubois, M. 2008. Outlook and trends 2008: Montana travel and recreation. Montana Business Quarterly 46(1):18-21.

73. Wilton, J.J. & Nickerson, N.P. 2006. Collecting and using visitor spending data. Journal of Travel Research 45(1):17-25.
http://dx.doi.org/10.1177/0047287506288875

74. Wright, M.V. & Sanyal, N. 1998. Differentiating motivations of guided versus unguided fly anglers. Human Dimensions of Wildlife 3(1):34-46. http://dx.doi.org/10.1080/10871209809359113

75. Fritz, H.W. 2002. Montana in the twenty-first century. In: Fritz, H.W., Murphy, M. & Swartout, R.R., Jr. (eds.), Montana Legacy: Essays on History, People, and Place. Montana Historical Society Press, Helena, MT, USA, pp. 341-358.

76. Malone, M.P., Roeder, R.B. & Lang, W.L. 1991. Montana: A History of Two Centuries. University of Washington Press, Seattle, WA, USA.

77. Utah Division of Wildlife Resources. 2013. Utah big game application guidebook. Retrieved on January 14, 2013. Available: http://wildlife.utah.gov/guidebooks/2013_pdfs/2013_biggameapp.pdf

78. Montana Fish, Wildlife & Parks. 2013. 2013 Available hunting licenses & permits. Retrieved on January 14, 2013. Available: http://fwp.mt.gov/hunting/licenses/availableLicenses.html

79. Mann, M.J. & Leahy, J.E. 2009. Connections: Integrated meanings of ATV riding among club members in Maine. Leisure Sciences 31(4):384-396. http://dx.doi.org/10.1080/01490400902988317

23

Hunter and Landowner Needs and Expectations and the Economic Impact of Hunting on Rural South Carolina Counties (USA)

B.J. Kendall[1], D.C. Guynn, Jr.[1], T.J. Straka[1,*] & G.K. Yarrow[1]

[1] School of Agricultural, Forest, and Environmental Sciences, Clemson University, Clemson, SC 29634, USA.

* Corresponding author address: School of Agricultural, Forest, and Environmental Sciences, P.O. Box 340310, Clemson University, Clemson, SC 29634-0310, USA; e-mail: tstraka@clemson.edu

Keywords
Economic impact;
Hunting;
Hunting expenditures;
Hunter characteristics.

Abstract

Hunting has a significant impact on South Carolina's economy. While state-level hunting related expenditure information is readily available, county level information is not. This study was designed to provide insight regarding the size and impact of the hunting industry on two South Carolina county economies, and additionally, to examine hunter and landowner needs and expectations. Over three-quarters of the private land in the two counties are open to some form of hunting. Nearly half of hunters come from out-of-state. Much of the land is leased for hunting. Landowner needs centered on technical and financial assistance, property access issues, and better hunter relations. Hunter needs were for enhanced habitat, improved roads, clearer property boundaries, and more prescribed burning. Total economic impact of hunting in the two counties was about $123 million per year. The current recession appeared to have little effect on this impact. Specific forest management practices could make hunting in the counties more attractive and changes in community structure could also serve to attract more hunters.

Introduction

In 2006, South Carolina residents and non-residents spent a total of $279 million on hunting related expenses [1]. The majority of these were related to fuel/auto, wildlife management, lease fees, license fees, food purchases, and purchasing hunting related equipment. In the period between 1999 and 2009, the number of licensed hunters in South Carolina, fell 9% [2]. Understanding the expenditures of hunters and those involved in hunting related activities is important to economic planning for rural counties. Additionally, most (71%) of South Carolina's hunting takes place exclusively on private land [3], providing a possible opportunity for an economic gain for the state's landowners and the businesses that serve them.

To insure that the hunting industry remains viable, it is important to understand the economic impacts of hunting activities and the behaviors and preferences of the parties involved [4]. National and state-level economic impacts of hunting are important, but these impacts also need to be assessed at the county-level in order for local communities to understand the magnitude and type of hunter expenditures on local economies [5].

County government, chambers of commerce and development groups may gain insight as to the economic importance of hunting, as well as possible revenue that hunting can provide. This information may help shape community planning efforts in supporting landowners that provide hunting opportunities and the hunters that use private land in the county. Additionally, landowners should be surveyed to assess trends in hunting related revenue and expenditures and willingness to provide access to their lands. Maximizing hunting's economic potential is vital to many rural county governments, businesses, development groups, landowners, sportsmen, land managers and the state as a whole [6].

An understanding of the benefits and problems associated with hunting will help ensure that it remains a part of South Carolina's culture and also aid communities, businesses and the state in determining the types of services and facilities needed to facilitate hunting activities. The contribution of hunting expenditures to the local economy and its impact on local economic conditions will highlight the importance of hunting at the county level. This study should help businesses to understand the needs of their landowner and hunter clientele, target them more effectively, provide them with the goods and services they desire, and help shape programs offered by Clemson University, South Carolina Department of Natural Resources, as well as other natural resources agencies and economic development groups.

Estimates of the economic impact of hunting at the state and national levels are in the research literature. The U.S. Fish and Wildlife publishes a national wildlife survey every five years that includes the economic impacts of hunting [1] and similar studies have looked at hunting impacts [7-10]. State-level studies on the economic impact of hunting in general and for specific species are not common, but exist for various states [6, 11-15].

A prior study focused on the magnitude and impact of hunting related expenditures on a county level in Jasper and McCormick Counties in South Carolina, as well as determining the problems and benefits associated with hunting and the needs of both landowners and hunters [7]. The current study focuses on Allendale and Jasper Counties. To allow findings for Jasper County to be compared with findings from the prior study, the survey methods and questions from that study were maintained. Some additional questions were added to the current study in order to enhance understanding of hunter and landowner groups. Jasper County is the closest South Carolina county to Savannah, Georgia, a large metropolitan area. McCormick and Allendale Counties are central South Carolina are considered to be "rural" counties. Jasper County was located near a large population center that consisted of out-of-state hunters. Differences in results might suggest influences of these variables on economic impact.

The study objectives were to (1) estimate the annual county-level economic impact of hunting in Allendale and Jasper Counties, (2) to determine the needs of landowners that provide hunting access to their lands in these counties, (3) determine the needs of hunters that utilize private lands in these counties, (4) determine the demographics of landowners and hunters in these counties, and (5) determine the impact the current economic recession has on hunting related expenditures in these counties.

Methods

Two counties were chosen for this study, Allendale and Jasper County, both located in the southeastern corner of South Carolina. Allendale County is in the coastal plain of South Carolina; it is 170,000 ha in size with a population of 10,500 and has experienced a population decline of about 7% between 1996 and 2006. It is about half forested [16, 17]. Jasper County is located on the coast of South Carolina between Savannah and Charleston. Currently it is feeling urban development pressure from Beaufort. The county has a population of 22,000 and encompasses 106,000 ha. It is nearly three-quarters forested [18, 19].

A two-part survey was required, a landowner component followed by a hunter component. This was necessary because there were no means to determine who hunts in these counties, based on hunting licensing or any other data. As part of the landowner survey, responding landowners were asked to provide the names and contact information for hunt club officers or the contact information for those that hunt their land. After collecting contact information from landowners, hunters were then surveyed to obtain their profile and expenditure information. Landowner and hunter data were then used to create county-level statistics. We recognize a potential bias in data collection as all hunters surveyed were known to the landowners, but some sort of county identification system was necessary to select hunter respondents.

Private landowners in Allendale and Jasper Counties were surveyed to determine a profile of landowners, the availability of private lands for hunting recreation, the amount of revenue generated by hunting activities, the amount landowners spent providing hunter services and facilities and managing their wildlife, and the needs, benefits and problems landowners associate with hunting in the county, plus provide mailing lists of hunters who use their lands. The contact information for landowners owning more than 20 ha was prepared using 2008 county tax records. This list contained a total population of 421 landowners in Allendale County and a total population of 422 landowners in Jasper County. In order to retain the ability to compare results of the prior study [7], the mailing list was divided into two categories; large landowners owning more than 160 ha, and small landowners owning between 20 and 160 ha. The large landowner group contained 101 and 98 individuals for Allendale and Jasper County respectively; all large landowners were surveyed. The small landowner category contained 320 and 324 landowners for Allendale and Jasper County respectively. A random sample of half the small landowners in both counties were surveyed.

The methods used were a modification of those used in the earlier survey [7]. All questions asked in the 1992 survey remained with the addition of questions regarding change in revenue between the 2007-2008 and 2008-2009 hunting seasons (to judge the impact of the economic downturn). An additional question regarding race of landowners and one regarding other natural resource enterprise activities that landowners were involved in was also asked. The questionnaire sent to landowners contained a cover letter, 13 major questions (some with sub-questions) and a section to provide names and contact information for hunters utilizing their property. Respondents were first asked demographic questions, then to identify relevant potential benefits, needs, problems, and involvement from provided lists, and finally were asked to add relevant items to any list. The lists correspond to the items in Table 1. Economic data were requested in terms of variables relative to economic impact; that is, revenue by source and expenditures for capital, labor, and profit. Economic data were reported as local (county-generated) and non-local (out-of-county and out-of-state). The landowner questionnaire was mailed to landowners in both counties in June of 2009, followed by a reminder postcard that emphasized the importance of the survey in September of 2009. The total response rate by landowners in Allendale and Jasper Counties was 27.2% (71 landowners) and 28.5% (74 landowners), respectively.

The list of hunters furnished by the landowner survey was used to determine hunter profiles and expenditures for the period of 30 June 2008 to 1 May 2009. The hunter survey mailing list included 101 hunters in Allendale County and included 31 hunters in Jasper County. The hunter questionnaire was a modification of the one used in the earlier survey [7]. All questions asked in that survey were used, plus questions regarding race, change in relative expenditures from the 2007-2008 to the 2008-2009 hunting season, and as well as a question regarding the composition of habitat management expenses. The questionnaire used contained a cover letter and 16 major questions (many with sub-questions). The questions asked were organized into the following categories: hunter profile and demographics, hunter needs, and hunter expenditures. Expenditures were organized as for the landowners for consistent reporting. The survey was mailed in September of 2009 to all hunters provided by landowners. The total response rate for the hunter survey was 45.5% (46 responses) and 41.9% (13 responses) for Allendale and Jasper Counties, respectively. These sample sizes may seem low, but the response rates for the survey are very high (for mail-based social science surveys). One problem of working at the county level is a small population, of landowners and especially reported hunters. The high response rates show a significant effort at increasing sample size; however, we acknowledge this is a weakness of the study.

Landowner and hunter questionnaires were entered into spreadsheets representing Allendale County landowners, Jasper County landowners, Allendale County hunters, and Jasper County hunters. Surveys were inspected for data quality and entered into Excel spreadsheets. Input-output analysis is a matrix algebra-based model, so expenditures organized into large tables are a normal input method. The input-output model converted these combined landowner and hunter expenditures by category into state-level economic impact. The model used was IMPLAN, the most wide-accepted and used input-output model in current practice. This model was used to determine the direct, indirect, and induced effect of landowner and hunter expenditures for four economic measures: output, value added, labor income, and employment [6, 20-22].

Landowner Profiles

Allendale County

There are 421 parcels larger than 20 ha in Allendale County, accounting for 90,823 ha (Allendale County Tax Assessor, personal communication, 4 June 2008). Survey respondents were mostly male (86%) and Caucasian (98%). The average age was 62 years. Average tract size for all respondents was 396 ha; additionally landowners owned an average of 2.3 tracts. Tracts owned by large landowners were an average of 824 ha; small landowners owned an average of 64 ha. About one-third of respondents lived in Allendale County (35%) and nearly a quarter lived out-of-state (24%). Forty-one percent of respondents lived in South Carolina, but not Allendale County. The majority indicated that their property was held in individual ownership (66%), while 20% was help in partnerships, 8% by corporations, and 6% in other forms of ownership.

Over three-quarters of respondents (78%) allowed some form of hunting activity on their property. Large landowners were more likely (90%) to allow hunting on their property than small landowners (70%). The average landowner that allowed hunting granted permission to hunt to 17.3 hunters, of these, 39% paid some type of fee to hunt. Expanding the survey results equates to 3,752 hunters on private lands in Allendale County in 2008.

Over two-thirds (69%) of this forest land (owned by 54% of landowners) was in some form of fee-hunting arrangement. Landowners that charged a fee held an average of 517 ha under the fee-hunting arrangement (representing 86% of their total land). Three-quarters of large landowners that allowed hunting charged a fee, while only 36% of small landowners charged one. The average landowner that leased hunting rights received $29.92/ha annually (weighted average by ha owned). Large landowners received slightly lower lease revenue ($29.87/ha) than small landowners ($31.83/ha). None of the landowners chose to charge hunter groups by hunting season, species being hunted or by any other method other than an annual per ha fee.

Over half (58%) of respondents belonged to an association that promotes wildlife management; this was 73% of large landowners and 43% of small landowners. Landowners that allowed hunting spent an average of $15.32 per ha to manage their lands for wildlife; large landowners averaged $15.30 and small landowners averaged $15.74. Landowners that charged a fee to hunt averaged $13.47/ha in wildlife-related expenditures, while those that allowed hunting without a fee averaged $24.56/ha. In order to provide services and facilities to hunters, landowners currently leasing their land spent an average of $14.65/ha. Large landowners spent more money providing hunter services and facilities ($18.36/ha) than did small landowners ($13.12/ha).

Average landowner expenditures on wildlife management and providing hunter services and facilities indicate that landowners were willing to reinvest the income they received into their property. Landowner expenditures (for both wildlife management and providing hunter services and facilities) were quite high in relation to the amount of revenue gained from leasing. Additionally, those that did not choose to lease their lands spent a great deal on these activities. This indicates that income is not the primary goal of many landowners, illustrating the non-monetary benefits that they receive from hunting.

Table 1 reports Allendale County landowners' needs, benefits, problems, and involvement with hunting. The most common needs were technical and financial assistance. Other needs concerned hunter relations dealing with property access and hunter relations. The main benefits of owning forested property were for personal enjoyment and wildlife habitat. Supplemental income was also important [1, 3]. Landowners reported small management problems, mainly related to hunter relations and wild dogs. Landowners enjoyed various outdoor activities like bird watching, wildlife viewing, and gathering pine straw. Landowners were also asked to report on changes in hunting lease revenue over the economic recession. Hunting lease revenue remained very stable over the recession.

Table 1: Needs, benefits, problems, and involvement expressed by Allendale and Jasper County landowners (2008). Allendale Co., n = 71; Jasper Co., n = 74.

Attributes	Allendale Co. Percent	Jasper Co. Percent
Needs		
Wildlife management technical assistance/expertise	28	33
Property management/maintenance assistance	28	30
No additional services needed	14	15
Other	14	19
More or improved community facilities	8	0
Assistance with controlling access	6	0
Improved hunter ethics/safety	3	4
Benefits		
Outdoor recreation/personal pleasure	33	38
Results in improved game populations/habitat	31	18
Supplemental income for landowner	14	15
Social aspects, family, friends, neighbors, meeting new people	9	9
Wildlife population control	7	6
Protection of the land	7	0
Improves local community/economy	0	9
Provides wild game meat (food)	0	6
Problems		
Other	39	30
No problems	16	22
Poaching/Trespassing	14	19
Neighbors' dogs/Wild dogs	14	15
Property damage	9	11
Lack of hunter ethics/education	7	4
Interferes with landowner's personal activities	3	0
Involvement		
Bird watching	8	7
Wildlife viewing	8	11
Pine straw harvesting	8	2
Horseback riding	3	7
Camping	3	3
Other	3	4
Mushroom collecting	0	1

Jasper County

There are 422 parcels larger than 20 ha in Jasper County (Jasper County Tax Assessor, personal communication, 4 June 2008). Survey respondents were mostly male (88%) and Caucasian (100%). The average age was 60 years. Average tract size for all respondents was 511 ha; additionally landowners owned an average of 2.4 tracts. Tracts owned by large landowners were an average of 1,112 ha; small landowners owned an average of 53 ha. About one-half of respondents lived in Jasper County (49%) and nearly a quarter lived out-of-state (24%). Twenty-seven percent of respondents lived in South Carolina, but not Jasper County. The majority indicated that their property was held in individual ownership (68%), while 9% was help in partnerships, 19% by corporations, and 4% in other forms of ownership.

Sixty-one percent of respondents allowed some form of hunting activity on their property. Large landowners were more likely (91%) to allow hunting on their property than small landowners (38%). Landowners allowed an average of 18.5 hunters to hunt their property. Small landowners allowed an average of 2 hunters to utilize their lands; large landowners granted access to an average of 40 hunters. Expanding the survey results equates to 4,596 hunters on private lands in Jasper County in 2008.

Over one-third (38%) of this forest land was in some form of fee-hunting arrangement. Landowners that charged a fee held an average of 768 ha under the fee-hunting arrangement (representing 62% of their total land). Forty-eight percent of large landowners that allowed hunting charged a fee, while only 13% of small landowners charged one. The average landowner that leased hunting rights received $21.52/ha annually (weighted average by ha owned). Large landowners received more lease revenue ($22.14/ha) than small landowners ($14.82/ha). None of the landowners chose to charge hunter groups by hunting season, species being hunted or by any other method other than an annual per ha fee. A quarter of the landowner exchanged hunting rights for labor from hunters rather than charge a monetary fee.

Over two-thirds (68%) of respondents belonged to an association that promotes wildlife management; this was 73% of large landowners and 43% of small landowners. Landowners that allowed hunting spent an average of $18.98 per ha to manage their lands for wildlife; large landowners averaged $17.47 and small landowners averaged $79.29. Landowners that charged a fee to hunt averaged $9.12/ha in wildlife-related expenditures, while those that allowed hunting without a fee averaged $32.37/ha. In order to provide services and facilities to hunters, landowners currently leasing their land spent an average of $3.11/ha. Large landowners spent more money providing hunter services and facilities ($4.74/ha) than did small landowners ($3.09/ha).

Jasper County landowners are willing to invest large amounts of money in managing their lands for wildlife and to provide hunters with services and facilities. Landowners that did not lease their lands spent more than those that did; indicating that landowners that lease expect hunters to assist with wildlife management and their facilities. Additionally, the fact that landowners were willing to spend more than their lease revenue (if they lease) indicates that non-monetary benefits play a key role in their decision making process.

The most common need expressed by Jasper County landowners was for technical and financial assistance in dealing with wildlife management (Table 1). Landowners cited the greatest benefits of their property as personal enjoyment, game habitat, and a source of additional income. The main problems cited were people-related ones like trespassing and property damage.Wild dogs were also mentioned. Landowners' main involvement with the property is personal activates like bird watching, wildlife viewing, horseback riding, and camping. Jasper County landowners' hunting fee revenue remained fairly stable over the economic recession.

Hunter Profiles

Allendale County

Based on the hunter survey, there are an estimated 3,752 hunters utilizing private lands in Allendale County. Most were male (97%). The average age was 56 years, the average educational level was 15.7 years, and the average family annual income was between $100,000 and $119,999.

Half of the hunters lived in South Carolina, but not Allendale County. Forty-five percent were from out-of-state and 5% resided in Allendale County. Hunters drove an average of 3.3 hours, 184 miles one way per hunting trip, and made an average of 30 hunting related trips per year to Allendale County. On these trips hunters traveled with an average of 1.8 hunters in their vehicle. Surveyed hunters in Allendale County spent 27 days hunting on private land, as opposed to an average of 2.3 days hunting public lands. When hunting on private land, hunters reported hunting on tracts averaging 1,341 ha.

White-tailed deer, wild turkey, wild hog, and mourning dove were the most popular game species hunted in Allendale County (Table 2). Other activities reported as being enjoyed while hunting were fishing (18%), sightseeing (9%), rest and relaxation (9%), camping (6%), socializing/eating out (9%), being outdoors (4%), and golfing (3%). Activities reported as substitutes for hunting were fishing (44%), golf (12%), other sports (6%), camping (6%), shooting (6%), water sports (4%), and photography (1%).

Table 2: Average number of days hunted per hunter annually and percent hunter involvement by game species, Allendale (n =46) and Jasper Counties (n =13), (2008).

	Allendale County		Jasper County	
Game Species	Average Days Hunted	Percent Involved	Average Days Hunted	Percent Involved
Deer	21.1	94.6	21.5	92.5
Turkey	9.1	56.8	8.2	67.5
Hog	16.6	37.8	7.5	25.0
Dove	9.2	20.3	6.1	40.0
Duck	1.9	14.9	5.6	42.5
Quail	11.8	8.1	0.8	5.0
Pen-raised game	6.0	6.8	11.0	20.0
Varmint	9.7	9.5	10.0	10.0
Small game	5.0	10.8	2.8	10.0
Other	--	--	2.3	5.0

Respondents reported spending an average of $9,772 annually on hunting related activities. Local expenditures were mainly received by community businesses (64%); landowners received 32% and licenses and permits accounted for 4%. During the recession 44% of hunters reported increased hunting-related expenditures and 12% reported less.

Hunter needs were established in terms of the landowners and the communities. In order to retain or attract new hunters suggested practices for landowners were to increase food plots for game (30%), lower hunting fees (7%), improved roads (5%), provide clearer property boundaries (5%), conduct more prescribed burning (3%), increase access (2%), and better attitude towards hunters (2%). Suggestions for communities were more restaurants (20%), more lodging (7%), increased game law enforcement (5%), better attitude towards hunters (4%), and respect for hunting dogs (2%).

Jasper County

Based on the hunter survey, there were an estimated 4,596 hunters utilizing private lands in Jasper County. Most were male (98%). The average age was 55 years, the average educational level was 15.4 years, and the average family annual income was between $140,000 and $519,999.

Forty-eight percent of the hunters came from out-of-state; 25% lived in South Carolina, but not Jasper County; and 27% were from Jasper County. Hunters drove an average of 1.1 hours, 46 miles one way per hunting trip, and made an average of 46 hunting related trips per year to Jasper County. On these trips hunters traveled with an average of 1.9 hunters in their vehicle. Surveyed hunters in Jasper County spent most of their hunting time on private lands; 27 days hunting on private land, as opposed to an average of 2.3 days hunting public lands in the county. When hunting on private land, hunters reported hunting on tracts averaging 895 ha.

White-tailed deer, wild turkey, mourning dove, and waterfowl were the most popular game species hunted in Jasper County (Table 2). Other activities reported as being enjoyed while hunting were fishing (14% socializing/eating out (19%),), sightseeing (10%), rest and relaxation (2%), camping (9%), being outdoors (9%), and golfing (7%). Activities reported as substitutes for hunting were fishing (35%), golf (13%), other sports (9%), shooting (4%), water sports (4%), photography (2%) and hiking (2%).

Respondents reported spending an average of $12,345 annually on hunting related activities. Local expenditures were mainly received by community businesses (75%); landowners received 14% and licenses and permits accounted for 11%. During the recession 38% reported increased hunting related expenditures and 23% reported less.

Hunter needs were established in terms of the landowners and the communities. In order to retain or attract new hunters suggested practices for landowners were increased food plots for game (36%), improved roads (9%), more prescribed burning (9%), lower hunting fees (5%), increased access (5%), and better attitude towards hunters (5%). Suggestions for communities were more restaurants (9%) and improved meat processing/storage (9%).

Economic Impact

Economic impacts involving hunting are especially difficult to estimate due to the wide variability of landowner and hunter expenditures [23]. IMPLAN is one of the most widely-used input-output models to estimate regional economic impact. An input-output model is used to define the relationships between industrial sectors of the economy. It shows individual sectors' purchases and outputs are distributed among all the sectors of the economy [21]. The key to the model is that all economic activity is linked. If a hunter makes a payment to a landowner, this payment might support a contractor's work on the forest property, a meal in a restaurant, or a new automobile. The money spent on the contractor, meal, or automobile would then be spent again by the recipient. So the impact of a dollar spent is expanded as it passes through the economy. This is called the multiplier effect [6]. That is, spending results in more spending via a multiplier effect. Spending impacts many other sectors and its impact is magnified as it passes through multiple industrial sectors and through time [24].

Economic multipliers are used to study the total impact of an economic event or action across an entire economy or a regional economy. In this study the state-level impacts were estimated. Due to a small sample size, large variability, and the problems associated with estimation of hunter expenditures, economic impact projections may be approximations. These types of problems are common in estimates of activities like sport fishing, hunting, and recreation [25, 26]. Even with some uncertainty with the estimates, much research in these areas has taken place and has provided useful tools in terms of planning, spending trends, and identification of areas that provide maximum economic impact [27, 28].

Input-output analysis captures the linkages in expenditures and outputs between industry sectors and multiplier effect of spending. IMPLAN recognizes 508 industrial sectors in South Carolina [6]. In input-output analysis, total economic impact is the sum of direct impact, indirect impact, and induced impact. *Direct impact* is the effect of expenditures on a specific industry. This is the initial impact of a change that impacts the industry's purchases or wages. For example, increased demand for hunting results in more land being leased for this activity. *Indirect impact* is the related effect on other industries. For example, increased hunting leads to increased ammunition purchases. *Induced impact* is the related effect of increased wages (household income) on the economy. For example, the landowner with increased hunting fee income spends more money at local restaurants. IMPLAN reports on all three impacts, for hunters and landowners combined, in order to determine total economic impact.

IMPLAN reports on economic value in terms of four criteria: value added, labor income, employment, and output. *Value added* is the difference between the value of an industry's total output and the cost of its intermediate inputs. It is the value of the outputs with the expense of producing them subtracted. That makes it a measure of the industry's net contribution to the state's gross domestic product (GDP). *Labor income* represents all employment income in an industry. *Employment* is the number of jobs resulting from the economic activity of an industry. *Output* or total impact is the value of an industry's production. Combined expenditure data for both landowners and hunters are necessary to obtain the total economic impact. To derive direct, indirect, and induced effects, the model inputs are expenditures by broad-based industries. Hunting and landowner services impact other industries like hunting equipment, lodging, restaurants, transportation, entertainment, and construction. Total expenditures are reported as a large matrix in a spreadsheet by type and industry and this is the input into the IMPLAN model. IMPLAN results for the hunting industry (hunters and landowners) in Allendale and Jasper Counties are reported in Table 3.

Table 3: State-level economic impact of hunting in Allendale and Jasper Counties, S.C.

	Direct	Indirect	Induced	Total
		Allendale County		
Value added	$8.952.084	$2.096.028	$2.773.385	$13.821.497
Labor income	$5.964.576	$1.234.425	$1.536.737	$8.735.738
Employment	226	35	48	309
Output	$27.437.350	$3.879.802	$4.930.455	$36.247.607
		Jasper County		
Value added	$26.479.469	$7.129.506	$9.315.770	$42.924.745
Labor income	$19.599.273	$4.582.094	$5.161.882	$29.343.249
Employment	486	122	161	769
Output	$56.736.126	$13.119.082	$16.561.347	$86.416.555

Allendale and Jasper Counties both have substantial economic contributions to their economies that are derived directly or indirectly from hunting. Direct impacts, expenditures made directly into the county economy, were $27.4 million and $56.7 million for Allendale and Jasper Counties, respectively; total impacts on Allendale and Jasper Counties were $36.2 million and $86.4 million, respectively. An understanding of the needs and expectations of these hunters can help landowners and community planners to grow local economies. Ninety-five percent of Allendale County's hunters and nearly three-quarters of Jasper County hunters live outside of the county. That means most of the economic impact of hunting originates from outside of the county; hunters from outside of the counties add money to the local economy that was not available without that sport.

Addressing hunter needs could enhance opportunities to increase that economic contribution. Plus about one-quarter of Allendale County hunting expenditures by local residents take place outside of the county and about one-fifth of Jasper County hunting expenditures by county residents take place outside of the county. Both counties may want to address ways to encourage spending to become local. Both counties could address additional goods and services desired by these hunters. Increases in the number and variety of retail and grocery stores that cater to hunters and increased entertainment opportunities would likely increase hunting and its economic impact.

It is significant that most of the private forest land in both counties that are in tracts greater than 20 ha in size are available for hunting and wildlife-related recreation, nearly 80% in Allendale County and just over 90% in Jasper County. The availability of private lands for hunting is a large asset to both counties as it serves to attract hunters to the area who add much to the overall economy. Community businesses receive most of that revenue (60 to 75%) and landowners have fee revenue as a source of additional income. License and permit fees were also significant. Landowners reinvested much of their fee revenue back into the property for wildlife habitat improvement, facilities, and services. In Allendale County weighted average hunting lease revenue was $29.92/ha and in Jasper County it was $18.85. Reinvestment was $25.64 in Allendale County and $11.32 in Jasper County, or 86% in Allendale County and 62% in Jasper County. This shows that landowners that lease private lands for hunting were committed to providing hunting opportunities and value the non-monetary benefits that hunting provides.

There has been meaningful research on the economic impact of hunting. Input-output analysis is the traditional analysis model of choice in economic impact estimation [28, 29]. However, little of it addresses county-level impacts. State-level impacts are widely reported [6, 12, 14, 15, 23, 27] and our estimated county-level economic impacts are proportionately consistent with these published reports [30]; plus, our statewide estimates are also consistent with these national-level studies that include state-level estimates [8, 23, 31-33]. Rather than a regional or statewide focus, many studies are game species specific [8, 14-17, 34, 35]; the magnitude of these estimates was well within the range of our study results. Overall, economic impact studies show regional impacts that followed the patterns of our study.

Results from this study indicate that hunting in both Allendale and Jasper Counties has a significant economic impact on the local and state economy. These benefits can be maximized by county governments by providing the services and encouraging the types of businesses hunters and landowners desire. Together with hunters and landowners, county governments, chambers of commerce, and natural resource managers can ensure that hunting remains a key component in rural South Carolina economies.

Jasper County was part of an earlier survey and consistent mythology was utilized to compare results over a fifteen year time period. IMPLAN was not used in the earlier study, so total economic impact could not be compared. However, basic results like hunting lease revenue and hunter expenditures were consistent across time. The inflation rate between the two time periods averaged 2.65% annually [36]. Based on these foundation results, it appears total economic impact increased at slightly more than the inflation rate over the fifteen year time period.

Conclusion

Direct expenditures from hunting were $27.4 million in Allendale County and $56.7 million in Jasper County for the 2008-2009 hunting season. Local community businesses received over 60% of these expenditures in Allendale County and three-quarters in Jasper County. Plus, landowners themselves spend $3.8 million in Allendale County and $6.0 million in Jasper County for wildlife management and hunter facilities. This makes hunting a considerable component of the regional economy.

Landowners showed an interest in investing in hunting as an enterprise, indicating a positive outlook for the industry by the business community. Hunting fared well over the economic recession, rewarding that investment. From the 1990-1991 to the 2008-2009 hunting season, the average hunting lease amount increased 6% annually. Hunters expressed satisfaction with the hunting experience in the two counties, but did suggest areas that could be strengthened to enhance the overall experience, mainly a broadening of the entertainment and eating opportunities in the counties. It is clear that the hunting opportunities are attractive in the region by the time and expense hunters expend to get to the counties and for the necessary supplies and licenses.

Economic impact is difficult to estimate and the wide variance in hunter expenditures and the small sample size make the estimates in this study somewhat conjectural. Economic multipliers are complex and no stronger than the data that go into calculating them. The variability in the data that went into developing the multipliers could lead to inflated expenditure estimates [25, 26]. But, the importance of hunting to the local economies is still significant. Hunting represents an important component of the region's economy.

This study and discussion has centered on small American regions, but implications for similar studies are international. Economic impact from wildlife and hunting is significant worldwide and can be a major economic engine in some countries [37, 38]. Issues like trophy hunting, subsistence hunting, and market hunting can be important in many regions of the world. Regional economic impact analysis is relevant in even the smallest economy and issues from this economic impact range across the globe.

Not all economic output is tangible products. Some impact is services. The hunting experience is certainly one of those services. It makes a major contribution to the regional economy and the product, intangible experiences and recreational opportunities, is one that hunters truly value. Those same hunters spend considerable money for the equipment needed for the sport. South Carolina Counties need to consider hunting as an industry that impacts their county. Its eventual impact is the same as a manufacturing plant in terms of growing the local economy. Often this is overlooked.

References

Five "key references", selected by the authors, are marked below (Three recommended (●) and two highly recommended (●●) papers).

1. ● U.S. Fish and Wildlife Service. 2007. 2006 National survey of fishing, hunting, and wildlife-related recreation. U.S. Department of the Interior, Fish and Wildlife Service and Department of Commerce, U.S. Census Bureau, Washington, DC, USA.
2. Phillips, N. 2009. The thrill of the hunt is still there, but hunting continues to fade. The State, Columbia, SC, USA. 1 November 2009.
3. Mozumder, P.C., Starbuck, C.M., Berrens, R.P. & Alexander S. 2007. Lease and fee hunting on private lands in the U.S.: A review of the economic and legal issues. Hum. Dimens. Wildl. 12(1):1-14. http://dx.doi.org/10.1080/10871200601107817
4. ●● International Association of Fish and Wildlife Agencies. 2002. Economic Importance of hunting in America. International Association of Fish and Wildlife Agencies, Washington, DC, USA.
5. Slee, B., Farr, H. & Snowdon, P. 1997. The economic impact of alternative types of rural tourism. J. Agr. Econ. 48(2):179-192. http://dx.doi.org/10.1111/j.1477-9552.1997.tb01144.x
6. ●● Moore School of Business. 2009. Underappreciated assets: The economic impact of South Carolina's natural resources. University of South Carolina, Moore School of Business, Division of Research, Columbia, SC, USA.
7. ● Richardson, C.L., Yarrow, G.K. & Smathers, W.M., Jr. 1992. Economic impact of hunting on rural communities: Jasper and McCormick Counties in South Carolina. Clemson University, Extension Wildlife Program, Clemson, SC, USA.
8. Carver, E. 2008. Economic impact of waterfowl hunting in the United States (Report 2006-2). U.S. Fish and Wildlife Service, Washington, DC, USA.
9. Maharaj, V. & Carpenter, J. 2000. The economic impacts of fishing, hunting and wildlife viewing on National Forests lands. The Theodore Roosevelt Conservation Alliance, Washington, DC, USA.
10. National Shooting Sports Foundation. 2011. The economic impact of Sunday hunting. National Shooting Sports Foundation, Newtown, CT, USA.
11. Pickton, T., & Sikorowski, L. 2008. The economic impact of hunting, fishing and wildlife watching in Colorado (final report). BBC Research & Consulting, Denver, CO, USA.
12. Southwick Associates, Inc. 2007. The 2006 economic benefits of hunting, fishing and wildlife watching in Texas. Southwick Associates, Inc., Fernandina Beach, FL, USA.
13. Burger, L.W., Miller, D.A. & Southwick, R.I. 1999. Economic impact of northern bobwhite hunting in the southeastern United States. Wildl. Soc. Bull. 27(4):1010-1018.

14. Grado, S.C., Hunt, K.M., Hutt, C.P., Santos, X.T. & Kaminski, R.M. 2011. Economic impacts of waterfowl hunting in Mississippi derived from a state-based mail survey. Hum. Dimens. Wildl. 16(2):100-113.
http://dx.doi.org/10.1080/10871209.2011.551175

15. Grado, S.C., Hunt, K.M. & Whiteside, M.W. 2008. Economic impacts of white-tailed deer hunting in Mississippi. Proc. Annu. Conf. Southeast. Assoc. Fish and Wildl. Agencies 61: 59-67.

16. U.S. Census Bureau. 2013. State and county quickfacts: Allendale County, South Carolina. U.S. Census Bureau, Washington, DC, USA. http://quickfacts.census.gov/qfd/states/45/45005.html. Cited 10 July 2013.

17. USDA National Agricultural Statistics Service. 2007. Census of agriculture, county profile, Allendale County, South Carolina. USDA National Agricultural Statistics Service, Washington, DC, USA.

18. U.S. Census Bureau. 2013. State and county quickfacts: Jasper County, South Carolina. U.S. Census Bureau, Washington, DC, USA. http://quickfacts.census.gov/qfd/states/45/45053.html. Cited 10 July 2013.

19. USDA National Agricultural Statistics Service. 2007. Census of agriculture, county profile, Jasper County, South Carolina. USDA National Agricultural Statistics Service, Washington, DC, USA.

20. Hussain, A., Munn, I.A., Holland, D.W., Armstrong, J.B. & Spurlock, S.R. 2012. Economic impact of wildlife-associated recreation expenditures in the southeastern United States: A general equilibrium analysis. J. Agr. Appl. Econ. 44(1):63-82.

21. Miernyk, W.H. 1965. The Elements of Input-Output Analysis. Random House, New York, NY, USA.

22. Munn, I.A., Hussain, A., Spurlock, S.R. & Henderson, J.E. 2012. Economic impact of fishing, hunting, and wildlife-associated recreation expenditures on the southeastern U.S. regional economy: An input-output analysis. Hum. Dimens.Wildl. 15(6):433-449.
http://dx.doi.org/10.1080/10871209.2010.508193

23. Shafer, E.L., Carline, R., Guldin, R.W. & Cordell, H.K. 1993. Economic amenity values of wildlife: Six case studies in Pennsylvania. Environ. Manage. 17(5):669-682.
http://dx.doi.org/10.1007/BF02393728

24. Hughes, D.W. 2003. Policy uses of economic multiplier and impact analysis. Choices 18(2):25-30.

25. Helgeson, J.G. & Beatty, S.E. 1987. Price expectations and price recall error: An empirical study. J. Consum. Res. 14(3):379-386.
http://dx.doi.org/10.1086/209121

26. Sudman, S. & Bradburn, N.M. 1973. Effects of time and memory factors on response in surveys. J. Am. Stat. Assoc. 68(344):805-815.
http://dx.doi.org/10.1080/01621459.1973.10481428

27. Grado, S.C., Kaminski, R.M., Munn, I.A. & Tullos, T.A. 2001. Economic impact of waterfowl hunting on public lands and private lodges in the Mississippi Delta. Wildl. Soc. Bull. 29(3):846-855.

28. ● Cooper, A.B., Stewart, F., Unsworth, J.W., Kuck, L., McAuthur, T.J. & Rachael, J.S. 2002. Incorporating economic impacts into wildlife management decisions. Wildl. Soc. Bull. 30(2):565-574.

29. English, B. K. & Bergstrom, J.C. 1994. The conceptual links between recreations and site development and regional economic impact. J. Regional Sci. 34(4):599-611. http://dx.doi.org/10.1111/j.1467-9787.1994.tb00885.x

30. Southwick, R.I. 1994. Economic impacts of hunting in the Southeast. Proc. 1994 Annu. Conf. Southeast. Asssoc. Fish and Wildl. Agencies 48: 88-98.

31. Steinback, S.R. 1999. Regional economic impact assessment of recreational fisheries: An application of the IMPLAN modeling system to marine party and charter boat fishing in Maine. N. Am. J. Fish. Sci. 19(3):724-736. http://dx.doi.org/10.1577/1548-8675(1999)019<0724:REIAOR>2.0.CO;2

32. Chulwon, K., Scott, D., Thigpen, J.F. & Kim, S. 1998. Economic impact of a bird festival. Festival Manage. Event Tourism 5(1-2):51-58.

33. Dudensing, R. and Hernandez, P. 2011. The economic impact of hunters on Sutton County. Texas A&M University, AgriLIFE Extension, College Station, TX, USA.

34. Pope, C.A., III. & Stoll, J.R. 1985.The market value of ingress rights for white-tailed deer hunting in Texas. Southern J. Agr. Econ. 17(1):177-182.

35. Gan, C.E.C., & Luzar, E.J. 1993. An economic analysis of waterfowl hunting in Louisiana. No.841. Louisiana State University, Agricultural Experiment Station, Baton Rouge, LA, USA.

36. United States Department of Labor, Bureau of Labor Statistics.b2013. Databases, Tables and Calculators by Subject: CPI Inflation Calculator. Bureau of Labor Statistics, Washington, DC, USA. http://www.bls.gov/data/inflation_calculator.htm. Cited 10 July 2013.

37. Taylor, M.S. and Brander, J. A. 1997. International trade and open-access renewable resources: The small open economy model. Can. J. Economics 30(3):526-552. http://dx.doi.org/10.2307/136232

38. Gibson, C.C. 1999. Politicians and Poachers: The Political Economy of Wildlife Policy in Africa. Cambridge University Press, Cambridge, UK. http://dx.doi.org/10.1017/CBO9780511625640

Permissions

All chapters in this book were first published in WBP, by Portuguese Wildlife Society; hereby published with permission under the Creative Commons Attribution License or equivalent. Every chapter published in this book has been scrutinized by our experts. Their significance has been extensively debated. The topics covered herein carry significant findings which will fuel the growth of the discipline. They may even be implemented as practical applications or may be referred to as a beginning point for another development.

The contributors of this book come from diverse backgrounds, making this book a truly international effort. This book will bring forth new frontiers with its revolutionizing research information and detailed analysis of the nascent developments around the world.

We would like to thank all the contributing authors for lending their expertise to make the book truly unique. They have played a crucial role in the development of this book. Without their invaluable contributions this book wouldn't have been possible. They have made vital efforts to compile up to date information on the varied aspects of this subject to make this book a valuable addition to the collection of many professionals and students.

This book was conceptualized with the vision of imparting up-to-date information and advanced data in this field. To ensure the same, a matchless editorial board was set up. Every individual on the board went through rigorous rounds of assessment to prove their worth. After which they invested a large part of their time researching and compiling the most relevant data for our readers.

The editorial board has been involved in producing this book since its inception. They have spent rigorous hours researching and exploring the diverse topics which have resulted in the successful publishing of this book. They have passed on their knowledge of decades through this book. To expedite this challenging task, the publisher supported the team at every step. A small team of assistant editors was also appointed to further simplify the editing procedure and attain best results for the readers.

Apart from the editorial board, the designing team has also invested a significant amount of their time in understanding the subject and creating the most relevant covers. They scrutinized every image to scout for the most suitable representation of the subject and create an appropriate cover for the book.

The publishing team has been an ardent support to the editorial, designing and production team. Their endless efforts to recruit the best for this project, has resulted in the accomplishment of this book. They are a veteran in the field of academics and their pool of knowledge is as vast as their experience in printing. Their expertise and guidance has proved useful at every step. Their uncompromising quality standards have made this book an exceptional effort. Their encouragement from time to time has been an inspiration for everyone.

The publisher and the editorial board hope that this book will prove to be a valuable piece of knowledge for researchers, students, practitioners and scholars across the globe.

List of Contributors

P. Méndez-Carvajal
Fundación Pro-Conservación de los Primates Panameños (FCPP), 0816-07905, Panamá, República de Panamá

I. Ruiz-Bernard
Fundación Pro-Conservación de los Primates Panameños (FCPP), 0816-07905, Panamá, República de Panamá

G. De León
Fundación Pro-Conservación de los Primates Panameños (FCPP), 0816-07905, Panamá, República de Panamá

Y. González
Fundación Pro-Conservación de los Primates Panameños (FCPP), 0816-07905, Panamá, República de Panamá

E. Miranda
Fundación Pro-Conservación de los Primates Panameños (FCPP), 0816-07905, Panamá, República de Panamá

L. Loría
Fundación Pro-Conservación de los Primates Panameños (FCPP), 0816-07905, Panamá, República de Panamá

G. Berguido
Fundación Pro-Conservación de los Primates Panameños (FCPP), 0816-07905, Panamá, República de Panamá

A. Cortes
Fundación Pro-Conservación de los Primates Panameños (FCPP), 0816-07905, Panamá, República de Panamá

E. Soto
Fundación Pro-Conservación de los Primates Panameños (FCPP), 0816-07905, Panamá, República de Panamá

C.M. Nagy
Mianus River Gorge Preserve, 167 Mianus River Rd., Bedford, NY 10506 USA; Sackler Institute of Comparative Genomics, American Museum of Natural History, Central Park West at 79th Street., New York 10024 USA

M.E. Weckel
Mianus River Gorge Preserve, 167 Mianus River Rd., Bedford, NY 10506 USA; Sackler Institute of Comparative Genomics, American Museum of Natural History, Central Park West at 79th Street., New York 10024 USA

A. Toomey
Lancaster Environment Centre, Lancaster University, United Kingdom LA1 4YQ.

C.E. Burns
402 Ranch View Rd., Santa Cruz, CA, 95064 USA

J. Peltz
College Internship in Suburban Ecology, Mianus River Gorge Preserve, Bedford, NY, 10506 USA

F. Santilli
Game & Wildlife Management Consulting, Campiglia Marittima (Li) Italy

G. Paci
Department of Veterinary Science, University of Pisa, Italy

M. Bagliacca
Department of Veterinary Science, University of Pisa, Italy

I.D. Parker
Department of Wildlife and Fisheries Sciences, Texas A&M University, College Station, TX 77801

R.R. Lopez
Department of Wildlife and Fisheries Sciences, Texas A&M University, College Station, TX 77801

N.J. Silvy
Department of Wildlife and Fisheries Sciences, Texas A&M University, College Station, TX 77801

D.S. Davis
Department of Veterinary Pathobiology, Texas A&M University, College Station, TX 77801

J.C. Cathey
Department of Wildlife and Fisheries Sciences, Texas A&M University, College Station, TX 77801

A. Majumder
Wildlife Institute of India, P.O. Box # 18, Chandrabani, Dehra Dun-248001, Uttarakhand, India

S. Basu
Wildlife Institute of India, P.O. Box # 18, Chandrabani, Dehra Dun-248001, Uttarakhand, India

K. Sankar
Wildlife Institute of India, P.O. Box # 18, Chandrabani, Dehra Dun-248001, Uttarakhand, India

Q. Qureshi
Wildlife Institute of India, P.O. Box # 18, Chandrabani, Dehra Dun-248001, Uttarakhand, India

Y.V. Jhala
Wildlife Institute of India, P.O. Box # 18, Chandrabani, Dehra Dun-248001, Uttarakhand, India

P. Nigam
Wildlife Institute of India, P.O. Box # 18, Chandrabani, Dehra Dun-248001, Uttarakhand, India

R. Gopal
Member Secretary, National Tiger Conservation Authority, Government of India, New Delhi

M. Papakosta
Aristotle University of Thessaloniki, Faculty of Forestry and Natural Environment, University Campus, 54124, Thessaloniki, Greece

K. Kitikidou
Democritus University of Thrace, Department of Forestry and Management of the Environment and Natural Resources, Pandazidou 193, 68200, Orestiada, Greece

D. Bakaloudis
Aristotle University of Thessaloniki, Faculty of Forestry and Natural Environment, University Campus, 54124, Thessaloniki, Greece

C. Vlachos
Aristotle University of Thessaloniki, Faculty of Forestry and Natural Environment, University Campus, 54124, Thessaloniki, Greece

L. Gorini
Department of Forestry and Wildlife Management, Faculty of Applied Ecology and Agricultural Sciences, Hedmark University College, Campus Evenstad, NO-2418 Elverum
Norwegian Institute for Nature Research, NO-7485 Trondheim, Norway

S.J. Wedul
Department of Forestry and Wildlife Management, Faculty of Applied Ecology and Agricultural Sciences, Hedmark University College, Campus Evenstad, NO-2418 Elverum

J.M. Arnemo
Department of Forestry and Wildlife Management, Faculty of Applied Ecology and Agricultural Sciences, Hedmark University College, Campus Evenstad, NO-2418 Elverum
Department of Wildlife, Fish and Environmental Studies, Faculty of Forest Sciences, Swedish University of Agricultural Sciences, SE-901 83 Umeå, Sweden

J.D.C. Linnell
Norwegian Institute for Nature Research, NO-7485 Trondheim, Norway

L. Boitani
University of Rome "La Sapienza", Department of Environmental Biology, V.le dell`Università 32, 00185 Rome, Italy

E.B. Nilsen
Norwegian Institute for Nature Research, NO-7485 Trondheim, Norway

P. Tizzani
Department of Veterinary Science, University of Turin, V. Leonardo da Vinci 44, 10095 Grugliasco (TO), Italy

M. Bessone
Department of Veterinary Science, University of Turin, V. Leonardo da Vinci 44, 10095 Grugliasco (TO), Italy

S. Lioy
Department of Veterinary Science, University of Turin, V. Leonardo da Vinci 44, 10095 Grugliasco (TO), Italy

P. Meneguz
Department of Veterinary Science, University of Turin, V. Leonardo da Vinci 44, 10095 Grugliasco (TO), Italy

H. Kafley
Department of Fisheries and Wildlife Sciences, University of Missouri, Columbia, MO 65211, USA
Nepal Nature Foundation, Koteshwor, Kathmandu-34, Nepal

M. E. Gompper
Department of Fisheries and Wildlife Sciences, University of Missouri, Columbia, MO 65211, USA

F. Spinelli
221 W, Grant St., Bozeman, MT 59715, USA

K. L. Poudel
Department of Agricultural Economics, University of Missouri, Columbia, MO 65211, USA

B. P. Thapaliya
Chitwan National Park, Kasara, Chitwan, Nepal

K. Narasimmarajan
Zoological Survey of India (ANRC), National Coral Reef Research Institute, Port Blair, Andaman -744102, India

C. Raghunathan
Zoological Survey of India (ANRC), National Coral Reef Research Institute, Port Blair, Andaman -744102, India

I. de la Casa-Resino
Toxicology Area, Fac. of Veterinary Medicine (UEX). 10003 Cáceres, Spain

D. Hernández-Moreno
Toxicology Area, Fac. of Veterinary Medicine (UEX). 10003 Cáceres, Spain
Universidad Autónoma de Chile

A. López-Beceiro
Veterinary Clinical Sciences, Dpt. Fac. of Veterinary Medicine (USC), Campus Universitario s/n. 27003 Lugo, Spain

L. Rigueira
Veterinary Clinical Sciences, Dpt. Fac. of Veterinary Medicine (USC), Campus Universitario s/n. 27003 Lugo, Spain

M. P. Míguez
Toxicology Area, Fac. of Veterinary Medicine (UEX). 10003 Cáceres, Spain

M. Pérez-López
Toxicology Area, Fac. of Veterinary Medicine (UEX). 10003 Cáceres, Spain

L. E. Fidalgo
Veterinary Clinical Sciences, Dpt. Fac. of Veterinary Medicine (USC), Campus Universitario s/n. 27003 Lugo, Spain

K. L. Gee
The Samuel Roberts Noble Foundation, 2510 Sam Noble Parkway, Ardmore, OK 73401, USA
Present address: 6438 Gene Autry Road, Ardmore, OK 73401, USA

S. L. Webb
The Samuel Roberts Noble Foundation, 2510 Sam Noble Parkway, Ardmore, OK 73401, USA

P. D. Jones
Department of Forest and Wildlife Ecology, University of Wisconsin-Madison, 1630 Linden Drive, Room 226, Madison, WI 53706, USA

J. C. Levesque
Geo-Marine, Inc., Environmental Resources Division, Marine Science Department, 2201 Avenue K, Suite A2, Plano, Texas 75074 USA
Present address: Environmental Resources Management, 10210 Highland Manor Drive, Suite 140, Tampa, Florida (USA)

J. A. Martin
Warnell School of Forestry and Natural Resources, Athens, GA 30602, USA

C. Marshall
Agricultural Ecology Lab, Department of Wildlife, Fisheries, and Aquaculture, Mississippi State University, Mississippi State, MS 39761, USA

J. L. Belant
Carnivore Ecology Lab, Mississippi State University, Mississippi State, MS 39761, USA

S. Cagle
Mississippi State Extension Service, Mississippi State University, Houston, MS 38851, USA

B. C. West
Extension Regional Director, University of Tennessee Instituted of Agriculture, Jackson, TN 38301, USA

E. Nájera-Hillman
COSTASALVAJE A.C. Blvd. Las Dunas 160-203. Fracc. Playa Ensenada, 22880 Ensenada, Baja California, México

S. Mandujano
Red de Biología y Conservación de Vertebradas, Instituto de Ecología A.C. (INECOL). Carretera antigua a Coatepec No. 351. El Haya, 91070 Xalapa, Veracruzn , México

P.D. Meek
NSW Department of Primary Industries and Invasive Animal CRC, PO Box 530, Coffs Harbour, NSW 2450, Australia.
Ecosystem Management, University of New England, Armidale, NSW 2351, Australia

K. Vernes
Ecosystem Management, University of New England, Armidale, NSW 2351, Australia

G. Falzon
C4D: Centre 4 Dimensions, University of New England, Armidale, NSW 2351, Australia

V. K. Koli
Wildlife Research Laboratory, Department of Zoology, University College of Science, Mohanlal Sukhadia University, Udaipur (Rajasthan), 313 001, India

C. Bhatnagar
Wildlife Research Laboratory, Department of Zoology, University College of Science, Mohanlal Sukhadia University, Udaipur (Rajasthan), 313 001, India

R. Z. Clardy
School of Agricultural, Forest, and Environmental Sciences, Clemson University, Clemson, SC 29634, USA

Jr. W. M. Smathers
Department of Economics, Clemson University, Clemson, SC 29634, USA

T. J. Straka
School of Agricultural, Forest, and Environmental Sciences, Clemson University, Clemson, SC 29634, USA

D.S. Jachowski
Department of Fisheries and Wildlife Sciences, Univ. of Missouri, 302 ABNR Building, Columbia, MO 65211, USA
Department of Fish and Wildlife Conservation, Virginia Tech, 106 Cheatham Hall, Blacksburg, VA 24061, USA

T.W. Mong
Department of Fisheries and Wildlife Sciences, Univ. of Missouri, 302 ABNR Building, Columbia, MO 65211, USA

B.E. Washburn
Department of Fisheries and Wildlife Sciences, Univ. of Missouri, 302 ABNR Building, Columbia, MO 65211, USA
USDA, Wildlife Services, National Wildlife Research Center, 6100 Columbus Avenue, Sandusky, OH 44870, USA

J.H. Schulz
Department of Fisheries and Wildlife Sciences, Univ. of Missouri, 302 ABNR Building, Columbia, MO 65211, USA
Missouri Department of Conservation, Conservation Research Center, 1110 South College Avenue, Columbia, MO 65201, USA

J.J. Millspaugh
Department of Fisheries and Wildlife Sciences, Univ. of Missouri, 302 ABNR Building, Columbia, MO 65211, USA

S. Anile
Dipartimento di Biologia Animale "Marcello La Greca", Università di Catania, Via Androne 81, 95124 Catania, Italy

C. Amico
Ripartizione Faunistico Venatoria di Catania,Via San Giuseppe alla Rena 30/b,95121 Catania, Italy

B. Ragni
Dipartimento di Biologia Cellulare ed Ambientale, Università degli Studi di Perugia, Via Elce di Sotto, 06123 Perugia, Italy

S. Barber-Meyer
U.S. Geological Survey, Northern Prairie Wildlife Research Center, 8711 - 37th St., SE, Jamestown, ND 58401-7317, USA

L. D. Mech
U.S. Geological Survey, Northern Prairie Wildlife Research Center, 8711 - 37th St., SE, Jamestown, ND 58401-7317, USA
Postal address: The Raptor Center, 1920 Fitch Avenue, University of Minnesota, St. Paul, 55108, USA

S. L. Eliason
Department of Social Sciences and Cultural Studies, Montana State University, Billings 1500 University Dr, Billings, Mt 59101, USA

B.J. Kendall
School of Agricultural, Forest, and Environmental Sciences, Clemson University, Clemson, SC 29634, USA

Jr. D.C. Guynn
School of Agricultural, Forest, and Environmental Sciences, Clemson University, Clemson, SC 29634, USA

T. J. Straka
School of Agricultural, Forest, and Environmental Sciences, Clemson University, Clemson, SC 29634, USA

G. K. Yarrow
School of Agricultural, Forest, and Environmental Sciences, Clemson University, Clemson, SC 29634, USA